Semiconducting Polymers
Controlled Synthesis and Microstructure

RSC Polymer Chemistry Series

Editor-in-Chief:
Professor Ben Zhong Tang, *The Hong Kong University of Science and Technology, Hong Kong, China*

Series Editors:
Professor Alaa S. Abd-El-Aziz, *University of Prince Edward Island, Canada*
Professor Stephen Craig, *Duke University, USA*
Professor Jianhua Dong, *National Natural Science Foundation of China, China*
Professor Toshio Masuda, *Shanghai University, China*
Professor Christoph Weder, *University of Fribourg, Switzerland*

Titles in the Series:
1: Renewable Resources for Functional Polymers and Biomaterials
2: Molecular Design and Applications of Photofunctional Polymers and Materials
3: Functional Polymers for Nanomedicine
4: Fundamentals of Controlled/Living Radical Polymerization
5: Healable Polymer Systems
6: Thiol-X Chemistries in Polymer and Materials Science
7: Natural Rubber Materials: Volume 1: Blends and IPNs
8: Natural Rubber Materials: Volume 2: Composites and Nanocomposites
9: Conjugated Polymers: A Practical Guide to Synthesis
10: Polymeric Materials with Antimicrobial Activity: From Synthesis to Applications
11: Phosphorus-Based Polymers: From Synthesis to Applications
12: Poly(lactic acid) Science and Technology: Processing, Properties, Additives and Applications
13: Cationic Polymers in Regenerative Medicine
14: Electrospinning: Principles, Practice and Possibilities
15: Glycopolymer Code: Synthesis of Glycopolymers and their Applications
16: Hyperbranched Polymers: Macromolecules in-between of Deterministic Linear Chains and Dendrimer Structures
17: Polymer Photovoltaics: Materials, Physics, and Device Engineering

18: Electrical Memory Materials and Devices
19: Nitroxide Mediated Polymerization: From Fundamentals to Applications
    in Materials Science
20: Polymers for Personal Care Products and Cosmetics
21: Semiconducting Polymers: Controlled Synthesis and Microstructure

How to obtain future titles on publication:
A standing order plan is available for this series. A standing order will bring
delivery of each new volume immediately on publication.

For further information please contact:
Book Sales Department, Royal Society of Chemistry, Thomas Graham House,
Science Park, Milton Road, Cambridge, CB4 0WF, UK
Telephone: +44 (0)1223 420066, Fax: +44 (0)1223 420247
Email: booksales@rsc.org
Visit our website at www.rsc.org/books

Semiconducting Polymers
Controlled Synthesis and Microstructure

Edited by

Christine Luscombe
University of Washington, Seattle, WA, USA
Email: luscombe@uw.edu

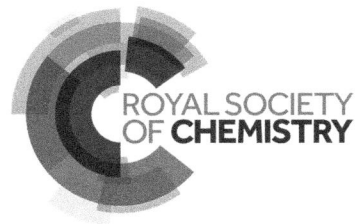

THE QUEEN'S AWARDS
FOR ENTERPRISE:
INTERNATIONAL TRADE
2013

RSC Polymer Chemistry Series No. 21

Print ISBN: 978-1-78262-034-1
PDF eISBN: 978-1-78262-400-4
EPUB eISBN: 978-1-78262-920-7
ISSN: 2044-0790

A catalogue record for this book is available from the British Library

Published by The Royal Society of Chemistry,
Thomas Graham House, Science Park, Milton Road,
Cambridge CB4 0WF, UK

Registered Charity Number 207890

For further information see our web site at www.rsc.org

Printed in the United Kingdom by CPI Group (UK) Ltd, Croydon, CR0 4YY, UK

Preface

The field of organic electronics has grown enormously since the seminal work on the conductivity of polyacetylene by Heeger, MacDiarmid, and Shirakawa was published in the 1970s. The ability of polymers to move charges and to absorb and emit visible light, while maintaining solution processability and flexibility, has given rise to a new generation of electronic devices. Flat panel displays consisting of organic light-emitting diodes (OLEDs) are now commercially available and work toward developing flexible and stretchable electronic devices is underway in numerous research groups around the world.

The progress made in organic electronics has been driven by the syntheses of π-conjugated polymers with increasingly complex structures. These complex structures have enabled us to reach a stage where the quantum efficiencies of OLEDs exceed those of inorganic LEDs. Charge mobilities >20 cm^2 V^{-1} s^{-1} have been reported for aligned π-conjugated polymers and organic photovoltaics now have power conversion efficiencies >10%.

Although significant advances have been made in the performance of these materials, their controlled synthesis has remained a bottleneck in research. These polymers are typically synthesized *via* polycondensation reactions (also called step-growth synthesis), which results in polymers with broad molecular weight distributions, random end-groups, and molecular weights that are difficult to control. All these characteristics become defects in the semiconducting material, which can adversely affect their performance in devices. Polycondensation reactions are also difficult to perform in a reproducible manner, giving rise to batch-to-batch variations when the synthesis is repeated. For π-conjugated polymers to be reliable in devices, it is important that they can be synthesized in a reproducible manner.

RSC Polymer Chemistry Series No. 21
Semiconducting Polymers: Controlled Synthesis and Microstructure
Edited by Christine Luscombe
© The Royal Society of Chemistry 2017
Published by the Royal Society of Chemistry, www.rsc.org

Having access to polymers synthesized in a controlled and reproducible manner allows the determination of detailed structure–property relationships. It is widely recognized that the morphology within the active layer is strongly dependent on the chemical (*e.g.* the chemistry of the monomer) and physical (*e.g.* molecular weight, alkyl chain, backbone flexibility) properties of the conjugated polymer. However, there are discrepancies within the literature about how the polymer structure affects its properties. Taking the ubiquitous poly(3-hexylthiophene) (P3HT) as an example, it is widely accepted that its charge carrier mobility is larger in the crystalline domains than in the amorphous regions. A number of studies have reported that the charge carrier mobility in organic field-effect transistors increases with molecular weight (MW). However, when studying P3HT in a diode configuration, it has been reported that the MW does not have a significant effect on the charge mobility, whereas a reversed dependence of MW on mobility has been observed in time-of-flight experiments. What may be giving rise to these contradictory reports (apart from the device structures used) is the presence of defects in a single polymer chain. Defects are rarely discussed in papers on π-conjugated semiconducting polymers, but can be minimized in a controlled polymerization.

This book, which is part of the RSC Polymer Chemistry Series, focuses on two aspects of π-conjugated polymers: (1) the controlled synthesis of these classes of materials; and (2) the microstructure that results from these polymers and their characterization and application. The first two chapters of this book focus on the most recent advances that have been made in the development of monomers and catalysts to achieve controlled polymerization from the perspective of two different research groups. The third and fourth chapters then explore new semiconducting polymer architectures that have been achieved with these controlled polymerizations, again from the perspective of two different research groups. The following chapters discuss the latest techniques are available for the characterization of polymer microstructures and the effect of the microstructure on device performance (in this case, OLEDs).

Christine Luscombe

Contents

Section I: Controlled Synthesis of Semiconducting Polymers

Chapter 1   **Controlled Synthesis of Conjugated Polymers in Catalyst-transfer Condensation Polymerization: Monomers and Catalysts**   3
T. Yokozawa and Y. Ohta

    1.1   Introduction   3
    1.2   Kumada–Tamao Coupling Polymerization of Grignard Monomers   4
        1.2.1   Background and Discovery   4
        1.2.2   Mechanistic Studies   6
        1.2.3   Monomers   9
        1.2.4   Catalysts   11
        1.2.5   Initiators   13
        1.2.6   Functionalization of Polymer Ends   16
    1.3   Suzuki–Miyaura Coupling Polymerization of Boronic Acid (Ester) Monomer   16
        1.3.1   Background and Discovery   16
        1.3.2   Monomers   17
        1.3.3   Initiators and Catalysts   20
        1.3.4   Functionalization of Polymer Ends   22
    1.4   Other Coupling Polymerization Reactions   24
        1.4.1   Negishi Coupling Polymerization of Zn-containing Monomers   24
        1.4.2   Stille Coupling Polymerization of Tin-containing Monomers   28

RSC Polymer Chemistry Series No. 21
Semiconducting Polymers: Controlled Synthesis and Microstructure
Edited by Christine Luscombe
© The Royal Society of Chemistry 2017
Published by the Royal Society of Chemistry, www.rsc.org

1.4.3   Murahashi Coupling Polymerization of
        Aryllithium Monomers   29
1.4.4   Mizoroki–Heck Coupling Polymerization of
        Non-metallated Monomers   30
1.4.5   Nucleophilic Substitution Polymerization of
        Silylated Monomers   30
1.5   Conclusions   31
Acknowledgements   32
References   32

**Chapter 2   Controlled Chain-growth Synthesis of Conjugated
              Polymers: Moving Beyond Thiophene   38**
            C. R. Bridges and D. S. Seferos

2.1   Introduction   38
2.2   Mechanism of Controlled Chain Growth in
      Conjugated Polymers   41
2.3   Chain-growth Synthesis of Conjugated Homopolymers   44
2.4   Chain-growth Synthesis of Alternating Copolymers   56
2.5   Mixed-mechanism Chain-growth Synthesis of Block
      Copolymers   63
2.6   Chain-growth Synthesis of Fully Conjugated Block
      Copolymers   66
2.7   Chain-growth Synthesis of Other Copolymers   71
2.8   Future Outlook   78
References   80

**Chapter 3   Application of Catalyst Transfer Polymerizations:
              From Conjugated Copolymers to Polymer
              Brushes   85**
            Yanhou Geng and Aiguo Sui

3.1   Introduction   85
3.2   Fully Conjugated Block Copolymers   86
      3.2.1   Donor–Donor Type Conjugated Block
              Copolymers   86
      3.2.2   Donor–Acceptor Type Fully Conjugated Block
              Copolymers   91
3.3   Conjugated Gradient Copolymers   97
3.4   Block Copolymers Consisting of Conjugated and
      Non-conjugated Blocks   99
      3.4.1   Synthesis *via* Macromolecular Coupling
              Reactions   99

3.4.2 Synthesis with Macromolecular Initiators 102
3.4.3 Synthesis *via* One-pot Polymerization 107
3.5 Conjugated Polymer Brushes 109
3.6 Other Applications 111
3.7 Summary and Outlook 113
References 114

Chapter 4 Controlled Synthesis of Chain End Functional, Block and Branched Polymers Containing Polythiophene Segments 121
Tomoya Higashihara and Mitsuru Ueda

4.1 Introduction 121
4.2 Controlled Synthesis of Polythiophenes 122
4.3 Regio-regularity 123
4.4 Control of Molecular Weight and Dispersity 124
4.5 Dehydrogenative Synthesis of Polythiophene 126
4.6 Chain End Functional Polythiophenes 128
4.7 Block Copolymers with Polythiophene Segments 134
    4.7.1 All-conjugated Block Copolythiophenes 134
    4.7.2 All-conjugated Block Copolymers Containing Polythiophene and Other Polymer Segments 137
    4.7.3 All-conjugated Donor–Acceptor Block Copolymers 139
    4.7.4 Semi-conjugated Block Copolymers Containing Polythiophene Segments 142
4.8 Graft Copolymers with Polythiophene Segments 146
4.9 Star-branched Polymers with Polythiophene Segments 149
4.10 Hyperbranched Polythiophene with a Controlled Degree of Branching 154
4.11 Conclusions 155
Acknowledgements 155
References 155

Section II: Microstructure of Semiconducting Polymers

Chapter 5 Characterization of Polymer Semiconductors by Neutron Scattering Techniques 165
Gregory M. Newbloom, Kiran Kanekal, Jeffrey J. Richards and Lilo D. Pozzo

5.1 Introduction 165

	5.2	Small-angle Neutron Scattering	166
		5.2.1 SANS of Dissolved Conjugated Polymers	167
		5.2.2 SANS of Colloidal Polymer Nanostructures	168
	5.3	Neutron Scattering in Thin Films	171
		5.3.1 Transmission SANS	172
		5.3.2 Neutron Reflectometry	174
		5.3.3 Grazing Incidence Small-angle Neutron Scattering	175
	5.4	Quasi-elastic Neutron Scattering	177
		5.4.1 Theory	178
		5.4.2 Instrumentation and Methods for Analysis	179
		5.4.3 QENS of Polymer Semiconductors	180
		References	184

Chapter 6 Structural Control in Polymeric Semiconductors: Application to the Manipulation of Light-emitting Properties   187
Ioan Botiz, Cosmin Leordean and Natalie Stingelin

	6.1	Introduction	187
	6.2	Approach 1: Chemical Design	189
	6.3	Approach 2: Physical and Physicochemical Methods	195
		6.3.1 Solvent Vapour and Solvent Quality	195
		6.3.2 Pressure	196
		6.3.3 Dewetting	196
		6.3.4 Chemical Cross-linking	198
		6.3.5 Controlled Aggregation	199
		6.3.6 Processing in Confined Spaces	199
		6.3.7 Stretchable Structures	200
	6.4	Approach 3: Blending	200
	6.5	Approach 4: Metal-enhanced Fluorescence	207
	6.6	Conclusions	212
		References	212

Chapter 7 Structure and Order in Organic Semiconductors   219
Chad R. Snyder, Dean M. DeLongchamp,
Ryan C. Nieuwendaal and Andrew A. Herzing

	7.1	Introduction	219
	7.2	Differential Scanning Calorimetry	220
		7.2.1 Introduction to Differential Scanning Calorimetry	220

7.2.2   Qualifying and Quantifying Crystallinity in Semiconducting Polymers   221

7.2.3   Determination of the Equilibrium Melting Temperature   225

7.2.4   Determination of the Enthalpy of Fusion Per Repeat Unit   228

7.2.5   Self-nucleation and Successive Annealing   231

7.2.6   Methods for Characterizing Thin Film Samples   232

7.3   Solid-state NMR Spectrometry   235

7.3.1   Introduction   235

7.3.2   Crystallinity and Order   236

7.3.3   Relaxation and Dynamics   238

7.3.4   Domain Sizes and Interfacial Structures in Donor/Acceptor Blends   248

7.3.5   Future Prospects   252

7.4   Transmission Electron Microscopy of Organic Semiconductors   252

7.4.1   Introduction to Transmission Electron Microscopy   252

7.4.2   Challenges for Characterizing Polymeric Materials in TEM   253

7.4.3   Methods for Characterizing Order and Morphology in Polymer Materials   254

7.5   Grazing Incidence Scattering   261

7.5.1   Introduction to X-ray Scattering   261

7.5.2   Grazing Incidence X-ray Diffraction   261

7.5.3   Grazing Incidence Small-angle X-ray Scattering   263

7.5.4   Polarized Resonant Soft X-ray Scattering   264

7.5.5   *In situ* X-ray Scattering Studies of Structure Evolution   265

References   266

Subject Index   275

Section I: Controlled Synthesis of Semiconducting Polymers

CHAPTER 1

Controlled Synthesis of Conjugated Polymers in Catalyst-transfer Condensation Polymerization: Monomers and Catalysts

T. YOKOZAWA* AND Y. OHTA

Kanagawa University, Department of Materials and Life Chemistry, Rokkakubashi, Kanagawa-ku, Yokohama 221-8686, Japan
*Email: yokozt01@kanagawa-u.ac.jp

1.1 Introduction

Semiconducting π-conjugated polymers were first synthesized by the chemical and electrochemical oxidative polymerization of electron-rich aromatic species such as pyrrole, thiophene, and aniline. This polymerization method is still widely used in, for example, the modification of electrodes.[1] Yamamoto[2] later reported organometallic polycondensation reactions involving cross-coupling reactions catalyzed by transition metals, such as the Kumada–Tamao coupling of organo-magnesium halides,[3] Negishi coupling of organo-zinc halides,[4] and Sonogashira coupling of acetylenic compounds,[5] in addition to the polycondensation of dibromoarenes with an equimolar Ni^0 complex (Yamamoto coupling).[6] The Pd-catalyzed polycondensations of organo-borons (Suzuki–Miyaura coupling)[7] and organo-stannanes (Stille coupling)[8] have also

RSC Polymer Chemistry Series No. 21
Semiconducting Polymers: Controlled Synthesis and Microstructure
Edited by Christine Luscombe
© The Royal Society of Chemistry 2017
Published by the Royal Society of Chemistry, www.rsc.org

been developed by other researchers. As these polymerizations proceeded in a non-chain reaction manner, research was focused on synthesizing high molecular weight π-conjugated polymers by changing the coupling reactions and polymerization conditions rather than on controlling molecular weight, polymer end-groups, and dispersity, which was believed to be achievable only by means of "living" chain polymerization using vinyl monomers and cyclic monomers.

In 2004, our research group[9] and McCullough and coworkers[10] independently found that bromothiophenemagnesium chloride and its zinc chloride counterpart[11] underwent chain polymerization with an Ni catalyst to afford poly(3-hexylthiophene) (P3HT) with a controlled molecular weight, low dispersity, and defined polymer ends. This finding was made during the course of research on the condensative chain polymerization of aromatic polyamides[12] and the synthesis of regioregular P3HT,[13] respectively. We proposed that chain polymerization involves the intramolecular transfer of the catalyst to the polymer C–Br ends after reductive elimination and we named this type of polymerization catalyst-transfer condensation polymerization (CTCP). This finding opened up a new area of research, aimed on extending CTCP to other coupling polymerizations and applying it to develop organic electronic devices, such as photovoltaics and field-effect transistors. This chapter focuses on the monomers and catalysts used for catalyst-transfer Kumada–Tamao coupling polymerization, Suzuki–Miyaura coupling polymerization, and other coupling polymerizations. π-Conjugated polymer architectures such as block copolymers, obtained by virtue of CTCP, are reviewed in Chapter 3.

1.2 Kumada–Tamao Coupling Polymerization of Grignard Monomers

1.2.1 Background and Discovery

We developed condensative chain polymerization by the activation of the polymer end-group based on a difference of substituent effects between the monomer and the polymer.[12,14] An example of condensative chain polymerization for the synthesis of a well-defined aromatic polyamide is shown in Figure 1.1a.[15] The amide anion of monomer **1** deactivates the ester moiety through its strong electron-donating resonance effect, which serves to suppress self-condensation. Monomer **1** then selectively reacts with the polymer end-group, which has a weaker electron-donating amide linkage at the *para* position, resulting in chain polymerization. Similar substituent effects are known in organometallic chemistry; for example, an electron-donating group on an aromatic halide retards oxidative addition with a zero-valent transition metal catalyst.[16–18] We utilized this chemistry for the Pd-catalyzed condensative chain polymerization of bromophenoxide **2** and carbon monoxide. Chain polymerization proceeded until the middle stage of polymerization and transesterification occurred in the final stage, resulting in a non-chain behavior (Figure 1.1b).[19]

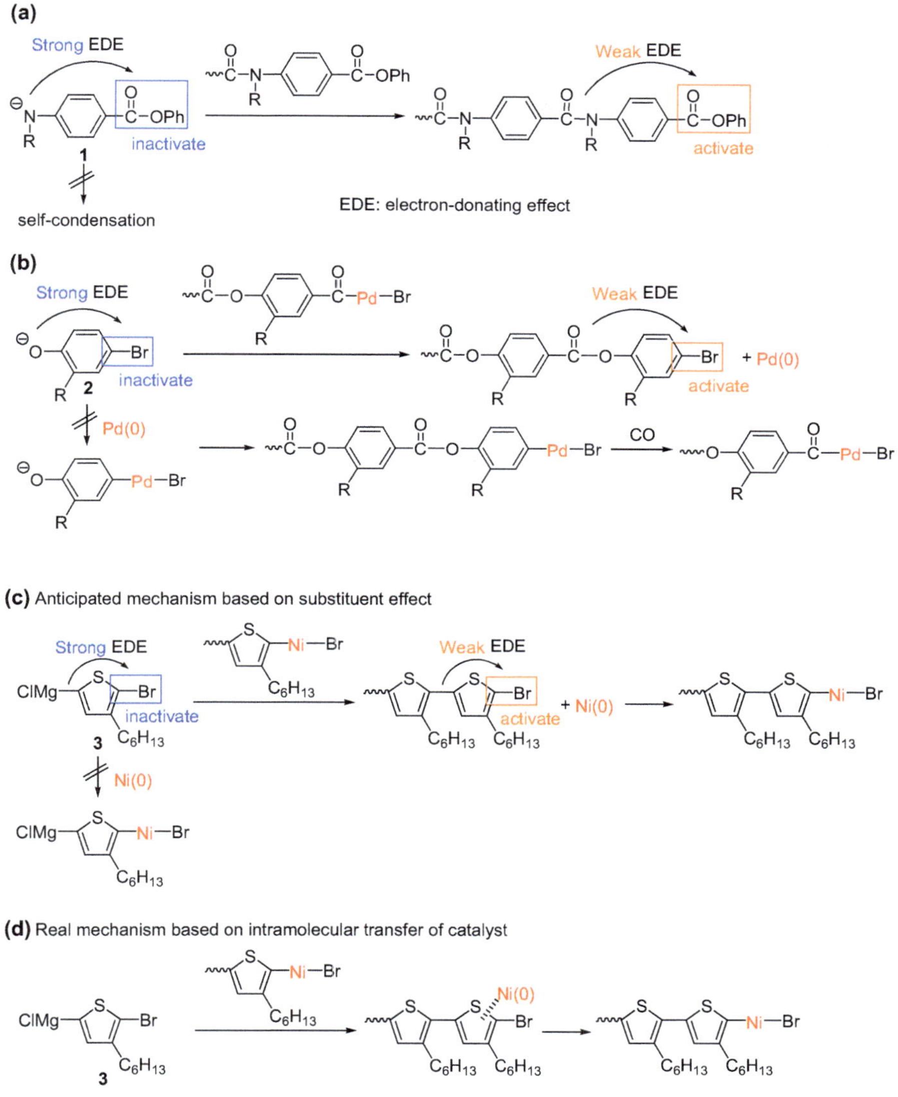

(a)

EDE: electron-donating effect

(b)

(c) Anticipated mechanism based on substituent effect

(d) Real mechanism based on intramolecular transfer of catalyst

Figure 1.1 Condensative chain polymerization: (a) polyamide synthesis based on the substituent effect; (b) polyester synthesis based on the substituent effect; (c) anticipated polythiophene synthesis based on the substituent effect; and (d) polythiophene synthesis based on intramolecular catalyst transfer.

We attempted to conduct a simpler Ni-catalyzed condensative chain polymerization of bromothiophenemagnesium chloride **3** on the basis of substituent effects, because bond-exchange reactions such as transesterification do not occur on the polythiophene backbone. Osaka and McCullough[13] reported that the Ni-catalyzed Kumada–Tamao coupling polymerization of

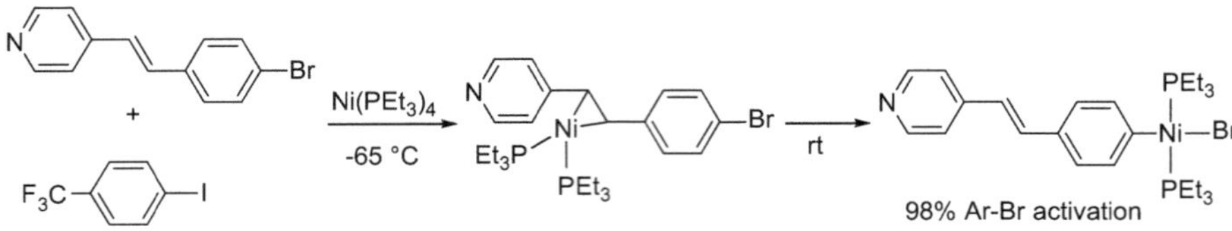

Figure 1.2   Intramolecular transfer of Ni⁰ in organic chemistry.

3 yielded regio-regular P3HT with superior electrical properties. We similarly anticipated that the Ni⁰ catalyst would insert selectively into the terminal C–Br bond of the polymer chain, rather than the C–Br bond of monomer **3**, because the strong electron-donating chloromagnesio moiety of **3** would deactivate the C–Br bond for oxidative addition (Figure 1.1c). The polymerization of **3** with Ni(dppp)Cl$_2$ (dppp = 1,3-bis(diphenylphosphino)propane) at room temperature showed chain polymerization behavior; the molecular weight increased in proportion to monomer conversion, and extension of the polymer chain took place upon further addition of the monomer to the reaction mixture.[9,20] At the same time, McCullough and coworkers reported the Ni-catalyzed chain polymerization of the zinc chloride counterpart[11] and **3**.[10]

However, molecular weight was not controlled by the addition of active aryl halides bearing an electron-withdrawing group, implying that the anticipated condensative chain polymerization based on a change in the substituent effect was not involved in this polymerization. After a detailed study of the mechanism, four important points were clarified: (1) the polymer end-groups are uniform among molecules – one end-group is Br and the other is H; (2) the propagating end-group is a polymer–Ni–Br complex; (3) one Ni molecule forms one polymer chain; and (4) the chain initiator is a dimer of **3** formed *in situ*. On the basis of these results, we proposed that the Ni⁰ catalyst is inserted into the intramolecular C–Br bond after transmetallation of the polymer–Ni–Br complex with **3** and reductive elimination during propagation (Figure 1.1d).[21] The intramolecular transfer of Ni catalysts has been reported in organic chemistry; van der Boom and coworkers[22] demonstrated that the Ni atom on an η²-C=C complex of bromostilbazole underwent intramolecular transfer to the aryl–Br bond followed by oxidative addition, even in the presence of reactive aryl iodide (Figure 1.2). Nakamura and coworkers,[23] in a study of kinetic isotope effects and theoretical calculations of Ni-catalyzed coupling reactions, showed that the π-complex of Ni⁰ and a haloarene does not dissociate and proceeds quickly to the oxidative addition step in an intramolecular manner.

1.2.2   Mechanistic Studies

The mechanism of the CTCP of **3** with an Ni catalyst has been studied in detail and the whole mechanism is shown in Figure 1.3. In the initiation step, two equivalents of **3** are reacted with Ni(dppp)Cl$_2$, followed by reductive elimination to form a tail-to-tail thiophene diad, accompanied by the

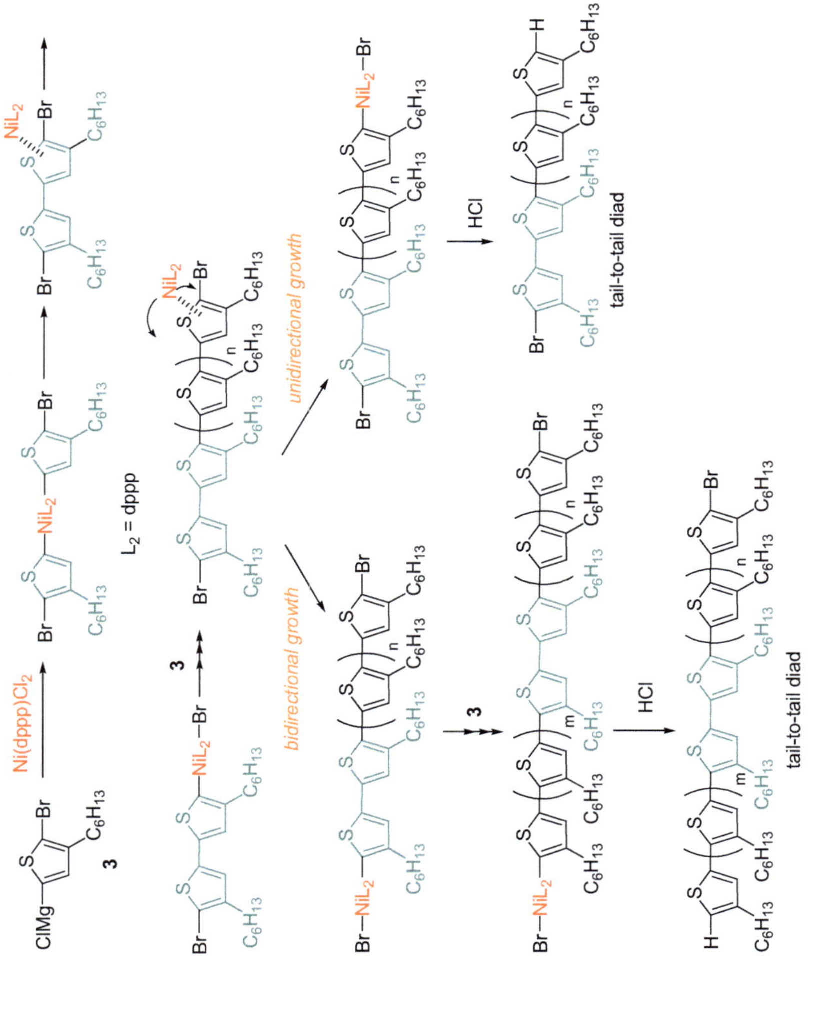

Figure 1.3 Mechanism of Kumada–Tamao CTCP of **3** with Ni(dppp)Cl$_2$.

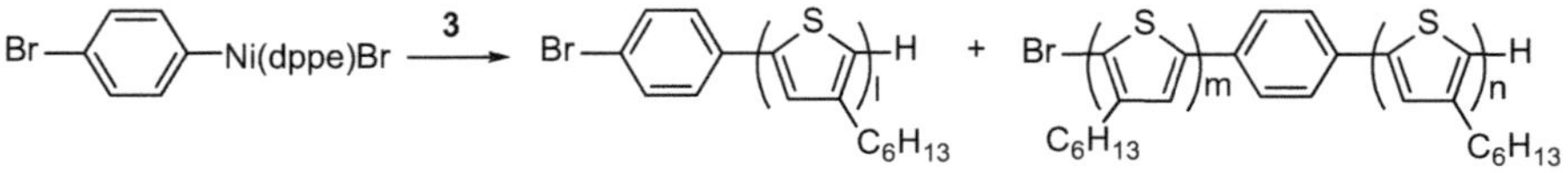

Figure 1.4   Bidirectional growth in the polymerization of **3** with BrC_6H_4-Ni(dppe)-Br.

generation of an Ni^0 complex, which inserts itself into the intramolecular C–Br bond of the dimer. Propagation involves the transmetallation of the polymer–Ni–Br complex with **3**, reductive elimination, intramolecular transfer to the terminal C–Br bond, and then oxidative addition (unidirectional growth). However, the C–Br bond of the tail-to-tail thiophene diad at the other terminal can also undergo intramolecular oxidative addition after ''long walking'' of the Ni^0 complex on the backbone. Similar propagation takes place to yield P3HT containing the tail-to-tail thiophene diad inside the backbone (bidirectional growth). In the termination step, the polymer–Ni–Br propagating end is hydrolyzed with hydrochloric acid to form the H-terminal. Therefore P3HTs containing the tail-to-tail thiophene diad at the terminal end and inside the backbone both bear H at one end and Br at the other. Koeckelberghs identified the signals of all possible end-groups in the ^1H NMR spectrum of P3HT obtained with Ni(dppp)Cl_2 and found that the ratio of Br-TT/(Br-HT + Br-TT) – where Br-TT represents a Br-terminal attached to the tail-to-tail diad and Br-HT represents a Br-terminal attached to the head-to-tail diad – decreased with an increasing degree of polymerization, indicating bidirectional growth.[24]

Before this finding, Kiriy and coworkers[25] had demonstrated bidirectional growth in the polymerization of **3** with Br–C_6H_4–Ni(dppe)Br (dppe = 1,2-bis(diphenylphosphino)ethane) as an initiator (Figure 1.4). The polymerization afforded not only bromophenyl-ended P3HT *via* unidirectional growth, but also P3HT containing the phenylene unit inside the backbone *via* directional growth. They also showed the presence of a single tail-to-tail defect distribution over the whole chain by means of an evaluation of the crystallinity of P3HT obtained with Ni(dppp)Cl_2.[26]

Lanni and McNeil conducted kinetic studies and showed that the rate-determining step is transmetallation in the polymerization[27] of **3** with Ni(dppp)Cl_2 and reductive elimination in the polymerization[28] with Ni(dppe)Cl_2. This means that an intermediate π-complex in the CTCP of **3** with the Ni catalyst could not be directly observed as a resting state in the rate-determining oxidative addition. However, McCullough and coworkers[11] reacted dibromothiophene with 0.5 equivalent of tolylmagnesium chloride in the presence of Ni(dppp)Cl_2 to exclusively obtain disubstituted thiophene. This result strongly supported the view that the second substitution would proceed *via* an intermediate π-complex between Ni^0 and thiophene (Figure 1.5). Bryan and McNeil[29] reacted the bromophenyl Ni complex **4** with an aryl Grignard reagent in the presence of active aryl bromide **5** with an electron-withdrawing CN group (Figure 1.6). If intramolecular transfer takes place *via* the π-complex, then the aryl-substituted Ni complex **6** would be

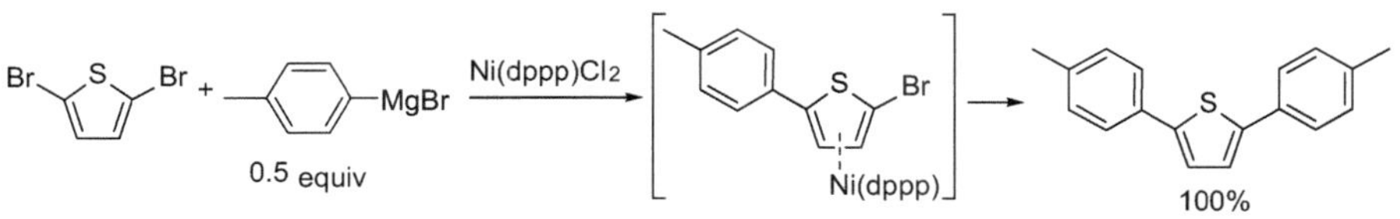

Figure 1.5   Reaction of dibromothiophene with tolylmagnesium bromide as a model reaction of the CTCP of **3** with Ni(dppp)Cl$_2$.

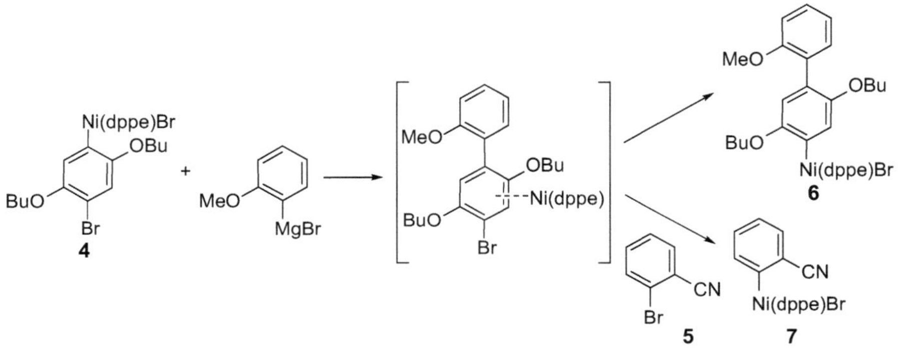

Figure 1.6   Competition experiment between intramolecular and intermolecular oxidative addition.

formed. However, if the π-complex does not form, then the resulting free Ni^0 should be selectively trapped by **5** to produce **7**. The experiment resulted in a 95:5 ratio of **6**:**7**, consistent with the involvement of an intermediate Ni^0 π-complex.

1.2.3 Monomers

Since the CTCP of **3** was first published, it has been reported that many monomers undergo Kumada–Tamao CTCP (Figure 1.7). Donor monomers were the first to be developed, including thiophenes,[12,30–33] selenophenes,[34,35] pyrroles,[36] phenylenes,[37,38] fluorenes,[39] cyclopentadithiophenes,[40] dithienosilole,[41] and non-conjugated bithienylmethylene.[42] Not only acceptor monomers such as pyridines,[43] benzotriazoles,[44] and thiazoles,[45,46] but also diaryl monomers,[47,48] have been reported. The CTCP of most of these monomers proceeds with phosphine-ligated Ni complexes such as Ni(dppp)Cl$_2$ or Ni(dppe)Cl$_2$, whereas the CTCP of the benzotriazole monomer proceeds with the Ni diimine complex, which bears electron-donating groups; Ni(dppp)Cl$_2$ and Ni(dppe)Cl$_2$ were ineffective (see Section 1.2.4).[44]

Several monomers have been reported not to undergo CTCP (Figure 1.8). Polymerization of the pyridine monomer was accompanied by disproportionation, probably due to the coordination of the pyridine N adjacent to the C–Ni–Br end to the Ni in another pyridine–Ni–Br end, yielding a polymer with broad molecular weight distribution and Br/Br ends.[49] The

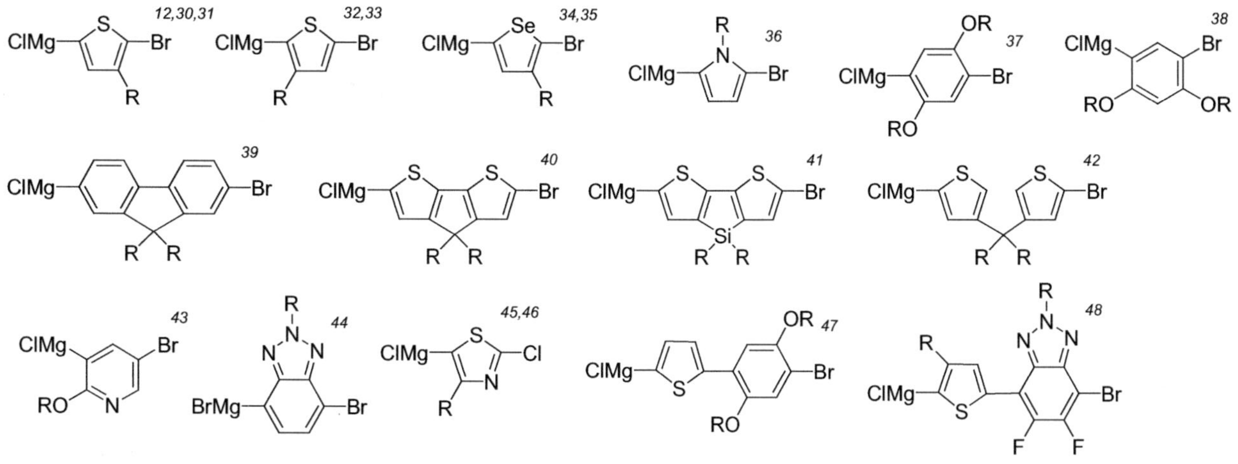

Figure 1.7 Grignard monomers for Kumada–Tamao CTCP.

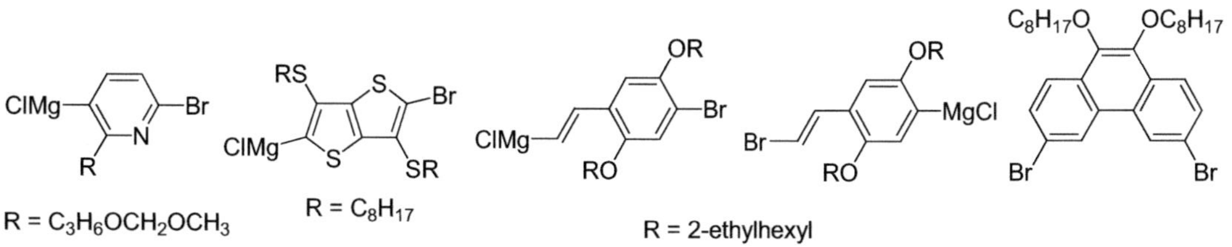

Figure 1.8 Unfavorable monomers for Kumada–Tamao CTCP.

thienothiophene monomer was not polymerized with an Ni^{II} catalyst because the dimer, formed by the reaction of the Ni^{II} catalyst with the monomer in the initiation step, strongly coordinated to the Ni^0 generated in this initiation step, blocking oxidative addition of the C–Br bond in the dimer.[50] Phenylenevinylene monomers did not efficiently undergo Kumada–Tamao coupling, affording only low molecular weight polymers.[51] Dibromophenanthrene was not quantitatively converted to the Grignard monomer with *i*-PrMgCl · LiCl.[52]

These Grignard monomers are generated *in situ via* a halogen–magnesium exchange reaction with *i*-PrMgCl or *t*-BuMgCl. The iodo–magnesium exchange reaction was fast enough to be practical; it was completed in 1 h at 0 °C.[9] However, the bromo-magnesium exchange reaction takes longer. For example, it took 20 h to generate the Grignard monomer when 2,5-dibromo-3-hexylthiophene was treated with *t*-BuMgCl at room temperature.[53] It is crucial to use an exactly equimolar amount of the Grignard reagent to generate the Grignard monomers.[21] Excess Grignard reagent reacts with the polymer–Ni–Br end-group to afford a polymer with a broad molecular weight distribution and unexpected end-groups.[53] Mori and coworkers[54] generated Grignard monomers by proton abstraction with TMPMgCl · LiCl (TMP = 2,2,6,6-tetramethylpiperidine) or by using a combination of a Grignard reagent and a catalytic amount of secondary amine.

1.2.4 Catalysts

The catalysts used for Kumada–Tamao CTCP are summarized in Figure 1.9. The first CTCP of **3** was achieved with Ni(dppp)Cl$_2$.[9] This catalyst is broadly effective for many monomers: selenophenes,[34,35] *m*-phenylenes,[38] cyclopentadithiophenes,[40] dithienosilole,[41] pyridines,[43] thiazoles,[45,46] and diaryl monomers.[47,48] The polymerization of the fluorene Grignard monomer with Ni(dppp)Cl$_2$ was poorly controlled,[55,56] whereas the Ni(acac)$_2$ (acac = acetylacetonate)/dppp system mediated well-controlled CTCP.[39] Ni(dppe)Cl$_2$ was effective for methoxyethoxyethoxymethylthiophene,[57,58] *p*-phenylene,[37] *N*-alkylpyrrole,[36] and bithienylmethylene.[42] McNeil and coworkers[59] studied CTCP with Ni(depe)Cl$_2$, anticipating that the more electron-donating phosphine ligand would increase the binding affinity of the polymers to the Ni catalyst and thus minimize the side-reactions. The polymerization of the phenylene monomer with Ni(depe)Cl$_2$ was significantly slower than that with Ni(dppp)Cl$_2$ at room temperature. The polymerization at 60 °C yielded a polymer with a low dispersity, as well as a small amount of oligomers, which resulted from the disproportionation of the polymer–Ni–Br end-group. The stronger binding affinity of Ni(depe)Cl$_2$ was demonstrated by means of small molecule competition experiments (Figure 1.6), as well as polymerization in the presence of a competing molecule, 2-bromobenzonitrile; the dispersity of the polymer obtained with Ni(depe)Cl$_2$ was narrower than that of the polymer obtained with Ni(dppp)Cl$_2$ under the same conditions.[29]

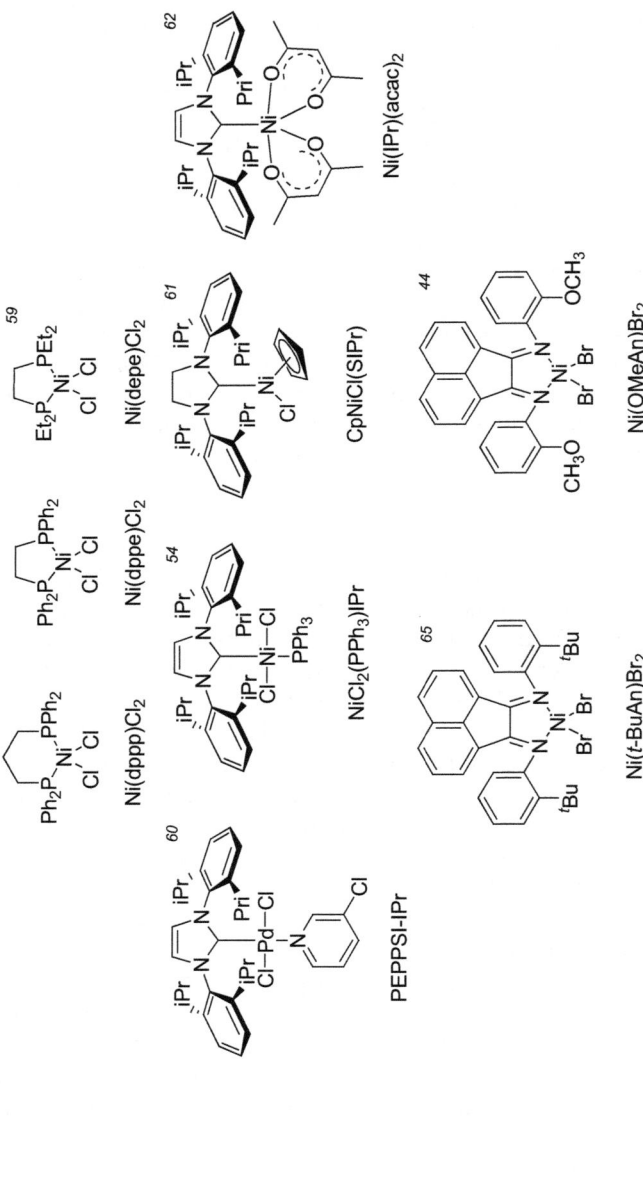

Figure 1.9 Catalysts for Kumada–Tamao CTCP.

N-Heterocyclic carbene (NHC)-ligated Pd and Ni catalysts were studied because NHC is a stronger σ-donor than phosphine. McNeil and coworkers[60] conducted the polymerization of thiophene, phenylene, and fluorene monomers with PEPPSI-IPr. Both the thiophene and phenylene monomers underwent CTCP, whereas the polymerization of the fluorene monomer was not well controlled. Mori and coworkers[54] used $NiCl_2(PPh_3)IPr$ for the polymerization of the chlorothiophene Grignard monomer, which was formed by means of proton abstraction of the corresponding chloro-thiophene with $TMPMgCl \cdot LiCl$. $Ni(dppp)Cl_2$ and $Ni(dppe)Cl_2$ were not effective in this polymerization. They also found that the polymerization of the bromothiophene monomer, similarly generated with $TMPMgCl \cdot LiCl$, with CpNiCl(SIPr) gave P3HT with $M_n > 220\,000$, although the dispersity was rather broad $(M_w/M_n = 1.85)$.[61] Geng and coworkers[62] reported that $Ni(IPr)(acac)_2$ enabled the CTCP of the thiophene monomer to yield P3HT with M_n up to 350 000; the M_n value increased in proportion to the feed ratio of the monomer to the catalyst, while M_w/M_n remained < 1.50.

Ni^{II} diimine complexes have been developed as catalysts for olefin poly-merization by Brookhart and coworkers.[63,64] The steric and electronic effects at the nickel center can be varied by changing the substituents of the diimine ligands. Stefan and coworkers[65] used $Ni(t\text{-}BuAn)Br_2$ for the CTCP of the thiophene monomer. A kinetic study indicated the presence of termination reactions, although the molecular weight increased with increasing con-version. Seferos and coworkers[44] investigated the CTCP of the benzotriazole monomer with several Ni catalysts. $Ni(dppe)Cl_2$ resulted in a polymer with lower M_n than expected and broad dispersity. Ni^{II} diimine complexes with bulky $(Ni(t\text{-}BuAn)Br_2)$, donor, and acceptor substituents were examined, and it was found that $Ni(OMeAn)Br_2$ mediated well-controlled CTCP. This catalyst also enabled the block copolymerization of the benzotriazole monomer and thiophene monomer by chain extension from either electron-deficient poly(benzotriazole) or electron-rich P3HT.[66]

1.2.5 Initiators

In CTCP with the described catalysts, the chain initiators are dimers formed *in situ* (see Section 1.2.2). The bromine of the dimers induces bidirectional growth, which becomes problematic in the one-pot synthesis of block copolymers of A and B, resulting in not only the AB diblock, but also BAB triblock copolymers.[24] If external initiators ArNiLX (L = dppp or dppe, X = Cl or Br) can be formed, only the AB diblock copolymer is synthesized and the functional group in Ar in the initiator is introduced at one end of the π-conjugated polymers. Three procedures have been reported for the gen-eration of ArNiLX initiators (Figure 1.10). (1) Bronstein and Luscombe[67] reacted ArCl with $Ni(PPh_3)_4$ to generate $ArNi(PPh_3)_2Cl$, the ligand of which was replaced with dppp. (2) Kiriy and coworkers[68] used $Et_2Ni(2,2'\text{-bipyridine})$ instead of $Ni(PPh_3)_4$ for the generation of the primary Ni^{II} complex and replaced the ligand. (3) Kiriy and coworkers[69] found that the reaction of

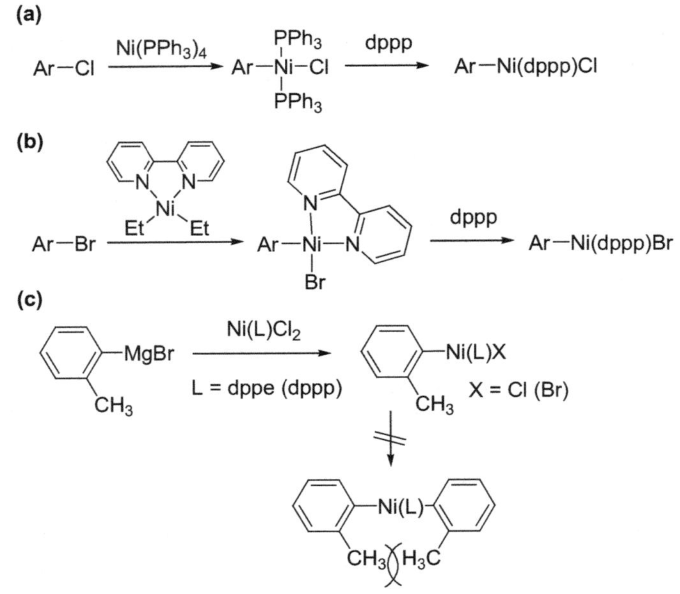
(a)
Ar—Cl
Ni(PPh3)4
PPh3
Ar—Ni—Cl
PPh3
dppp
Ar—Ni(dppp)Cl
(b)
Ar—Br
N
N
Ni
Et
Et
Ar—Ni—N
N
Br
dppp
Ar—Ni(dppp)Br
(c)
MgBr
CH3
Ni(L)Cl2
L = dppe (dppp)
Ni(L)X
CH3
X = Cl (Br)
Ni(L)
CH3
H3C

Figure 1.10 Synthesis of Ni initiators for Kumada–Tamao CTCP: (a) Luscombe method; (b) Kiriy method; and (c) convenient Kiriy method.

o-tolymagnesium bromide or (3-hexylthiophene-2-yl)magnesium chloride with Ni(dppe)Cl_2 or Ni(dppp)Cl_2 yielded ArNiLCl. The steric hindrance of the *ortho*-substituent of Grignard reagents is responsible for the monotransmetallation.

Many initiators have been formed by procedure 1 (Figure 1.11a). Protected hydroxyl and ethynyl groups were deprotected after CTCP and used for block copolymerization.[70,71] P3HT obtained with the pyridine initiator contained 18% H/Br-terminated polymer, presumably because of the coordination of the pyridine nitrogen to Ni^0, which might reinitiate the polymerization of **3**.[72] This pyridine-terminated P3HT was used to synthesize hybrid materials consisting of quantum dots (CdSe/ZnS) anchored with P3HT. The phosphonic ester terminal function was used to decorate Fe_3O_4 nanoparticles, and the phenol and thiol functions enabled the synthesis of hybrid Au nanoparticles.[72] Procedure 1 was also used to synthesize star P3HT[73] and donor–acceptor conjugated polymers grafted with P3HT.[74] The Ph–Ni–Br initiator was formed by procedure 2, as was an orthogonal difunctional initiator containing nitroxide for the preparation of a block copolymer of P3HT and polystyrene (Figure 1.11b).[75] The generation of *ortho*-substituted aromatic Ni initiators was performed by procedure 3 (Figure 1.11c).[69]

McNeil and coworkers investigated the substituent effects at the ligand (L)[76] and the aryl group[77] of ArNi(L)Br initiators in the CTCP of phenylene monomers (Figure 1.12). These initiators were synthesized from $Ni(COD)_2$ (COD = 1,5-cyclooctadiene) and the corresponding aryl bromide and then

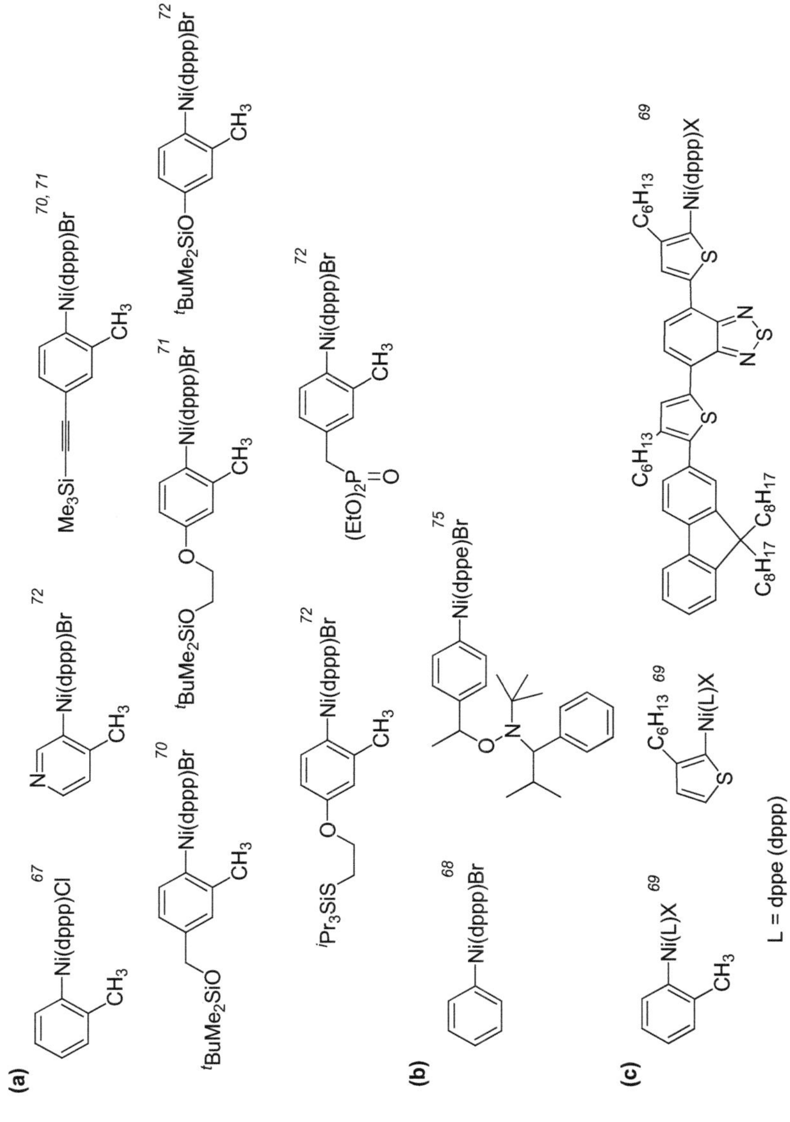

Figure 1.11 Range of Ni initiators available for Kumada–Tamao CTCP obtained by: (a) Luscombe method; (b) Kiriy method; and (c) convenient Kiriy method.

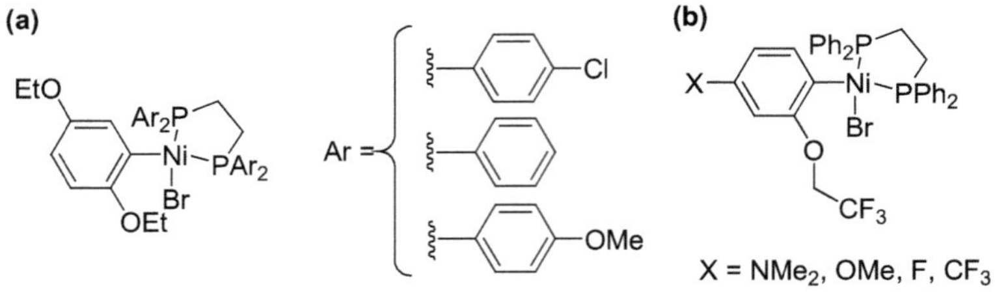

(a) **(b)**

EtO

Ar$_2$P

Ni–PAr$_2$

Br

OEt

Ar = Cl, OMe

Ph$_2$P

X–Ni–PPh$_2$

Br

O

CF$_3$

X = NMe$_2$, OMe, F, CF$_3$

Figure 1.12 (a) Effect of ligand electronic properties of the initiator. (b) Substituent effects of the aryl group on the Ni of the catalyst.

isolated. An electron-withdrawing Cl substituent on the phosphine ligand increased the rates of transmetallation and reductive elimination, and the molecular weight distribution became broader. However, the electron-donating MeO group made the transmetallation and reductive elimination slower and provided polymers with a narrower molecular weight distribution (Figure 1.12a). These results suggest that an electron-donating ligand stabilizes an intermediate Ni^0–polymer π-complex, suppressing competing side-reactions. With respect to the substituent effects in the aryl group on the Ni of the initiators, electron-donating Me$_2$N-substituted aryl–Ni–Br underwent the fastest initiation, at a rate similar to that of propagation, resulting in the narrowest molecular weight distribution (Figure 1.12b). Model reactions confirmed that fast initiation was due to fast reductive elimination.

1.2.6 Functionalization of Polymer Ends

Even when Kumada–Tamao CTCP is carried out with a catalyst, not an initiator, functional groups can be introduced onto one or both ends of the π-conjugated polymers by the addition of Grignard reagents to the polymerization mixture.[78] Allyl, ethynyl, and vinyl Grignard reagents react at one end, whereas aryl and alkyl Grignard reagents react at both ends. Exceptionally, the dimethylsilylaminophenyl Grignard reagent afforded a monofunctional product.[78,79] End-group functionality was 80–99%. 2-(4-Bromobutyl)-5-thienylmagnesium chloride was used for the introduction of a terminal bromoalkyl group on both ends of P3HT.[80] Many functional groups have been introduced by post-polymerization modification.[81]

1.3 Suzuki–Miyaura Coupling Polymerization of Boronic Acid (Ester) Monomer

1.3.1 Background and Discovery

We aimed to extend the range of CTCP to include not only Kumada–Tamao coupling polymerization with a Ni catalyst, but also on Suzuki–Miyaura coupling polymerization with a Pd catalyst. We first searched for a Pd catalyst with a high propensity for intramolecular transfer by using

Suzuki–Miyaura coupling reaction of dibromophenylene with phenylboronic acid ester as a model reaction (Figure 1.13a).[82,83] If the catalyst undergoes intramolecular transfer after the first substitution, then disubstituted phenylene is preferentially obtained; this would be similar to the exclusive formation of disubstituted thiophene in the Kumada–Tamao coupling reaction of dibromothiophene with tolylmagnesium bromide as a model reaction of Ni CTCP (Figure 1.5). During this work, it was reported that *m*- or *p*-iodobenzene underwent selective double coupling reactions with aryl-boronic acids and esters, whereas the dibromo counterparts underwent single couplings with high selectivity when Pd(PPh$_3$)$_4$ was used.[84] However, the mechanism of catalyst-transfer was not addressed. We found that Pd(PPh$_3$)$_4$ and Pd$_2$(dba)$_3$ (dba = dibenzylideneacetone)/bidentate phosphine ligands such as dppp and dppe gave disubstituted phenylene with 50–60% selectivity, whereas Pd$_2$(dba)$_3$/*t*-Bu$_3$P afforded the product with 93–96% select-ivity. When this catalyst system was used for the reaction of dibromofluorene with phenylboronic acid ester, the selectivity was increased to 99% (Figure 1.13b).[83,85] Dong and Hu[86] and Scherf and coworkers[87] independently conducted similar experiments and found that Pd$_2$(dba)$_3$/*t*-Bu$_3$P had a high propensity for intramolecular transfer in the Suzuki–Miyaura coupling reaction.

On the basis of our results for model reactions, we carried out polymer-ization of bromofluoreneboronic acid ester with a known reagent, ^{t}Bu$_3$PPd(Ph)Br, and isolated the complex,[88] which was expected to serve as an initiator (Figure 1.14a).[89] At that time, Ni initiators had not been utilized in Kumada–Tamao CTCP, and this Suzuki–Miyaura coupling polymerization was the first example of CTCP with an external initiator. The polymerization showed chain polymerization behavior and each polymer molecule had the phenyl group from the initiator at one end. The polymerization was carried out in the presence of a bromofluorene **8** (Figure 1.14b). If the polymer-ization involved intermolecular transfer of the Pd catalyst, then **8** would be consumed and the molecular weight of the obtained polymer should decrease, in contrast with the case of polymerization without **8**. The polymer-ization proceeded without the consumption of **8** to yield a polymer with almost the same molecular weight as in the polymerization without **8**. This result supported the intramolecular transfer of the *t*-Bu$_3$P-ligated Pd^0 catalyst. Kiriy and coworkers[90] applied this chemistry to the surface-initiated polymerization of the fluorene monomer by the formation of a ^{t}Bu$_3$PPd(Ph)Br complex from the bromophenyl group on the surface and (^{t}Bu$_3$P)$_2$Pd.

1.3.2 Monomers

Since the CTCP of fluorene was found, Suzuki–Miyaura CTCP of thio-phene,[91–93] phenylene,[94] phenanthrene,[52] and fluorene-benzothiadiazole[95] monomers has been reported (Figure 1.15). The dispersity of the obtained polymers is greater ($M_w/M_n = 1.3$) than that of polymers obtained by Kumada–Tamao CTCP. Different polymer end-groups are easily introduced

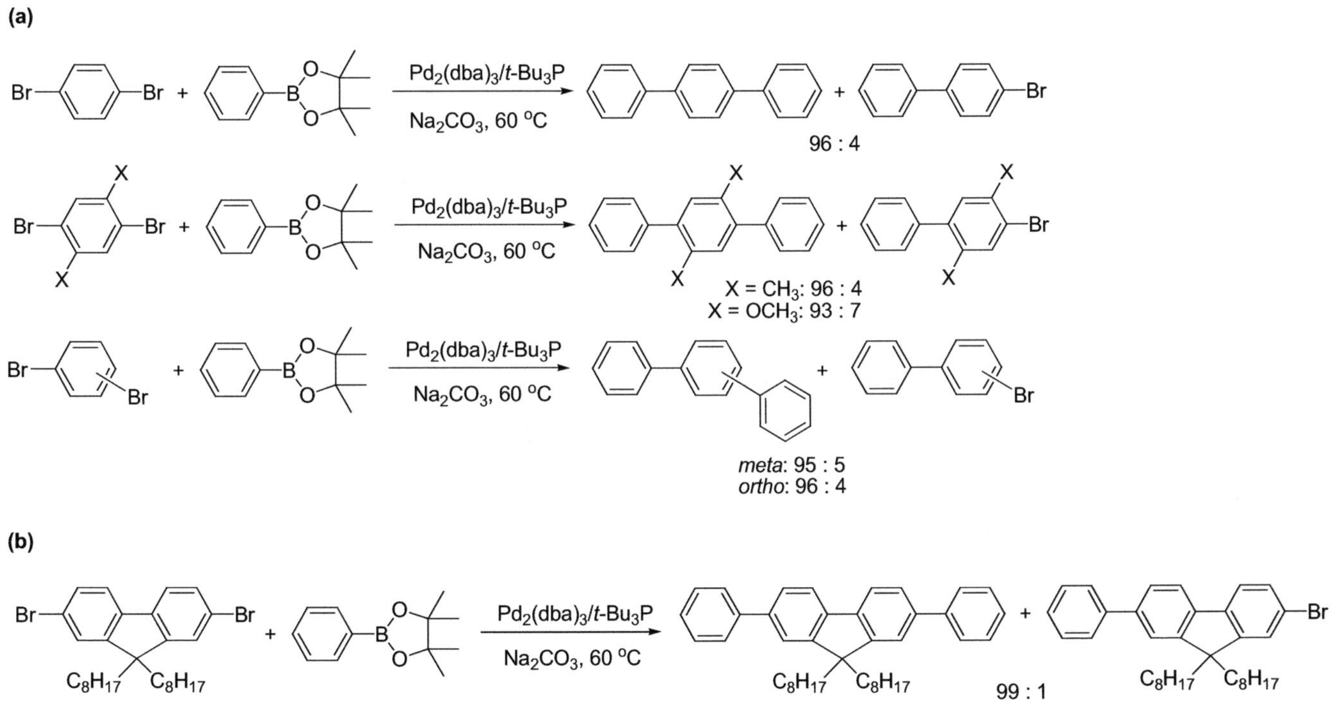

Figure 1.13 Model reactions of Suzuki–Miyaura CTCP for (a) polyphenylene and (b) polyfluorene.

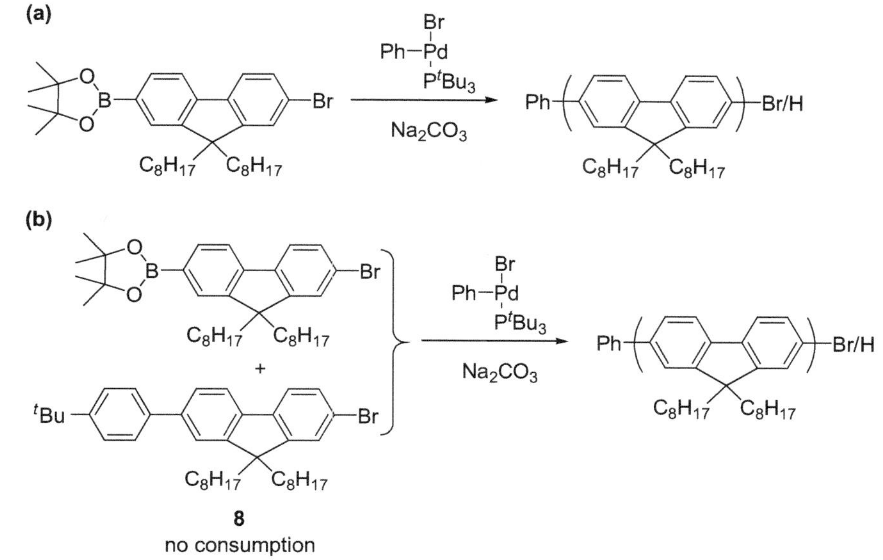

(a)

(b)

8
no consumption

Figure 1.14 (a) Suzuki–Miyaura CTCP for polyfluorene. (b) Suzuki–Miyaura CTCP in the presence of bromofluorene **8**.

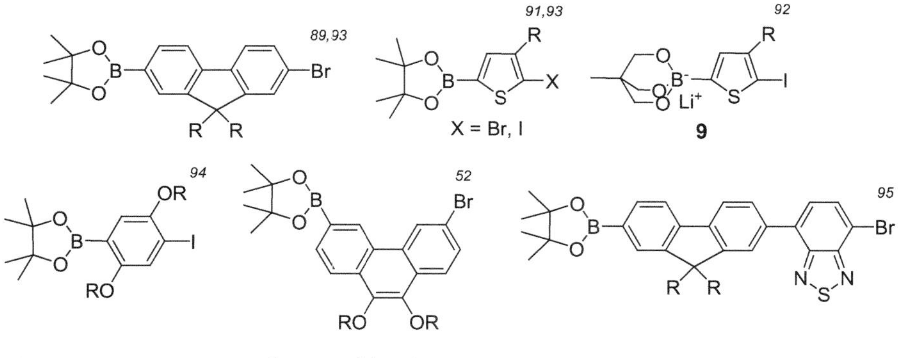

Figure 1.15 Monomers for Suzuki–Miyaura CTCP.

into the polymer by initiation with Ar–Pd–X complexes and termination can be performed with many kinds of boronic acids or esters (see Section 1.3.4).

We attempted the condensative chain polymerization of the triolborate monomer **9**,[92] which can be polymerized in dry solvent without the addition of a base, because deboronation and dehalogenation are caused by water and a base in the reaction mixture. However, the P3HT obtained had a broad molecular weight distribution and many kinds of polymer ends, whereas polymerization in THF/water yielded P3HT with a narrow molecular weight distribution and controlled polymer ends. This surprising result indicated that a small amount of water is necessary for the intramolecular transfer of the Pd catalyst in Suzuki–Miyaura CTCP.

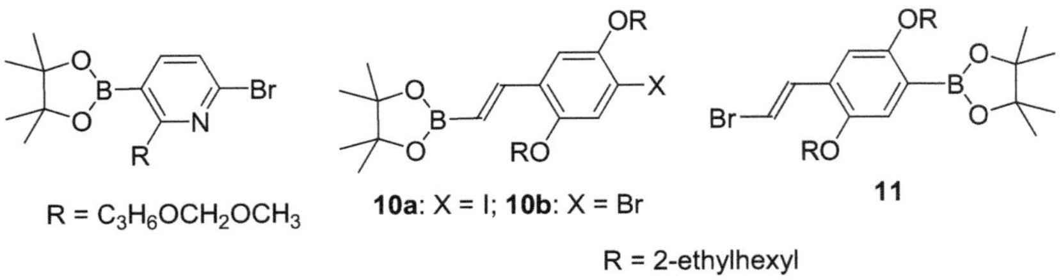

R = C$_3$H$_6$OCH$_2$OCH$_3$ **10a**: X = I; **10b**: X = Br   **11**

R = 2-ethylhexyl

Figure 1.16 Unfavorable monomers for Suzuki–Miyaura CTCP.

Figure 1.16 summarizes monomers that have been reported to be unfavorable. The pyridine monomer underwent disproportionation for the same reason as in the case of the Grignard counterpart mentioned in Section 1.2.3.[49] The polymerization of the phenylenevinylene monomer **10a** with ^{t}Bu$_3$PPd(Ph)Br in the presence of KOH/18-crown-6 proceeded quickly, even at 0 °C. The polymer obtained had a high molecular weight, in contrast with the result of the Kumada–Tamao coupling polymerization of the Grignard counterpart (Figure 1.8), but the molecular weight distribution was broad. Many kinds of end-groups – such as Ph/I, Ph/H, Bpin/I, and Bpin/H – were observed in the initial stage, but the end-groups uniformly became Ph/H in the final stage, implying that the catalyst underwent intermolecular transfer to the monomer and that dehalogenation took place as a side-reaction. To suppress the dehalogenation, polymerization of the bromide monomer **10b** was attempted, but the same result was observed. Monomer **11**, in which the boronate moiety and bromine in **10a** were exchanged, also gave a high molecular weight polymer, but the polymerization did not proceed *via* the CTCP mechanism.[51]

We then investigated why the CTCP of monomers containing a C=C bond did not occur by examining model reactions of dibromostilbenes with an arylboronic acid ester in the presence and absence of additives containing C=C (Figure 1.17a).[96] It was found that Pd^0 on the C=C bond of the substrate was readily trapped by the C=C bond of another substrate or additive and that this intermolecular transfer of the catalyst could be suppressed by the introduction of alkoxy groups at the *ortho* positions to the C=C bond. Consequently, the failure of the Suzuki–Miyaura CTCP of **10** and **11** is probably due to the intermolecular transfer of Pd^0 to the C=C bond of another monomer (Figure 1.17b).

1.3.3 Initiators and Catalysts

Many kinds of ^{t}Bu$_3$PPd(Ar)Br initiators have been synthesized by the reaction of ArBr and (^{t}Bu$_3$P)$_2$Pd[88] and used for Suzuki–Miyaura CTCP (Figure 1.18).[49,97–99] A dimeric Pd^{II} complex with P(*o*-Tol)$_3$ was used for the CTCP of the fluorene monomer and a triarylamino group was introduced at one end.[100] Hu and coworkers[101,102] developed the *in situ* generation of ^{t}Bu$_3$PPd(Ar)Br initiators by the reaction of Pd$_2$(dba)$_3$, *t*-Bu$_3$P, and

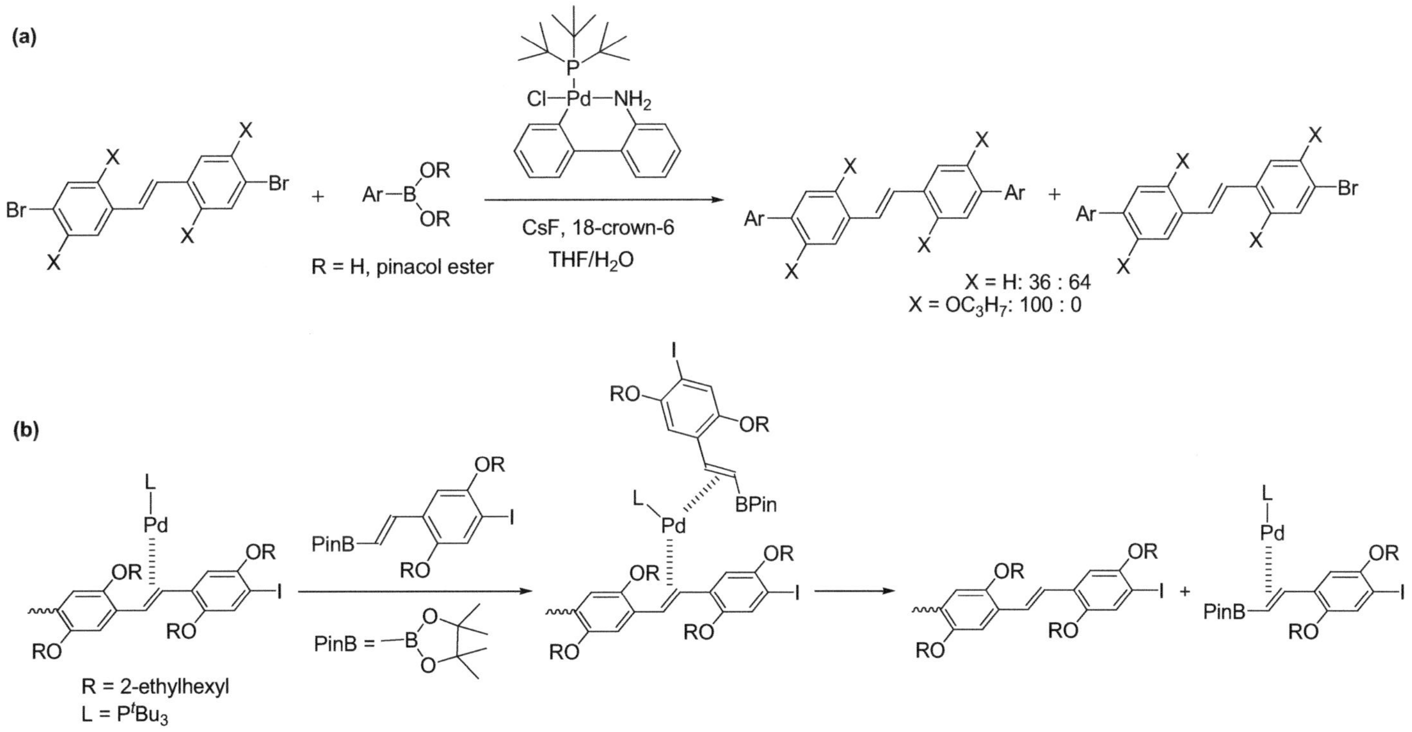

Figure 1.17 (a) Substituent effect of dibromostilbene on intramolecular transfer of Pd⁰. (b) Proposed mechanism of intermolecular transfer of Pd⁰ to phenylenevinylene boronic acid ester monomer.

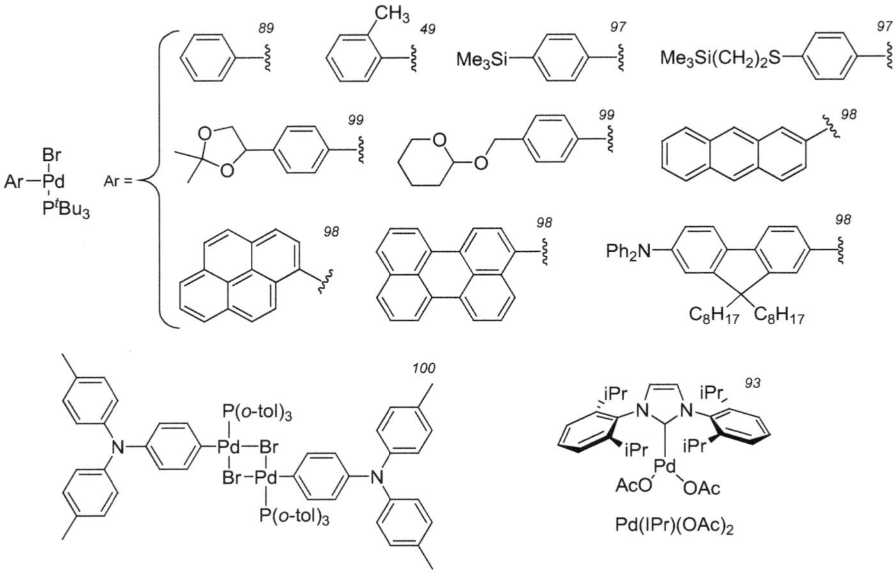

Figure 1.18 Pd initiators and catalyst for Suzuki–Miyaura CTCP.

ArX (X = I or Br). ArX with various substituents (Cl, Br, F, NO_2, CN, COPh, CO_2Et, OMe, CH_2OH) at the *para* position can be used. The *in situ* generated $^tBu_3PPd(Ar)Br$ initiators generally afforded polymer with a narrower molecular weight distribution than that obtained with the isolated Pd initiators.

A PdX_2 complex is not often used as a catalyst for the Suzuki–Miyaura coupling reaction because reduction to Pd^0 with boronic acid or ester is slow, in contrast with the fast reduction of NiX_2 to Ni^0 with a Grignard reagent in the Kumada–Tamao coupling reaction. However, Geng and coworkers[93] found that the NHC-ligated Pd complex $Pd(IPr)(OAc)_2$ was effective for the CTCP of fluorene and thiophene monomers. Chain polymerization behavior was observed, but the molecular weight distribution was rather broad ($M_w/M_n \approx 1.6$) due to the slow generation of dimer–Pd–Br initiator species.

1.3.4 Functionalization of Polymer Ends

Suzuki–Miyaura CTCP with Pd initiators (Figure 1.18), followed by end-capping with boronic acids or esters, can easily afford hetero end-functionalized π-conjugated polymers. Several examples are shown in Figure 1.19. Bao and coworkers[97] synthesized azide-functionalized polythiophene, which can be connected to DNA *via* a click reaction. Hu and coworkers carried out the CTCP of fluorene monomers with initiators generated *in situ* from $Pd_2(dba)_3$, t-Bu_3P and ArX and then used tolyl or phenylboronic acid for end-capping.[101,102] A variety of functional groups can

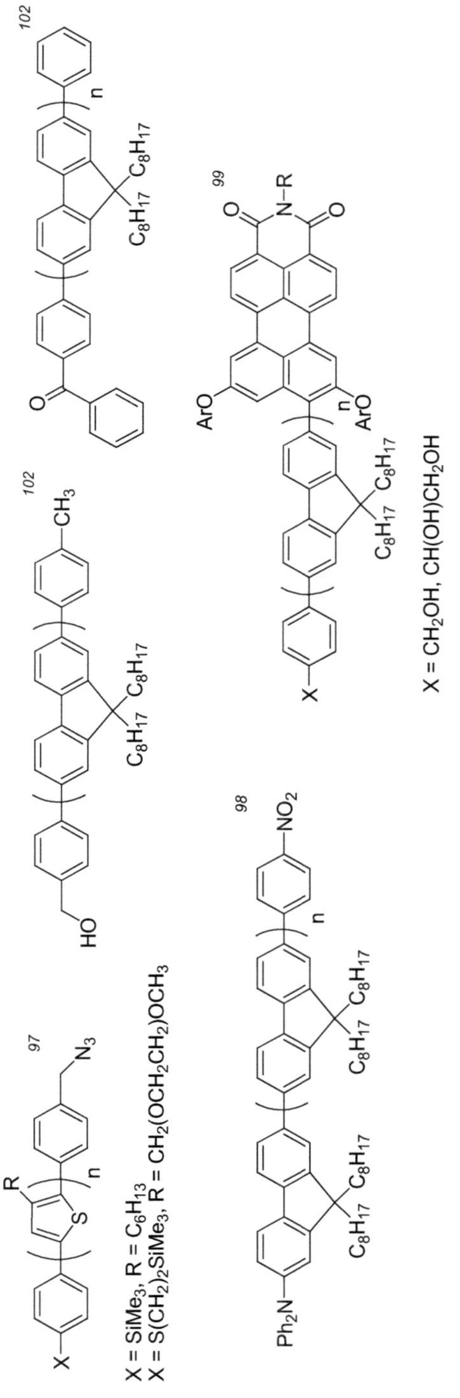

Figure 1.19 Hetero end-functionalized π-conjugated polymers.

be easily introduced to the initiator units in this method because the isolation of ^{t}Bu$_3$PPd(Ar)Br is sometimes difficult, depending on the substituent in the Ar group. Huck and coworkers[98] introduced a donor group at one end and an acceptor group at the other in polyfluorene using isolated ^{t}Bu$_3$PPd(Ar)Br initiators and end-capping boronate, and investigated charge transfer and energy transfer through the polyfluorene backbone. Mecking and coworkers[99] synthesized polyfluorene bearing one or two hydroxyl groups at one end and red-emitting perylenemonoimide at the other by using a hydroxyl-protected Pd initiator (Figure 1.18). The hydroxyl group was connected to poly(ethylene glycol) to yield an amphiphilic diblock co-polymer, which afforded stable micelles in water. These micelles exhibited a bright fluorescence emission with a quantum yield as high as $\Phi = 84\%$.

1.4 Other Coupling Polymerization Reactions

1.4.1 Negishi Coupling Polymerization of Zn-containing Monomers

Zinc-containing monomers undergo CTCP, although some polymerizations involve chain termination and do not show "living" polymerization behavior. Monomers and catalysts are shown in Figure 1.20. The first example of the CTCP of zinc-containing monomers was conducted by McCullough and coworkers[11] using a thiophene monomer and Ni(dppp)Cl$_2$, which showed "living" polymerization behavior. However, Koeckelberghs and coworkers[103] reported that polymerization of the same monomer with thienyl-Pd(Ruphos)Br proceeded through a catalyst diffusion mechanism, not *via* the CTCP mechanism, although "living" polymerization behavior was similarly observed. Thus Pd^0[Ruphos] is diffused from the propagating end, but selectively undergoes oxidative reinsertion into the C–Br bond at the end of a polymer chain, rather than the C–Br bond of the monomer, which is deactivated due to the electron-donating chlorozincio moiety (Figure 1.21). This mechanism was proposed on the basis of the fact that the molecular weight decreased with increasing amounts of bromothiophene, which has a non-activated C–Br bond. If the catalyst remains at the propagating chain end, the amount of bromothiophene should have no effect. As the polymerization of the reversed thiophene monomer with thienyl-Pd(Ruphos)Br proceeded according to the same mechanism, control of the regio-regularity of P3HT was achieved by the copolymerization of normal and reversed thiophene monomers.[104]

Higashihara and coworkers reported the polymerization of the zincate thiophene monomer, which was prepared by treatment of 2-bromo-3-hexyl-5-iodothiophene with ^{t}Bu$_4$ZnLi$_2$[105] and Ni(dppe)Cl$_2$ at 60 °C to afford well-defined P3HT.[106] An important feature of this polymerization is its tolerance to protonic impurities; as-received tetrahydrofuran can be used as the reaction solvent. Thiophene, having a hydroxyalkyl group, was directly polymerized by this method. The thienothiophene Grignard monomer was

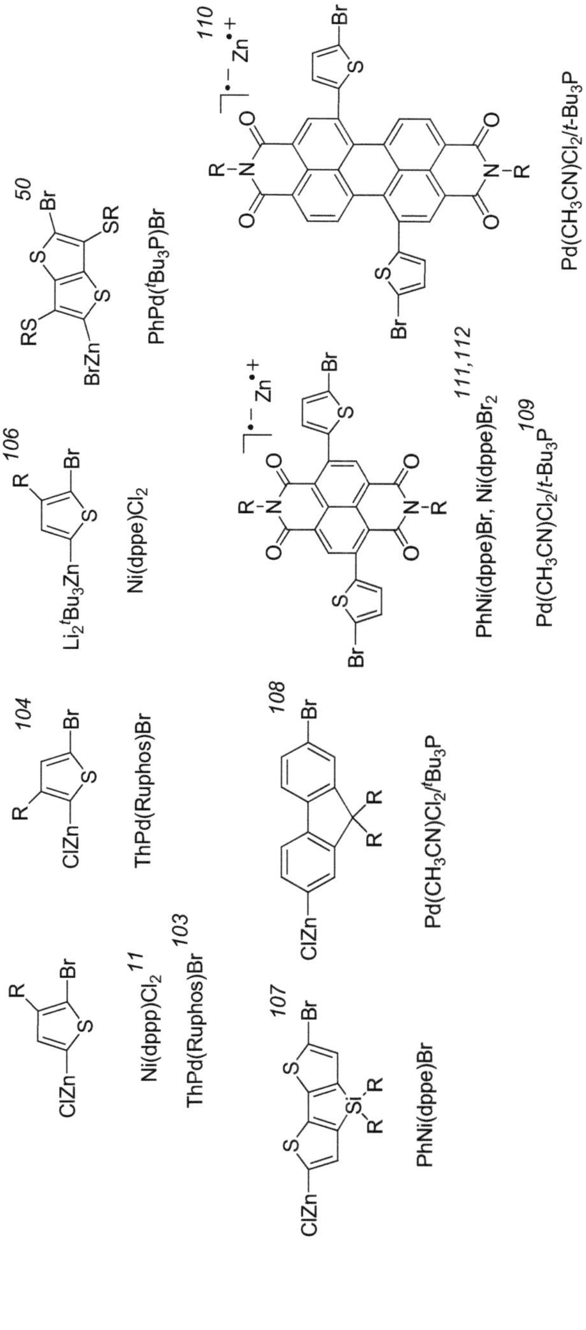

Figure 1.20 Zinc-containing monomers for CTCP.

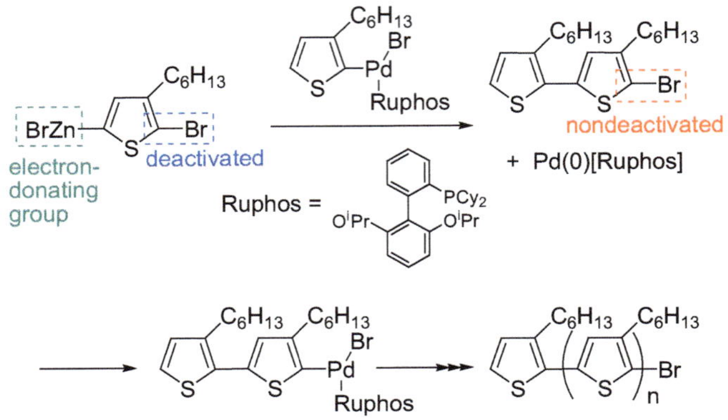

Figure 1.21 Negishi CTCP through the substituent effect.

not polymerized with an Ni catalyst (Figure 1.8), whereas the bromozinc counterpart was polymerized with a ^{t}Bu$_3$P-ligated Pd catalyst. The polymerization behavior indicated chain polymerization, but it did not show a "living" polymerization nature; termination may occur during polymerization.[50]

Kiriy and coworkers[107] investigated the polymerization of the four monomers in the second row of Figure 1.20. The polymerization of the dithienosilol Grignard monomer (Figure 1.7) with an Ni catalyst was first attempted and yielded a polymer with a broad molecular weight distribution and a lower molecular weight than that expected based on the feed ratio of the monomer to the catalyst. The conversion of dibromodithienosilol to the Grignard monomer with i-PrMgCl·LiCl did not exceed 90% and the remaining i-PrMgCl presumably terminated the polymerization. However, the zinc-containing monomer was prepared with a purity >99% by treatment of the dibromide with n-BuLi, followed by the addition of ZnCl$_2$. The polymerization proceeded in a "living" polymerization manner and the successive polymerization of the thiophene Grignard monomer and the dithienosilolzinc chloride monomer yielded a block copolymer. The polymerization of the fluorene monomer with Pd(CH$_3$CN)$_2$/t-Bu$_3$P proceeded very rapidly at room temperature, even with only a small amount of catalyst (turnover numbers >100 000; turnover frequencies ≤280 s^{-1}). However, the polymerization did not show a "living" polymerization nature; the polymerization proceeded *via* a chain polymerization mechanism accompanied by chain termination.[108] A similar behavior was observed in the polymerization of the naphthalenediimide-dithiophene monomer[109] and the perylenediimide-dithiophene monomer[110] with the same Pd catalyst.

In contrast, the naphthalenediimide-dithiophene monomer underwent chain polymerization in a "living" polymerization manner with the Ni catalyst.[111] This unusual anion radical monomer was generated by electron transfer from the activated Zn to the electron-deficient precursor.

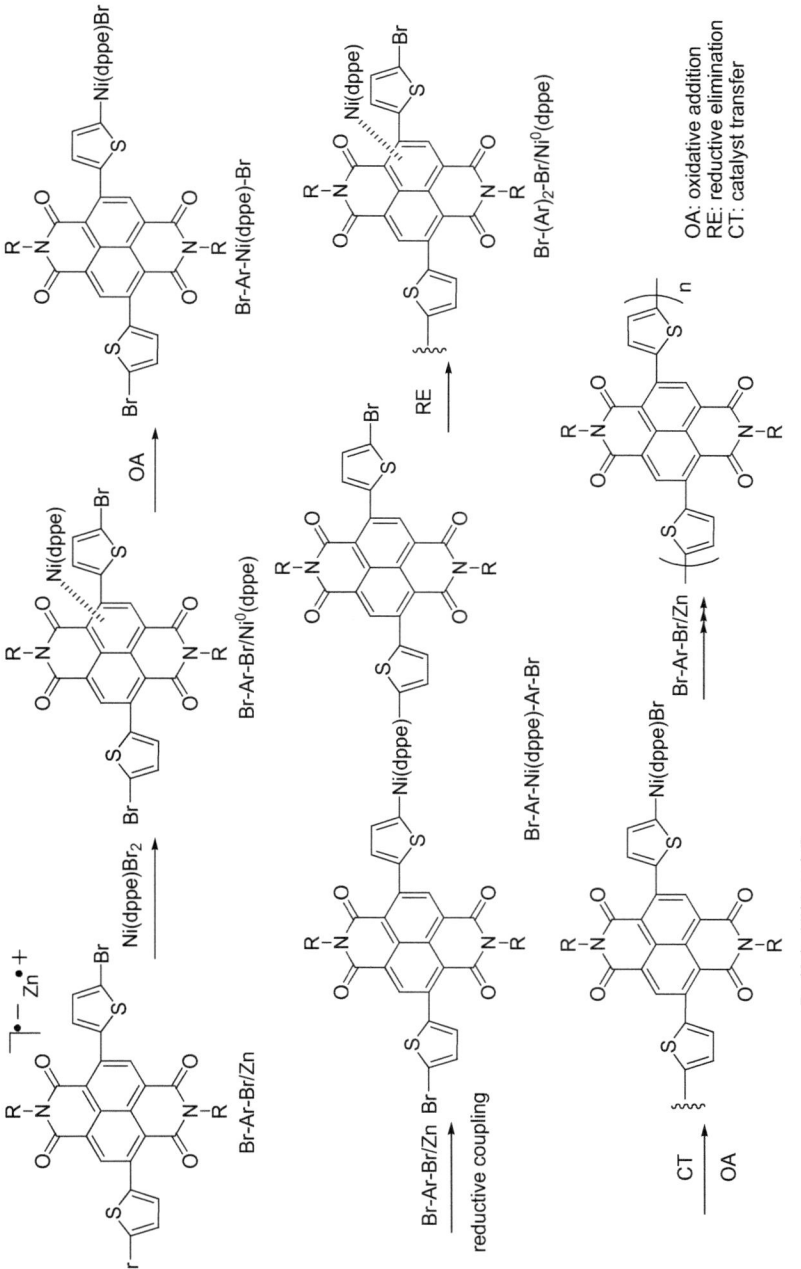

Figure 1.22 Proposed mechanism of CTCP of radical anion monomer with Ni(dppe)Br$_2$.

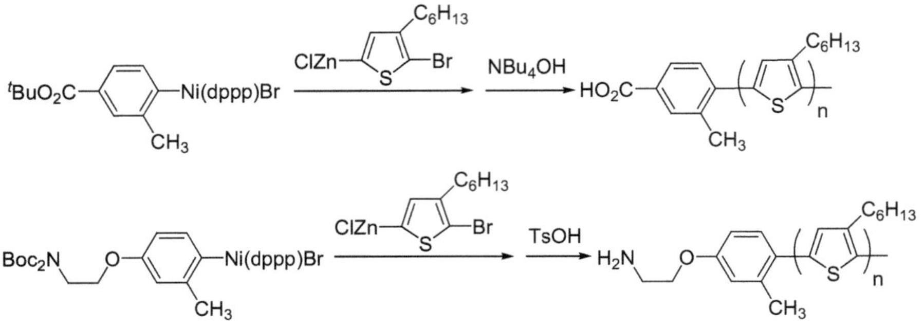

Figure 1.23   Initiators for Negishi CTCP.

The following mechanism has been proposed on the basis of NMR studies and model reactions (Figure 1.22).[112] When Ni(dppe)Br$_2$ is used as a catalyst, it is reduced by the anion radical monomer (Br–Ar–Br/Zn) in a two-electron transfer process to give the coordination complex of Br–Ar–Br and Ni^0, which rearranges *via* oxidative addition into Br–Ar–Ni(dppe)–Br. The latter species acts as an initiator. Chain propagation starts with a single-electron redox process between the chain end, (Ar)$_n$–Ni(dppe)–Br (or the initiator Br–Ar–Ni(dppe)–Br), and the anion radical monomer. Presumably, reduction with Zn$^{\bullet+}$ leads to an Ni^I species, (Ar)$_n$–Ni(dppe), which oxidatively adds to the C–Br bond of the anion radical monomer, followed by the elimination of the bromide anion, leading to (Ar)$_n$–Ni(dppe)–Ar–Br. This species undergoes reductive elimination followed by intramolecular oxidative addition to afford (Ar)$_{n+1}$–Ni(dppe)–Br.

Koeckelberghs and coworkers[71] synthesized external initiators for the polymerization of a zinc-containing thiophene monomer to yield P3HT with a carboxyl group and an amino group (Figure 1.23).

1.4.2   Stille Coupling Polymerization of Tin-containing Monomers

Stille coupling polymerization is widely used for the preparation of donor–acceptor alternating π-conjugated polymers by the polycondensation of AA and BB monomers. However, there are few examples of Stille CTCP. It is rather surprising that the Stille CTCP of P3HT was only reported in 2015. Noonan and coworkers[113] conducted polymerization of the Me$_3$Sn-containing thiophene monomer with PEPPSI-IPr (Figure 1.9) at 40 °C (Figure 1.24a). The polymerization showed a "living" chain polymerization nature. The obtained polymer had a tail-to-tail diad inside the backbone, indicating that the polymerization mainly proceeded *via* bidirectional growth (see Section 1.2.2).

Three years earlier, Bielawski and coworkers[114] had investigated Stille CTCP for the synthesis of well-defined poly(phenylene ethynylene) (PPE). They first attempted Sonogashira coupling polymerization of the non-stannylated monomer. The yield of polymer was not high, although PPE with

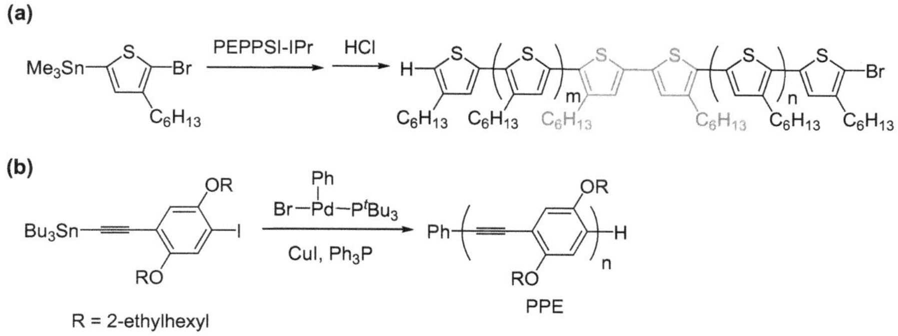

(a)

(b)

R = 2-ethylhexyl

Figure 1.24 Stille CTCP for the synthesis of (a) P3HT and (b) PPE.

low dispersity was obtained. They studied the polymerization conditions and found that the stannylated monomer quantitatively underwent CTCP with ^{t}Bu$_3$PPd(Ph)Br in the presence of PPh$_3$ and CuI (Figure 1.24b). Taking advantage of the "living" polymerization nature, they obtained a block co-polymer of PPE and poly(fluorenyl ethynylene), as well as PPE brushes from silica nanoparticles.

1.4.3 Murahashi Coupling Polymerization of Aryllithium Monomers

Aryllithium can be formed faster than arylmagnesium halides by deprotonation or a halogen–lithium exchange reaction with the alkyllithium reagent. Mori and coworkers[115] investigated the Murahashi CTCP of aryl-lithium monomers, which were generated by three methods: deprotonation, a halogen–lithium exchange reaction, and a halogen dance mechanism[116] (Figure 1.25). Chlorothiophene was selectively deprotonated with *n*-BuLi, although bromothiophene underwent a bromo–lithium exchange reaction. The thienyllithium monomer was polymerized with NiCl$_2$(PPh$_3$)IPr in cyclopentyl methyl ether at 0 °C for 10 min to afford P3HT with M_w/M_n of 1.35 (Figure 1.25a). Polyphenylene was obtained from the corresponding phenylenelithium monomer generated by a halogen–lithium exchange reaction (Figure 1.25b). These polymerizations showed a "living" polymerization nature and therefore chain extension of the thienyllithium monomer from polyphenylene was performed to obtain a block copolymer. When 2,5-dibromo-3-hexylthiophene was treated with lithium diisopropylamide at −78 °C to 0 °C, a halogen dance mechanism took place to generate 2,4-dibromo-3-hexyl-5-lithiothiophene, which was successively polymerized with NiCl$_2$(PPh$_3$)IPr to afford a bromine-containing polythiophene (Figure 1.25c). It should be noted that 2,5-dibromo-3-hexylthiophene is often used for the CTCP of Grignard monomers generated by *i*-PrMgCl or *t*-BuMgCl, yielding P3HT containing no bromine.

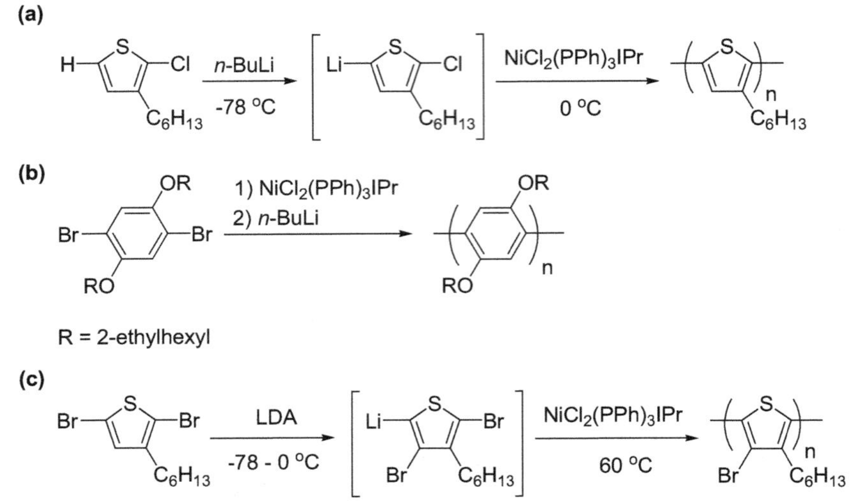

(a)

(b)

R = 2-ethylhexyl

(c)

Figure 1.25 Murahashi CTCP of aryllithium monomers generated by (a) proton abstraction, (b) halogen–lithium exchange, and (c) a halogen dance mechanism.

1.4.4 Mizoroki–Heck Coupling Polymerization of Non-metallated Monomers

We attempted Mizoroki–Heck CTCP of p-iodostyrene with ^{t}Bu$_3$PPd(tolyl)Br for the synthesis of well-defined poly(phenylenevinylene) (Figure 1.26)[117] because ^{t}Bu$_3$PPd^0 was reported to promote the Mizoroki–Heck coupling reaction even at room temperature.[118] The molecular weight was not increased until monomer conversion reached about 90%, and then sharply increased in the final stage. The molecular weight distribution was broad. These results indicated that the polymerization proceeded in a polycondensation manner. The polymer end-groups were H at one end and I at the other until the middle stage; the polymerization was not initiated from the tolyl group of the Pd initiator. In the final stage, the polymer end-groups were converted into tolyl and hydrogen, probably because the polymer with H/I ends reacted would react with the Pd initiator and the terminal polymer–Pd–I was hydrolyzed to afford an H end-group. The occurrence of a polycondensation is accounted for by the intermolecular transfer of the Pd catalyst induced by the C=C bond of the monomer, in a similar manner to the Suzuki–Miyaura coupling polymerization of the phenylenevinyleneboronic acid ester shown in Figure 1.17.

1.4.5 Nucleophilic Substitution Polymerization of Silylated Monomers

Iyoda and coworkers[119,120] reported the chain polymerization of monomers containing pentafluorophenyl and trimethylsilyl groups with a fluoride ion

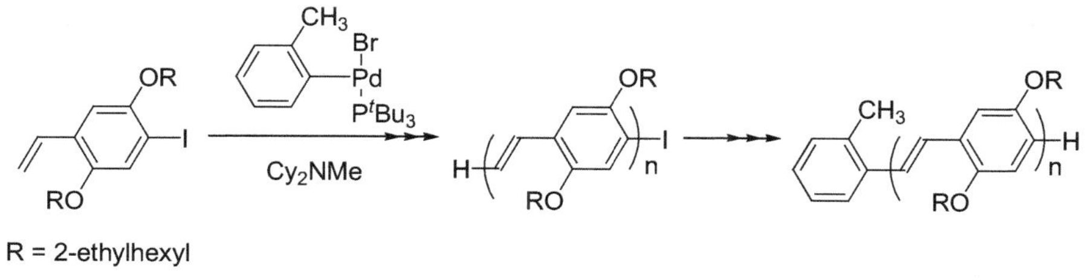

Figure 1.26 Attempted Mizoroki–Heck CTCP for poly(phenylenevinylene).

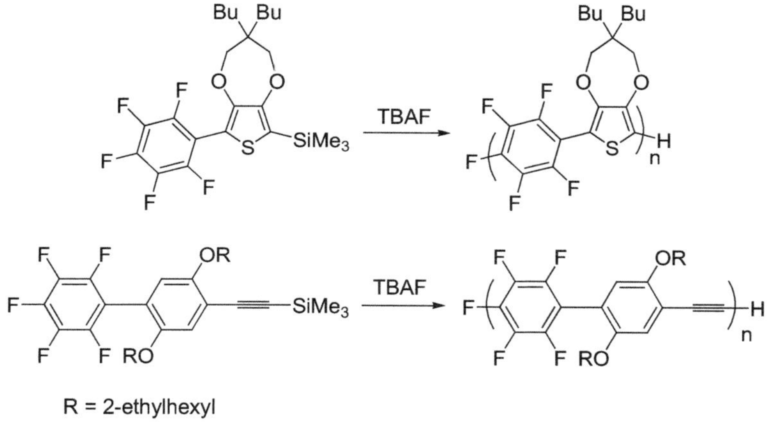

R = 2-ethylhexyl

Figure 1.27 Condensative chain polymerization based on nucleophilic substitution of silylated monomers.

catalyst on the basis of aromatic nucleophilic substitution (Figure 1.27). Tetrabutylammonium fluoride was the best source of fluoride ions and the molecular weight was controlled by the feed ratio of the monomer to tetrabutylammonium fluoride. The proposed mechanism of chain polymerization was due to the intramolecular transfer of fluoride ions based on anion–π interactions or a change in the substituent effect.

1.5 Conclusions

CTCP has been extensively investigated since the Kumada–Tamao coupling polymerization of the bromothiophene Grignard monomer with Ni(dppp)Cl$_2$ was discovered to proceed in a "living" chain polymerization manner as a result of catalyst transfer on the conjugated backbone. Many monomers have been found to undergo Kumada–Tamao CTCP with Ni(dppp)Cl$_2$ or Ni(dppe)Cl$_2$. Catalysts for Kumada–Tamao CTCP have been also developed from bidentate phosphine-ligated Ni catalysts to NHC- and diimine-ligated Ni catalysts; the NHC catalysts enable the synthesis of high molecular weight P3HT in a controlled manner and the diimine catalysts yield well-defined, acceptor polybenzotriazole. Functional chain ends can be introduced by using external Ni initiators and/or by the addition of Grignard end-capping reagents to the polymerization mixture.

CTCP has been extended to Suzuki–Miyaura coupling polymerization with ^{t}Bu$_3$P-ligated Pd initiators. However, there are fewer monomers available than for Kumada–Tamao CTCP. Acceptor monomers with functional groups that are not available for Grignard reagents should be further examined, because the formation of boronic acids or esters and the Suzuki–Miyaura coupling reaction can occur under less basic conditions. Pd initiators other than ^{t}Bu$_3$P-ligated initiators should also be further explored. The CTCP of zinc-containing monomers has been extended from donor monomers to acceptor monomers. The suppression of chain termination is still to be solved for the polymerization of some monomers. The Stille CTCP of Sn-containing monomers and Murahashi CTCP of aryllithium monomers are at an early stage, and are likely to be further extended in the future.

Acknowledgements

This study was supported by a Grant in Aid (No. 24550141) for Scientific Research from the Japan Society for the Promotion of Science (JSPS) and by the MEXT-Supported Program for the Strategic Research Foundation at Private Universities, 2013–2018.

References

1. *Handbook of Conducting Polymers*, ed. T. A. Skotheim, R. L. Elsenbaumer and J. R. Reynolds, 2nd edn, Revised and Expanded, 1997.
2. T. Yamamoto, *Bull. Chem. Soc. Jpn.*, 2010, **83**, 431.
3. T. Yamamoto and A. Yamamoto, *Chem. Lett.*, 1977, **6**, 353.
4. T. Yamamoto, K. Osakada, T. Wakabayashi and A. Yamamoto, *Makromol. Chem., Rapid Commun.*, 1985, **6**, 671.
5. K. Sanechika, T. Yamamoto and A. Yamamoto, *Bull. Chem. Soc. Jpn.*, 1984, **57**, 752.
6. T. Yamamoto, T. Ito and K. Kubota, *Chem. Lett.*, 1988, **17**, 153.
7. M. Rehahn, A.-D. Schlüter, G. Wegner and W. J. Feast, *Polymer*, 1989, **30**, 1060.
8. M. Bochmann and J. Lu, *J. Polym. Sci., Part A: Polym. Chem.*, 1994, **32**, 2493.
9. A. Yokoyama, R. Miyakoshi and T. Yokozawa, *Macromolecules*, 2004, **37**, 1169.
10. M. C. Iovu, E. E. Sheina, R. R. Gil and R. D. McCullough, *Macromolecules*, 2005, **38**, 8649.
11. E. E. Sheina, J. S. Liu, M. C. Iovu, D. W. Laird and R. D. McCullough, *Macromolecules*, 2004, **37**, 3526.
12. T. Yokozawa and A. Yokoyama, *Chem. Rev.*, 2009, **109**, 5595.

13. I. Osaka and R. D. McCullough, *Acc. Chem. Res.*, 2008, **41**, 1202.
14. T. Yokozawa and Y. Ohta, *Chem. Commun.*, 2013, **49**, 8281.
15. T. Yokozawa, T. Asai, R. Sugi, S. Ishigooka and S. Hiraoka, *J. Am. Chem. Soc.*, 2000, **122**, 8313.
16. A. Schoenberg, I. Bartoletti and R. F. Heck, *J. Org. Chem.*, 1974, **39**, 3318.
17. A. Schoenberg and R. F. Heck, *J. Org. Chem.*, 1974, **39**, 3327.
18. A. Jutand and A. Mosleh, *Organometallics*, 1995, **14**, 1810.
19. T. Yokozawa and H. Shimura, *J. Polym. Sci., Part A: Polym. Chem.*, 1999, **37**, 2607.
20. R. Miyakoshi, A. Yokoyama and T. Yokozawa, *Macromol. Rapid Commun.*, 2004, **25**, 1663.
21. R. Miyakoshi, A. Yokoyama and T. Yokozawa, *J. Am. Chem. Soc.*, 2005, **127**, 17542.
22. O. V. Zenkina, A. Karton, D. Freeman, L. J. W. Shimon, J. M. L. Martin and M. E. van der Boom, *Inorg. Chem.*, 2008, **47**, 5114.
23. N. Yoshikai, H. Matsuda and E. Nakamura, *J. Am. Chem. Soc.*, 2008, **130**, 15258.
24. M. Verswyvel, F. Monnaie and G. Koeckelberghs, *Macromolecules*, 2011, **44**, 9489.
25. R. Tkachov, V. Senkovskyy, H. Komber, J.-U. Sommer and A. Kiriy, *J. Am. Chem. Soc.*, 2010, **132**, 7803.
26. P. Kohn, S. Huettner, H. Komber, V. Senkovskyy, R. Tkachov, A. Kiriy, R. H. Friend, U. Steiner, W. T. S. Huck, J.-U. Sommer and M. Sommer, *J. Am. Chem. Soc.*, 2012, **134**, 4790.
27. E. L. Lanni and A. J. McNeil, *Macromolecules*, 2010, **43**, 8039.
28. E. L. Lanni and A. J. McNeil, *J. Am. Chem. Soc.*, 2009, **131**, 16573.
29. Z. J. Bryan and A. J. McNeil, *Chem. Sci.*, 2013, **4**, 1620.
30. A. Mori, K. Ide, S. Tamba, S. Tsuji, Y. Toyomori and T. Yasuda, *Chem. Lett.*, 2014, **43**, 640.
31. C. Pan, K. Sugiyasu, J. Aimi, A. Sato and M. Takeuchi, *Angew. Chem., Int. Ed.*, 2014, **53**, 8870.
32. S. Wu, L. Huang, H. Tian, Y. Geng and F. Wang, *Macromolecules*, 2011, **44**, 7558.
33. J. P. Lamps and J. M. Catala, *Macromolecules*, 2011, **44**, 7962.
34. J. Hollinger, A. A. Jahnke, N. Coombs and D. S. Seferos, *J. Am. Chem. Soc.*, 2010, **132**, 8546.
35. J. Hollinger and D. S. Seferos, *Macromolecules*, 2014, **47**, 5002.
36. A. Yokoyama, A. Kato, R. Miyakoshi and T. Yokozawa, *Macromolecules*, 2008, **41**, 7271.
37. R. Miyakoshi, K. Shimono, A. Yokoyama and T. Yokozawa, *J. Am. Chem. Soc.*, 2006, **128**, 16012.
38. K. Ohshimizu, A. Takahashi, T. Higashihara and M. Ueda, *J. Polym. Sci., Part A: Polym. Chem.*, 2011, **49**, 2709.
39. A. Sui, X. Shi, S. Wu, H. Tian, Y. Geng and F. Wang, *Macromolecules*, 2012, **45**, 5436.

40. P. Willot, S. Govaerts and G. Koeckelberghs, *Macromolecules*, 2013, **46**, 8888.
41. F. Boon, N. Hergue, G. Deshayes, D. Moerman, S. Desbief, J. De Winter, P. Gerbaux, Y. H. Geerts, R. Lazzaroni and P. Dubois, *Polym. Chem.*, 2013, **4**, 4303.
42. S. Wu, Y. Sun, L. Huang, J. Wang, Y. Zhou, Y. Geng and F. Wang, *Macromolecules*, 2010, **43**, 4438.
43. Y. Nanashima, A. Yokoyama and T. Yokozawa, *Macromolecules*, 2012, **45**, 2609.
44. C. R. Bridges, T. M. McCormick, G. L. Gibson, J. Hollinger and D. S. Seferos, *J. Am. Chem. Soc.*, 2013, **135**, 13212.
45. F. Pammer and U. Passlack, *ACS Macro Lett.*, 2014, **3**, 170.
46. F. Pammer, J. Jäger, B. Rudolf and Y. Sun, *Macromolecules*, 2014, **47**, 5904.
47. R. J. Ono, S. Kang and C. W. Bielawski, *Macromolecules*, 2012, **45**, 2321.
48. A. D. Todd and C. W. Bielawski, *ACS Macro Lett.*, 2015, **4**, 1254.
49. Y. Nanashima, R. Shibata, R. Miyakoshi, A. Yokoyama and T. Yokozawa, *J. Polym. Sci., Part A: Polym. Chem.*, 2012, **50**, 3628.
50. P. Willot and G. Koeckelberghs, *Macromolecules*, 2014, **47**, 8548.
51. M. Nojima, Y. Ohta and T. Yokozawa, *J. Polym. Sci., Part A: Polym. Chem.*, 2014, **52**, 2643.
52. M. Verswyvel, C. Hoebers, J. De Winter, P. Gerbaux and G. Koeckelberghs, *J. Polym. Sci., Part A: Polym. Chem.*, 2013, **51**, 5067.
53. R. H. Lohwasser and M. Thelakkat, *Macromolecules*, 2011, **44**, 3388.
54. S. Tamba, K. Shono, A. Sugie and A. Mori, *J. Am. Chem. Soc.*, 2011, **133**, 9700.
55. L. Huang, S. P. Wu, Y. Qu, Y. H. Geng and F. S. Wang, *Macromolecules*, 2008, **41**, 8944.
56. A. E. Javier, S. R. Varshney and R. D. McCullough, *Macromolecules*, 2010, **43**, 3233.
57. I. Adachi, R. Miyakoshi, A. Yokoyama and T. Yokozawa, *Macromolecules*, 2006, **39**, 7793.
58. T. Yokozawa, I. Adachi, R. Miyakoshi and A. Yokoyama, *High Perform. Polym.*, 2007, **19**, 684.
59. E. L. Lanni, J. R. Locke, C. M. Gleave and A. J. McNeil, *Macromolecules*, 2011, **44**, 5136.
60. Z. J. Bryan, M. L. Smith and A. J. McNeil, *Macromol. Rapid Commun.*, 2012, **33**, 842.
61. S. Tamba, K. Fuji, H. Meguro, S. Okamoto, T. Tendo, R. Komobuchi, A. Sugie, T. Nishino and A. Mori, *Chem. Lett.*, 2013, **42**, 281.
62. X. Shi, A. Sui, Y. Wang, Y. Li, Y. Geng and F. Wang, *Chem. Commun.*, 2015, **51**, 2138.
63. S. D. Ittel, L. K. Johnson and M. Brookhart, *Chem. Rev.*, 2000, **100**, 1169.
64. G. J. Domski, J. M. Rose, G. W. Coates, A. D. Bolig and M. Brookhart, *Prog. Polym. Sci.*, 2007, **32**, 30.

65. H. D. Magurudeniya, P. Sista, J. K. Westbrook, T. E. Ourso, K. Nguyen, M. C. Maher, M. G. Alemseghed, M. C. Biewer and M. C. Stefan, *Macromol. Rapid Commun.*, 2011, **32**, 1748.
66. C. R. Bridges, H. Yan, A. A. Pollit and D. S. Seferos, *ACS Macro Lett.*, 2014, **3**, 671.
67. H. A. Bronstein and C. K. Luscombe, *J. Am. Chem. Soc.*, 2009, **131**, 12894.
68. V. Senkovskyy, R. Tkachov, T. Beryozkina, H. Komber, U. Oertel, M. Horecha, V. Bocharova, M. Stamm, S. A. Gevorgyan, F. C. Krebs and A. Kiriy, *J. Am. Chem. Soc.*, 2009, **131**, 16445.
69. V. Senkovskyy, M. Sommer, R. Tkachov, H. Komber, W. T. S. Huck and A. Kiriy, *Macromolecules*, 2010, **43**, 10157.
70. A. Smeets, K. Van den Bergh, J. De Winter, P. Gerbaux, T. Verbiest and G. Koeckelberghs, *Macromolecules*, 2009, **42**, 7638.
71. A. Smeets, P. Willot, J. De Winter, P. Gerbaux, T. Verbiest and G. Koeckelberghs, *Macromolecules*, 2011, **44**, 6017.
72. F. Monnaie, W. Brullot, T. Verbiest, J. De Winter, P. Gerbaux, A. Smeets and G. Koeckelberghs, *Macromolecules*, 2013, **46**, 8500.
73. M. Yuan, K. Okamoto, H. A. Bronstein and C. K. Luscombe, *ACS Macro Lett.*, 2012, **1**, 392.
74. J. Wang, C. Lu, T. Mizobe, M. Ueda, W.-C. Chen and T. Higashihara, *Macromolecules*, 2013, **46**, 1783.
75. E. Kaul, V. Senkovskyy, R. Tkachov, V. Bocharova, H. Komber, M. Stamm and A. Kiriy, *Macromolecules*, 2010, **43**, 77.
76. S. R. Lee, Z. J. Bryan, A. M. Wagner and A. J. McNeil, *Chem. Sci.*, 2012, **3**, 1562.
77. S. R. Lee, J. W. G. Bloom, S. E. Wheeler and A. J. McNeil, *Dalton Trans.*, 2013, **42**, 4218.
78. M. Jeffries-El, G. Sauve and R. D. McCullough, *Macromolecules*, 2005, **38**, 10346.
79. A. Takahashi, Y. Rho, T. Higashihara, B. Ahn, M. Ree and M. Ueda, *Macromolecules*, 2010, **43**, 4843.
80. H. Fujita, T. Michinobu, M. Tokita, M. Ueda and T. Higashihara, *Macromolecules*, 2012, **45**, 9643.
81. N. V. Handa, A. V. Serrano, M. J. Robb and C. J. Hawker, *J. Polym. Sci., Part A: Polym. Chem.*, 2015, **53**, 831.
82. H. Suzuki, Bachelor Thesis, Kanagawa University, 2005.
83. T. Yokozawa and H. Higashimura, *Japan Pat.*, Kokai 2005-038853, 2005.
84. D. J. Sinclair and M. S. Sherburn, *J. Org. Chem.*, 2005, **70**, 3730.
85. T. Yokozawa, K. Okuchi, Y. Kubota and H. Higashimura, *Japan Pat.*, Kohkai 2006-039065, 2006.
86. C.-G. Dong and Q.-S. Hu, *J. Am. Chem. Soc.*, 2005, **127**, 10006.
87. S. K. Weber, F. Galbrecht and U. Scherf, *Org. Lett.*, 2006, **8**, 4039.
88. J. P. Stambuli, C. D. Incarvito, M. Buhl and J. F. Hartwig, *J. Am. Chem. Soc.*, 2004, **126**, 1184.

89. A. Yokoyama, H. Suzuki, Y. Kubota, K. Ohuchi, H. Higashimura and T. Yokozawa, *J. Am. Chem. Soc.*, 2007, **129**, 7236.

90. T. Beryozkina, K. Boyko, N. Khanduyeva, V. Senkovskyy, M. Horecha, U. Oertel, F. Simon, M. Stamm and A. Kiriy, *Angew. Chem., Int. Ed.*, 2009, **48**, 2695.

91. T. Yokozawa, R. Suzuki, M. Nojima, Y. Ohta and A. Yokoyama, *Macromol. Rapid Commun.*, 2011, **32**, 801.

92. K. Kosaka, Y. Ohta and T. Yokozawa, *Macromol. Rapid Commun.*, 2015, **36**, 373.

93. A. Sui, X. Shi, H. Tian, Y. Geng and F. Wang, *Polym. Chem.*, 2014, **5**, 7072.

94. T. Yokozawa, H. Kohno, Y. Ohta and A. Yokoyama, *Macromolecules*, 2010, **43**, 7095.

95. E. Elmalem, A. Kiriy and W. T. S. Huck, *Macromolecules*, 2011, **44**, 9057.

96. M. Nojima, Y. Ohta and T. Yokozawa, *J. Am. Chem. Soc.*, 2015, **137**, 5682.

97. J. K. Lee, S. Ko and Z. Bao, *Macromol. Rapid Commun.*, 2012, **33**, 938.

98. E. Elmalem, F. Biedermann, K. Johnson, R. H. Friend and W. T. S. Huck, *J. Am. Chem. Soc.*, 2012, **134**, 17769.

99. C. S. Fischer, M. C. Baier and S. Mecking, *J. Am. Chem. Soc.*, 2013, **135**, 1148.

100. Z. Zhang, P. Hu, X. Li, H. Zhan and Y. Cheng, *J. Polym. Sci., Part A: Polym. Chem.*, 2015, **53**, 1457.

101. H.-H. Zhang, C.-H. Xing and Q.-S. Hu, *J. Am. Chem. Soc.*, 2012, **134**, 13156.

102. H.-H. Zhang, C.-H. Xing, Q.-S. Hu and K. Hong, *Macromolecules*, 2015, **48**, 967.

103. M. Verswyvel, P. Verstappen, L. De Cremer, T. Verbiest and G. Koeckelberghs, *J. Polym. Sci., Part A: Polym. Chem.*, 2011, **49**, 5339.

104. P. Willot, J. Steverlynck, D. Moerman, P. Leclere, R. Lazzaroni and G. Koeckelberghs, *Polym. Chem.*, 2013, **4**, 2662.

105. M. Uchiyama, T. Furuyama, M. Kobayashi, Y. Matsumoto and K. Tanaka, *J. Am. Chem. Soc.*, 2006, **128**, 8404.

106. T. Higashihara, E. Goto and M. Ueda, *ACS Macro Lett.*, 2012, **1**, 167.

107. T. Erdmann, J. Back, R. Tkachov, A. Ruff, B. Voit, S. Ludwigs and A. Kiriy, *Polym. Chem.*, 2014, **5**, 5383.

108. R. Tkachov, V. Senkovskyy, T. Beryozkina, K. Boyko, V. Bakulev, A. Lederer, K. Sahre, B. Voit and A. Kiriy, *Angew. Chem., Int. Ed.*, 2014, **53**, 2402.

109. R. Tkachov, Y. Karpov, V. Senkovskyy, I. Raguzin, J. Zessin, A. Lederer, M. Stamm, B. Voit, T. Beryozkina, V. Bakulev, W. Zhao, A. Facchetti and A. Kiriy, *Macromolecules*, 2014, **47**, 3845.

110. W. Liu, R. Tkachov, H. Komber, V. Senkovskyy, M. Schubert, Z. Wei, A. Facchetti, D. Neher and A. Kiriy, *Polym. Chem.*, 2014, **5**, 3404.

111. V. Senkovskyy, R. Tkachov, H. Komber, M. Sommer, M. Heuken, B. Voit, W. T. S. Huck, V. Kataev, A. Petr and A. Kiriy, *J. Am. Chem. Soc.*, 2011, **133**, 19966.

112. V. Senkovskyy, R. Tkachov, H. Komber, A. John, J.-U. Sommer and A. Kiriy, *Macromolecules*, 2012, **45**, 7770.
113. Y. Qiu, J. Mohin, C.-H. Tsai, S. Tristram-Nagle, R. R. Gil, T. Kowalewski and K. J. T. Noonan, *Macromol. Rapid Commun.*, 2015, **36**, 840.
114. S. Kang, R. J. Ono and C. W. Bielawski, *J. Am. Chem. Soc.*, 2013, **135**, 4984.
115. K. Fuji, S. Tamba, K. Shono, A. Sugie and A. Mori, *J. Am. Chem. Soc.*, 2013, **135**, 12208.
116. M. Schnurch, M. Spina, A. F. Khan, M. D. Mihovilovic and P. Stanetty, *Chem. Soc. Rev.*, 2007, **36**, 1046.
117. M. Nojima, R. Saito, Y. Ohta and T. Yokozawa, *J. Polym. Sci., Part A: Polym. Chem.*, 2015, **53**, 543.
118. A. F. Littke and G. C. Fu, *J. Am. Chem. Soc.*, 2001, **123**, 6989.
119. T. Sanji and T. Iyoda, *J. Am. Chem. Soc.*, 2014, **136**, 10238.
120. T. Sanji, A. Motoshige, H. Komiyama, J. Kakinuma, R. Ushikubo, S. Watanabe and T. Iyoda, *Chem. Sci.*, 2015, **6**, 492.

CHAPTER 2

Controlled Chain-growth Synthesis of Conjugated Polymers: Moving Beyond Thiophene

C. R. BRIDGES AND D. S. SEFEROS*

Department of Chemistry, University of Toronto, Toronto, Ontario, Canada
*Email: dseferos@chem.utoronto.ca

2.1 Introduction

The controlled chain-growth polymerization of 3-hexylthiophene was in-dependently reported[1-6] in 2004 by the research groups of Yokozawa and McCullough, who showed that narrow dispersity, regio-regular poly(3-hexylthiophene) (P3HT) could be synthesized with good molecular weight control. This reaction proceeds by a catalyst-transfer polymerization (CTP) mechanism, which gives these polymerizations their controlled chain-growth nature. The evidence for controlled polymerization includes a narrow dispersity, a single distribution of predetermined end-groups, and a degree of polymerization that correlates with the initiator or catalyst to monomer ratio. These observations are rather superficial, however, and in-depth studies on the reaction kinetics are required to confirm controlled polymerization. In controlled polymerization, the monomer consumption will adhere to a first-order rate law with respect to the monomer concentration. Because the number of actively propagating chains remains constant, the

RSC Polymer Chemistry Series No. 21
Semiconducting Polymers: Controlled Synthesis and Microstructure
Edited by Christine Luscombe

molecular weight of the polymer will progress linearly with monomer consumption throughout. As a result of the "living" nature of the growing chains, chain extension will occur or block copolymers will form on subsequent additions of monomer (Figure 2.1).[7,8] Because of these characteristics, developing a controlled synthesis of P3HT allowed researchers to synthesize for the first time well-defined conjugated polymers with precise control over the degree of polymerization, end-groups, and copolymer composition.[1-13] Since then, astounding progress has been made in controlling the chain architecture, composition, morphology, and properties of conjugated polymers consisting of thiophene and thiophene analogs.[9-18]

Developing controlled polymerization methods for conjugated polymers has allowed researchers to extend polythiophenes from tools for fundamental studies to advanced materials for electronic applications.[19-22]

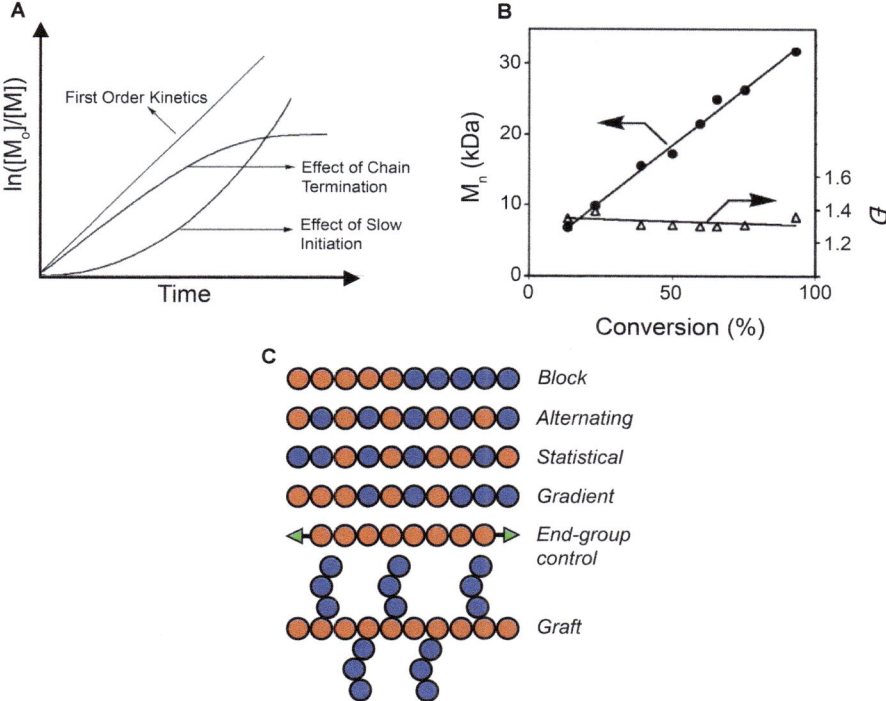

Figure 2.1 (A) Example of the first-order kinetics with respect to monomer consumption ([M_0]/[M]) observed in controlled chain polymerization, adapted with permission from K. Matyjaszewski and J. Xia, *Chem. Rev.*, 2001, **101**, 2921–2990.[8] Copyright 2001 American Chemical Society. (B) M_n *versus* monomer consumption for the controlled polymerization of 2,5-dibromo-3-hexylthiophene, adapted with permission from A. Yokoyama, R. Miyakoshi and T. Yokozawa, *Macromolecules*, 2004, **37**, 1169–1171.[1] Copyright 2004 American Chemical Society. (C) Some possible chain architectures of conjugated copolymers synthesized using controlled chain polymerization methods.

However, applying CTP methods to monomers that are either electronically or structurally different from thiophene has proved difficult. Until recently, only small, electron-rich heterocycles such as thiophenes, selenophenes, pyrroles, and phenylenes were amenable to CTP (Figure 2.2). This highlights the main limitation of this polymerization method.[23–30]

Despite the clear advantages of using controlled polymerization methodologies, they have rarely been used in the synthesis of high-performance conjugated polymers. This is because the majority of high-performance conjugated polymers used in transistors and solar cells are copolymers consisting of alternating electron-rich and electron-deficient monomers. Common examples of these so called "donor–acceptor" copolymers are poly[4,8-bis(5-(2-ethylhexyl)thiophen-2-yl)benzo[1,2-*b*;4,5-*b'*]dithiophene-2,6-diyl-*alt*-(4-(2-ethylhexyl)-3-fluorothieno[3,4-*b*]thiophene-)-2-carboxylate-2-6-diyl)] and poly(2,6-bisthien-5-yl)naphthalene-1,4,5,8-tetracarboxylic-*N,N'*-bis(2-alkyl) diimide (Figure 2.3).[31–35] The presence of electron-deficient moieties and the large size of donor–acceptor monomers are thought to disrupt the ring-walking mechanism that is critical to CTP.

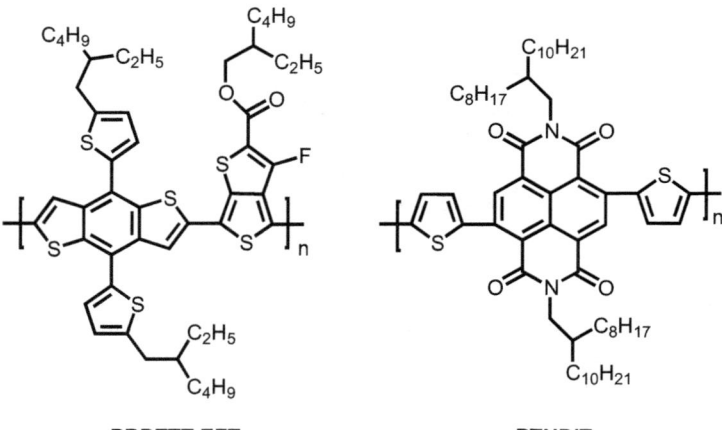

R = alkyl, thioalkyl, phenyl, carboxylate, etc.

Figure 2.2   Examples of conjugated polymers that can be synthesized using controlled chain polymerizations.

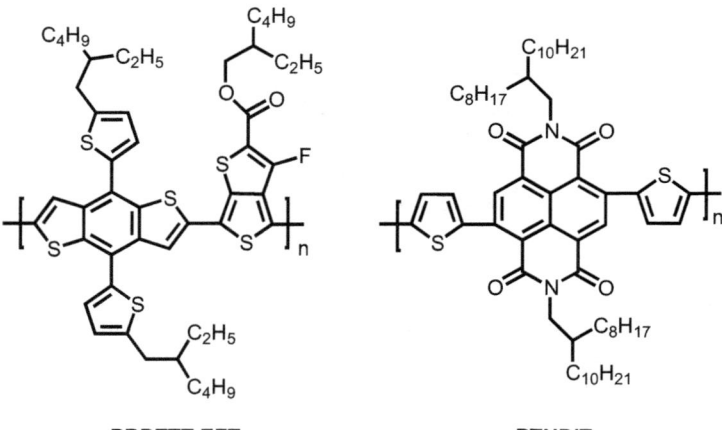

PBDTTT-EFT   PTNDIT

Figure 2.3   Structures of two common donor–acceptor alternating copolymers used in solar cells and transistors.

Donor–acceptor conjugated copolymers are most commonly synthesized using condensation polymerization, which gives little control over the degree of polymerization, dispersity, or end-groups. By using complex monomers and macroinitiators, alternating copolymers, semi-random copolymers, and block copolymers have been synthesized using polycondensation methods.[36–38] These procedures are more complicated than CTP, which achieves similar results. The ability to synthesize polymers with backbone structures similar to these high-performance donor–acceptor type copolymers, but with increased control over their sequence and composition, remains an important goal.

This chapter begins with a brief introduction to the mechanism of controlled chain growth in the synthesis of conjugated polymers, followed by an overview of some strategies that have succeeded in extending CTP to other monomer types. These include the controlled synthesis of homopolymers, block, alternating, gradient, and statistical conjugated copolymers containing both larger and electron-deficient monomers. The current limitations of, and future outlook for, the controlled synthesis of conjugated macromolecules is discussed.

2.2 Mechanism of Controlled Chain Growth in Conjugated Polymers

It is important to understand the controlled chain polymerization mechanism to identify potential reasons that could result in failure in certain cases. The most common method to achieve CTP is a Grignard metathesis polymerization. The catalytic mechanism behind this polymerization has been discussed previously[39–45] in several publications and will not be presented in depth here.

A dihalogenated heterocycle is first activated with a single equivalent of an alkyl Grignard reagent. The polymerization is then initiated using a nickel catalyst, most commonly dichloro[1,3-bis(diphenylphosphino)propane]-nickel (Ni(dppp)Cl$_2$). The activated monomers go thorough successive oxidative addition, transmetallation, and reductive elimination steps to form a polymer (Figure 2.4). An important intermediate is the transition state in which the nickel catalyst becomes coordinated to the conjugated π-system of the growing polymer chain. This transition state makes intramolecular oxidative addition (*i.e.* ring-walking) across the terminal repeat unit on a propagating polymer chain the preferred reaction pathway. This is an important contrast to polycondensation reactions in which the catalyst dissociates from the chain and forms polymers *via* intermolecular pathways. This transition state has been indirectly observed by Bryan and McNeil,[46] who showed that, even in the presence of a large excess of competing reagents, the preferential reaction pathway is intramolecular (Figure 2.5). The result is that each catalyst initiates and allows the propagation of a

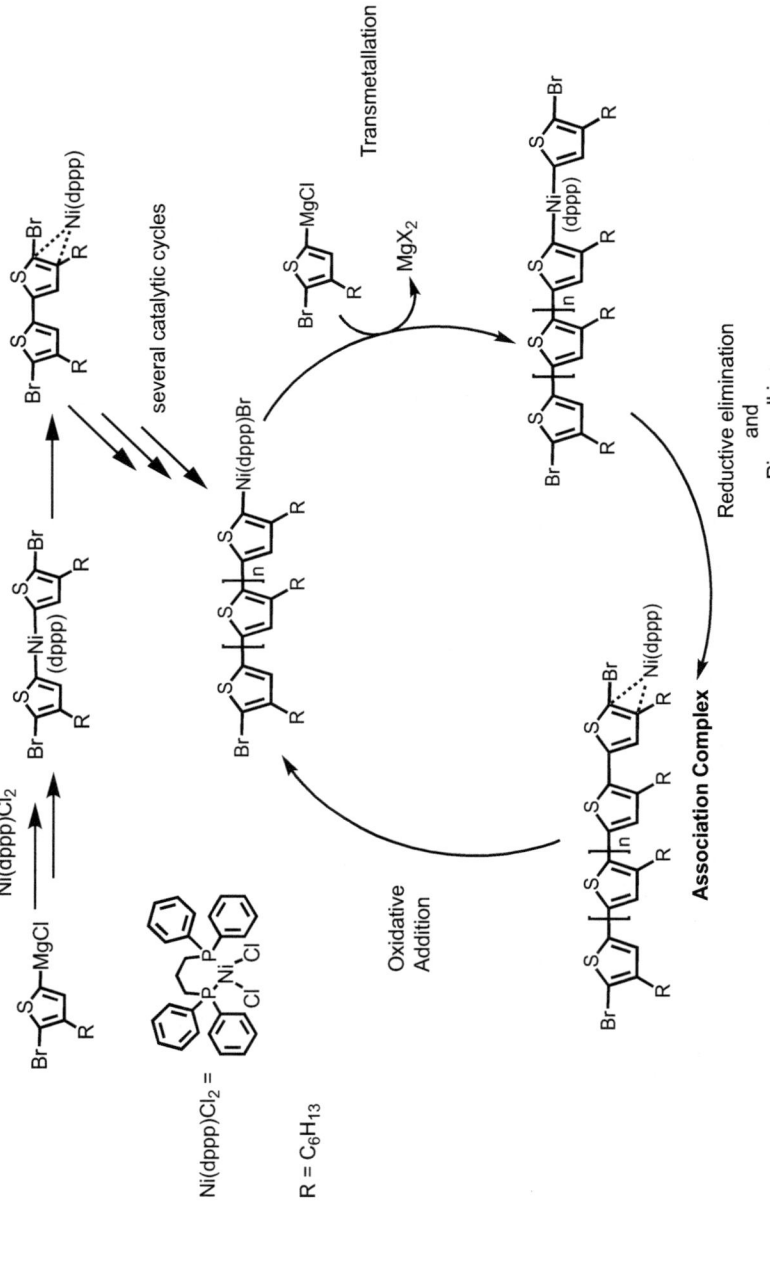

Figure 2.4 Mechanism for CTP.

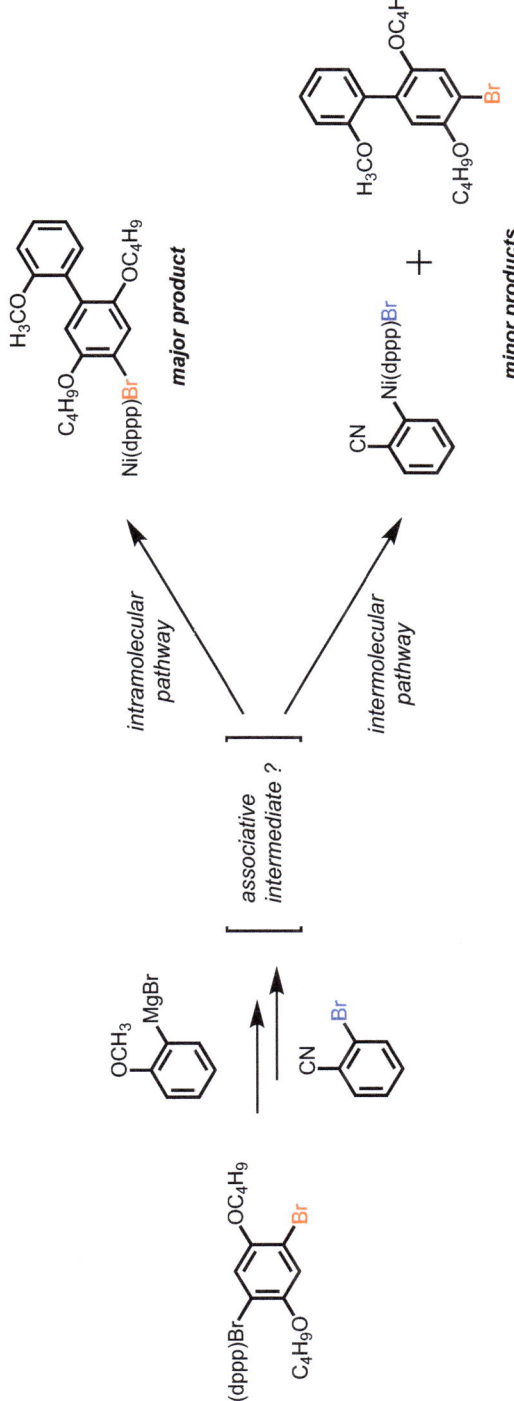

Figure 2.5 Proposed preferential intramolecular oxidative addition displayed in CTP. Adapted with permission from Z. J. Bryan and A. J. McNeil, *Macromolecules*, 2013, **46**, 8395–8405.[41] Copyright 2013 American Chemical Society.

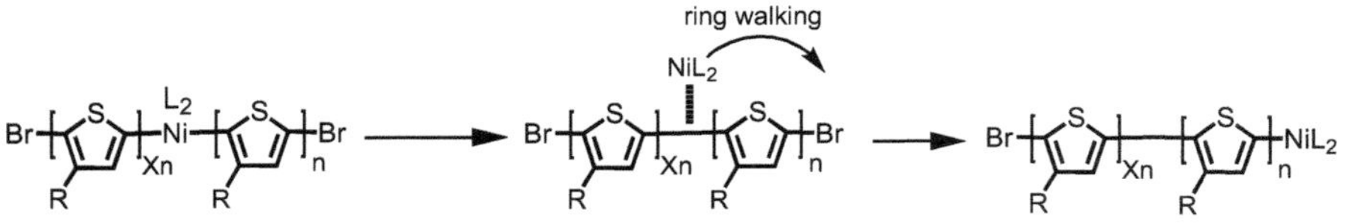

Xn = degree of polymerization   n = 1, 2, 3

Figure 2.6   Ring-walking on thiophene oligomers up to $n = 3$ repeat units.

single polymer chain rather than dissociating and initiating new chains, resulting in the controlled chain growth of CTP reactions.

The robustness of the association complex is an important factor in the success of CTP. If the association complex is too weak, then the catalyst will dissociate from the growing chain. If the monomer is too large, then the catalyst cannot reliably ring-walk across the terminal repeat unit prior to oxidative addition. Controlled chain growth will not occur in either of these two cases. When polymerizing thiophene oligomers, it has been shown that this ring-walking mechanism proceeds with monomers that are as large as three arene units.[47] As the monomer becomes larger, intramolecular oxidative addition becomes less favored and control over the polymerization is gradually lost (Figure 2.6).

The association complex between Ni-diphosphine catalysts and thiophene is sufficiently strong that intramolecular oxidative addition will occur even when the thiophene-based repeat units are large (*c.* 1 nm), but for other monomers this may not be true.[48] Several research groups have identified that the strength of the association complex strength is a key factor in extending CTP to larger and more diverse conjugated monomers. Some studies have succeeded in using this concept as a guiding principle for designing catalyst and monomer systems that display a controlled polymerization character.[48,65,97,104]

2.3   Chain-growth Synthesis of Conjugated Homopolymers

Several years after the original reports of the CTP of thiophene, other electron-deficient monomers were tested, but generally showed less controlled chain growth. One of the first examples of an electron-deficient conjugated polymer synthesized by chain-growth methods was poly(2,3-dialkyl-thieno[3,4-*b*]pyrazine) (PTPz). PTPz is considered to be an n-type, narrow HOMO–LUMO gap version of P3HT and, for these reasons, it is interesting as a photovoltaic material. When PTPz was synthesized using standard Grignard metathesis procedures using a Ni(dppp)Cl$_2$ catalyst (Figure 2.7) the resulting polymers had narrow dispersities ($Đ = 1.2$–1.5), as expected in controlled polymerization.[49–51]

PTPz is sparingly soluble and only low molecular weight polymers could be synthesized ($M_n = 4900$ Da), with some portion of the polymeric material

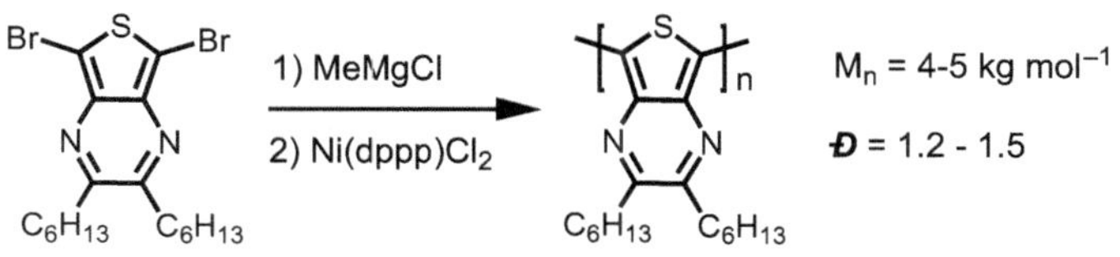

Figure 2.7 Synthesis, molecular weight range and dispersity values for PTPz.

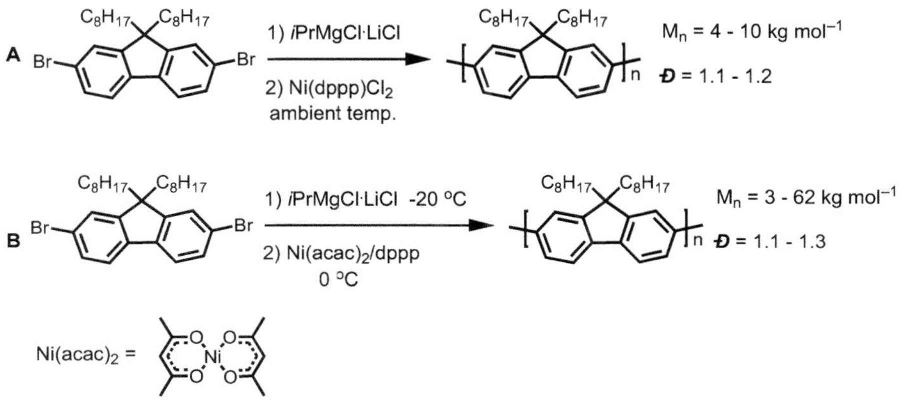

Figure 2.8 Synthesis, molecular weight range and dispersity values for PF from (A) Javier *et al.*[54] and (B) Sui *et al.*[55]

remaining insoluble. Because of this, the effect of catalyst loading on the molecular weight could not be studied and the dispersity of the polymer sample may have been low as a result of only measuring chains with low degrees of polymerization. Although these initial results may indicate that chain growth is occurring in this system, it is clear that electron-deficient monomers do not behave in the same way as electron-rich monomers such as thiophene. Studies of the effect of catalyst loading on the molecular weight and reaction kinetics are required to conclusively prove that this polymer was synthesized by CTP, or if it can indeed be controlled.

Poly(9,9-dialkylfluorene) (PF) is widely used in organic electronics for its brilliant emissive properties. Synthesizing PF using CTP would allow for more control over the solid-state morphology, which is important for device optimization. Huang *et al.*[52] and Stefan *et al.*[53] first studied the controlled polymerization of fluorene in 2008 and 2009, respectively.

Soon after these reports, Javier *et al.*[54] and Sui *et al.*[55] presented even more compelling results, showing that standard CTP conditions led to PFs with high yields, fast reaction kinetics, and narrow dispersities (Figure 2.8).

The manner by which molecular weight increases as a function of monomer conversion indicates the extent to which the CTP mechanism occurs. Ideally, a linear correlation between molecular weight and monomer conversion should be observed throughout the polymerization. For PF, it

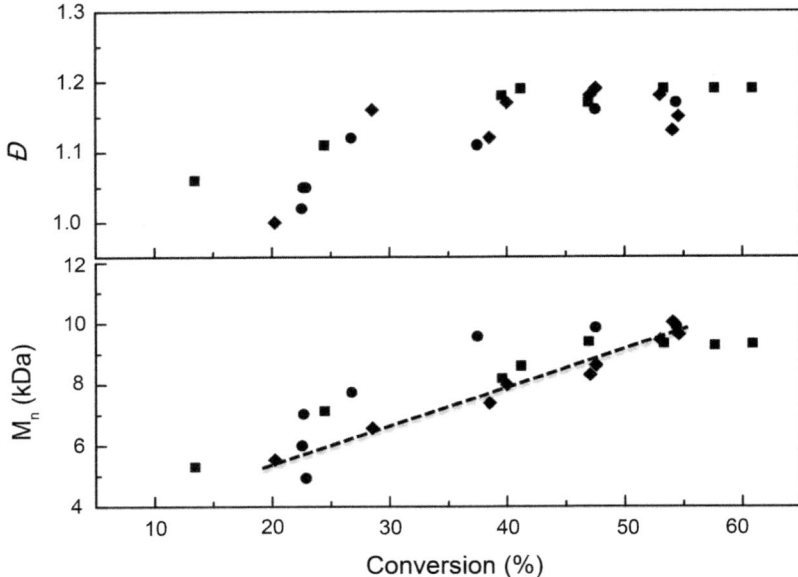

Figure 2.9 Molecular weight and dispersity as a function of monomer conversion during a typical synthesis of poly(9,9-dialkylfluorene).
Adapted with permission from A. E. Javier, S. R. Varshney and R. D. McCullough, *Macromolecules*, 2010, **43**, 3233–3237.[54] Copyright 2010 American Chemical Society.

appeared that the CTP mechanism was valid until 55% monomer conversion. At this point, the polymer reached its peak M_n and, in spite of further monomer consumption, the polymer chains were not extended (Figure 2.9).

Although the dispersity remained fairly low throughout the polymerization, the fact that the molecular weight stopped increasing with monomer conversion shows that the CTP mechanism is limited to low monomer conversion and low molecular weight polymers.

The fraction of polymer chains with active chain ends indicates how often chain termination occurs during polymerization. In the case of PFs synthesized by CTP, active chains can be end-capped with a Grignard reagent. When a CTP of PF was terminated with end-capping reagents and analyzed using matrix-assisted laser desorption/ionization mass spectrometry (MALDI), the end-capping efficiency decreased as polymerization progressed. In general, only moderate control over the end-groups was achieved. This is the result of a failure in the intramolecular oxidative addition reaction pathway, causing some polymer chains to terminate early in the polymerization. Sui *et al.*[55] screened a number of catalyst/ligand systems and found that, when using nickel(acac)$_2$/dppp, CTP occurred in fluorene up to $M_n = 62$ kg mol^{-1}. The conditions for the CTP of PF are very specific and several other groups have reported polymers with broad dispersities or lower than expected molecular weights when attempting to synthesize PF using different polymerization conditions.[56–58] Together, these results indicate that the intramolecular oxidative addition step for the fluorene monomer may not be as robust as that for thiophene, and that early chain termination results from catalyst diffusion from the growing polymer chains prior to oxidative addition. The aromatic backbone of fluorene is much larger than that of thiophene, which increases the ring-walking distance. Fluorene is also less electron-rich. These are probably the main reasons for the observed lack of the controlled chain growth in PF.

Thiazoles are similar in size to thiophene, but are significantly more electron-deficient. Polymers containing the thiazole unit are becoming more commonly used in organic photovoltaics;[59] however, poly(4-alkylthiazole) (PTz) homopolymers are not widely reported. Pammer and Passlack[60] studied the first chain-growth synthesis of PTz in 2014. Their initial attempts to synthesize PTz using CTP resulted in insoluble polymers. The kinetics of the polymerization could not be studied, but the dispersities and molecular weights of the polymers did not indicate a controlled polymerization.

To improve on this, a larger solubilizing side-chain and an asymmetrical monomer were used to synthesize two external initiator catalysts with electron-donating ligands. The result of this modification was soluble, high molecular weight PTz (Figure 2.10).[61]

These researchers showed that as the catalyst loading decreases, there is a correlated increase in M_n for both catalysts, which is consistent with controlled polymerization. It should be noted that in these studies the size-exclusion chromatography (SEC) elution curves had bimodal distributions and the SEC-measured molecular weight was much higher than expected given the catalyst loading. These two observations were explained by chain–chain coupling and the presence of impurities in the catalyst, respectively (Figure 2.11). The bulky side-chain may also contribute to the larger than expected M_n.

The narrow dispersity and the fact that the molecular weight of PTz can be controlled up to $M_n = 100$ kg mol^{-1} are good indications that the polymerization proceeds by the CTP mechanism. The electron-donating external initiator forces unidirectional chain propagation and may have improved the otherwise sluggish initiation that can occur with electron-deficient monomers. This probably facilitates the controlled polymerization behavior observed with this system.

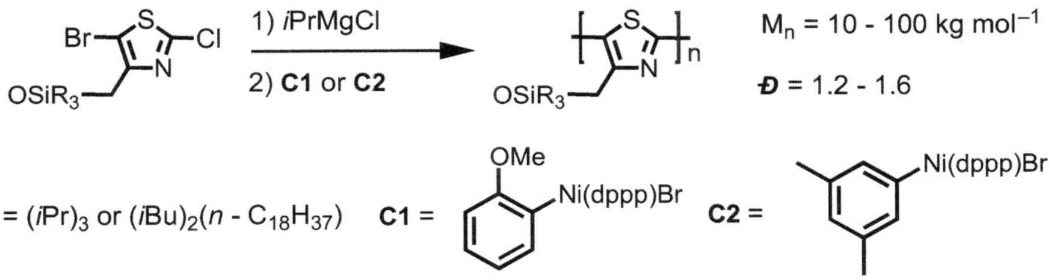

Figure 2.10   Synthesis, molecular weight range and dispersity values for soluble PTz.

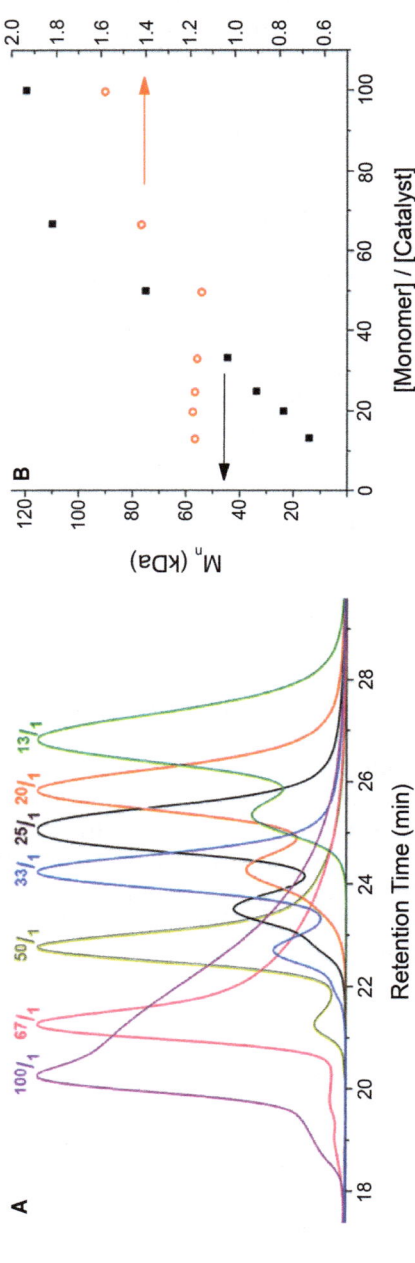

Figure 2.11 (A) Size-exclusion chromatography elution curves of PTz as a function of the monomer/C2 ratio. (B) Molecular weight and dispersity of PTz as a function of C2 loading. Adapted with permission from F. Pammer, J. Jäger, B. Rudolf and Y. Sun, *Macromolecules*, 2014, **17**, 5904–5912.[61] Copyright 2014 American Chemical Society.

Nanashima and co-workers[62–64] conducted in-depth studies on the chain-growth synthesis of n-type poly(2-alkoxypyridine-3,6-diyl), a *para*-linked conjugated polymer, and poly(2-alkoxypyridine-3,5-diyl), a *meta*-linked conjugated polymer. The authors first used a model reaction between 2,5-dibromopyridine and 0.5 equiv. phenylmagnesium bromide to show that intramolecular oxidative addition is preferred for these electron-deficient pyridine monomers. The major product was the intramolecular addition product 2,5-diphenylpyridine. The intermolecular reaction products, 2-bromo-5-phenylpyridine and 5-bromo-2-phenylpyridine, were not observed (Figure 2.12).

These results are similar to those observed by Bryan *et al.* using phenylene monomers.[41,46,56] It appears that intramolecular oxidative addition is also the favorable reaction pathway for pyridine. Although the preferred intramolecular oxidative addition is a good indication that CTP can occur, this was not the case for *para*-linked poly(2-alkoxypyridine-3,6-diyl). Significant optimization of the polymerization conditions, catalyst, and initiation reaction still resulted in low molecular weight polymers with broad dispersities. In contrast, *meta*-linked poly(2-alkoxypyridine-3,5-diyl) displayed some characteristics of controlled polymerization under similar reaction conditions (Figure 2.13). The authors showed that the *meta* version can be synthesized with narrow dispersity and a molecular weight that reflected the catalyst loading. They also observed a linear progression of molecular weight as a function of monomer conversion up to $M_n = 10$ kg mol^{-1} (Figure 2.14).

MALDI confirmed that the polymers consisted predominantly of a single distribution of Br–H terminated chains, which is expected for CTP in this instance. The polymer end-groups could be functionalized very efficiently by treating the active polymer with a Grignard reagent. These observations indicated that the CTP mechanism proceeded throughout the polymerization and that the chain ends remained active.

It is interesting to note that these two alkoxypyridine monomers have similar electronic properties, yet exhibit drastically different behaviors during polymerization. This suggests that the regio-chemistry of the linkages is important in the polymerization mechanism. It was speculated that, in *para*-linked poly(2-alkoxypyridine-3,6-diyl), the interactions between the catalyst and nitrogen disrupt the ring-walking mechanism, whereas these interactions do not occur in *meta*-linked poly(2-alkoxypyridine-3,5-diyl). It is also important to note that the ring-walking step only needs to proceed across two bonds in the *meta*-linked monomer. This may make intramolecular oxidative addition more likely in poly(2-alkoxypyridine-3,5-diyl) than in poly(2-alkoxypyridine-3,6-diyl).

It has been speculated that the reason why the CTP mechanism is so much more successful for electron-rich monomers is that the association between the catalyst and the monomer is much stronger, which allows more reliable intramolecular oxidative addition. We have studied this systematically by varying the electron-donating ability of a series of Ni(II)diimine catalysts using 2-alkyl-2*H*-benzo[*d*]-[1,2,3]triazole (BTz) as a model electron-deficient

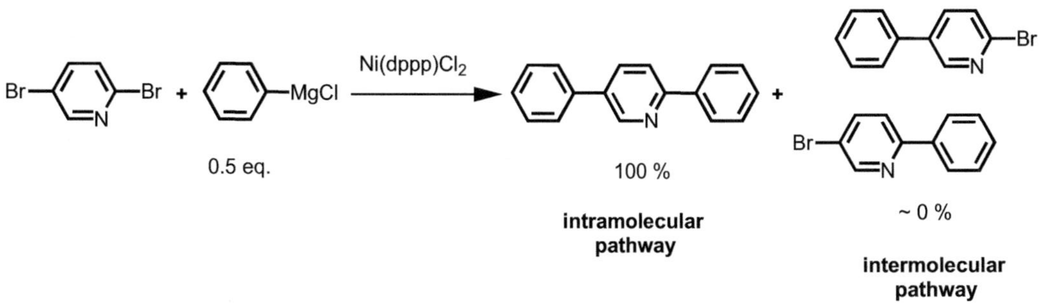

Figure 2.12 Preferred intramolecular oxidative addition with pyridine.

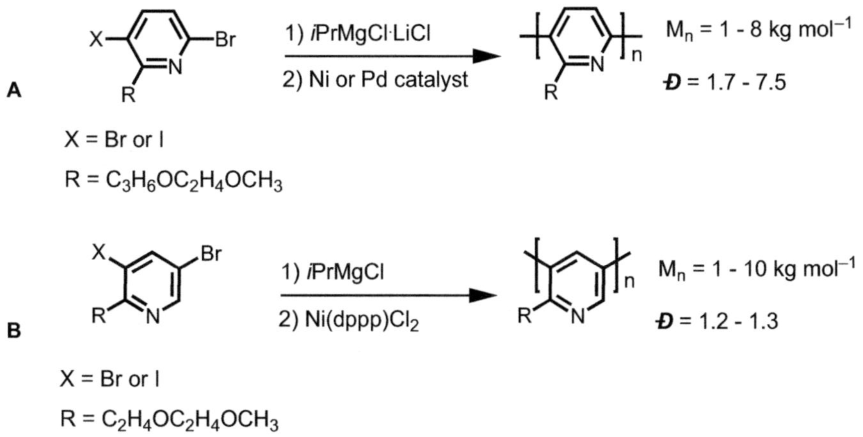

Figure 2.13 Synthesis, molecular weight range and dispersity values for (A) poly(2-alkoxypyridine-3,6-diyl) and (B) poly(2-alkoxypyridine-3,5-diyl).

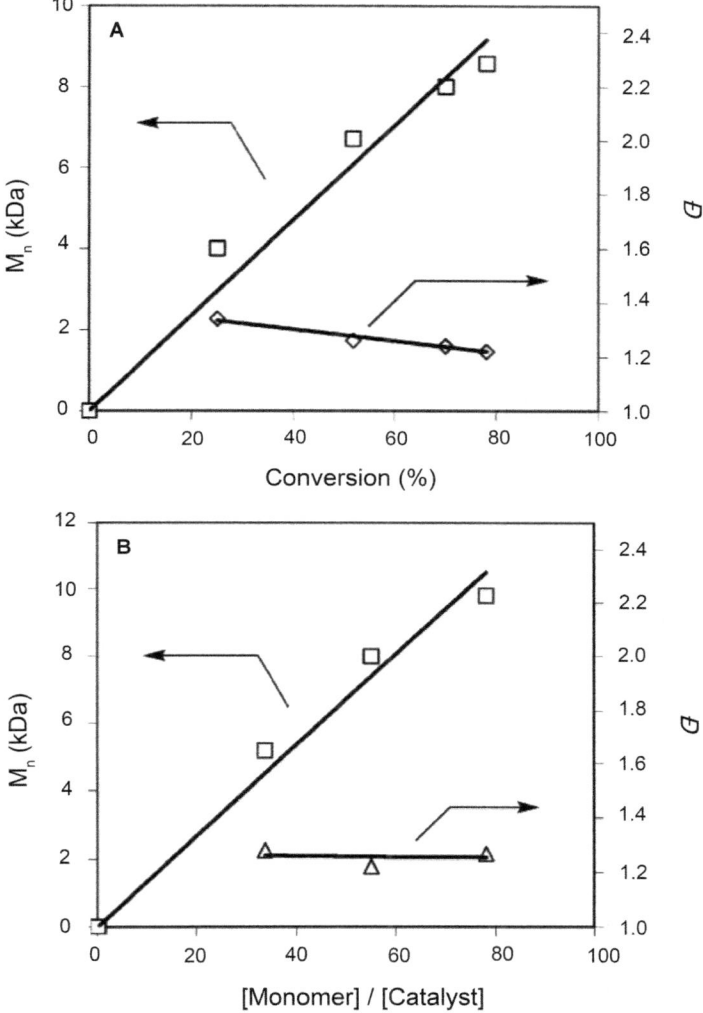

Figure 2.14 (A) Molecular weight and dispersity as a function of monomer conversion and (B) molecular weight as a function of catalyst loading for poly(2-alkoxypyridine-3,5-diyl).
Adapted with permission from Y. Nanashima, A. Yokoyama and T. Yokozawa, *Macromolecules*, 2012, **45**, 2609–2613.[63] Copyright 2012 American Chemical Society.

monomer (Figure 2.15).[65] We chose acenaphthylene-(1,2-diylidene)bis(2-*t*-butylaniline) nickel(ɪɪ) bromide (*t*-BuAn)as our parent catalyst, acenaphthylene-(1,2-diylidene)bis(2-methoxyaniline nickel(ɪɪ) bromide (OMeAn) as our electron-donating catalyst, and acenaphthylene-(1,2-diylidene)bis(2-trifluoromethyl) nickel(ɪɪ) bromide (CF3An) as our electron-deficient catalyst.

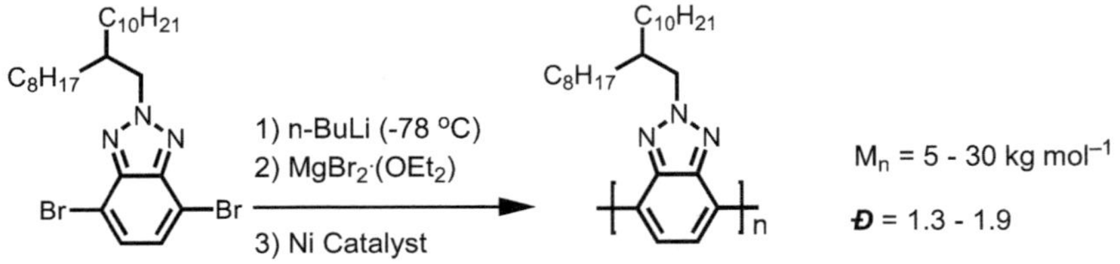

Figure 2.15 Synthesis, molecular weight range and dispersity values for PBTz.

We used density functional theory (DFT) to model the association complex between BTz and each catalyst. The resulting optimized geometry was used to calculate the association strength. A stronger predicted association should decrease the occurrence of catalyst dissociation from the polymer chain and make intramolecular oxidative addition more favorable (Figure 2.16).

We synthesized poly(2-(2-octyldodecyl)-2*H*-benzo[*d*][1,2,3]triazole) (PBTz) using each catalyst at several catalyst loadings. Catalysts predicted to associate more strongly with the monomer produce polymers with narrower dispersities and allow for a greater control over molecular weight as a function of catalyst loading. We thus used semi-logarithmic kinetic plots to gain a better understanding of the extent of control during each polymerization. In this analysis, a linear semi-logarithmic kinetic plot indicates that the number of growing chains is constant, whereas a deviation indicates a loss of control by chain termination, or slow initiation.

We observed that the more strongly associating catalysts controlled the polymerization to a higher monomer consumption and a higher degree of polymerization, resulting in a more linear plot (Figure 2.17). The best-performing catalyst (OMeAn) exhibited chain growth up to $M_n = 24$ kg mol^{-1} (from a 100 : 1 monomer : catalyst loading) while maintaining moderate dispersities throughout the polymerization. To prove that the OMeAn catalyst is indeed associated with the polymer chain and the chain is "living", we conducted a chain-extension polymerization. In this experiment, a PBTz macroinitaitor was extended by the subsequent addition of activated monomer. SEC showed that the final polymer chain length distribution had shifted entirely to longer chain lengths compared with the macroinitiator. This is the first reported example of the chain extension of an electron-deficient conjugated polymer (Figure 2.18).

Using the same computational chemistry methods, we later improved our synthesis of PBTz by using a more strongly associating catalyst [N,N'-dimesityl-2-3-(1,8-naphthyl)-1,4-diazabutadiene]dibromonickel (MesAn). MesAn was able to control the polymerization of BTz beyond 80% monomer conversion, maintain an M_n that reflected the catalyst loading up to a ratio of 1 : 100, and the resulting polymers had dispersities of <1.3.[97]

Koeckelberghs and co-workers[66–70] studied the universality of CTP using a variety of monomers such as thiophenes, selenophenes, fluorenes, cyclopentadithiophenes, and phenanthrenes. Although they observed CTP in most cases, only Ni(ii) phosphine catalysts were tested and all these

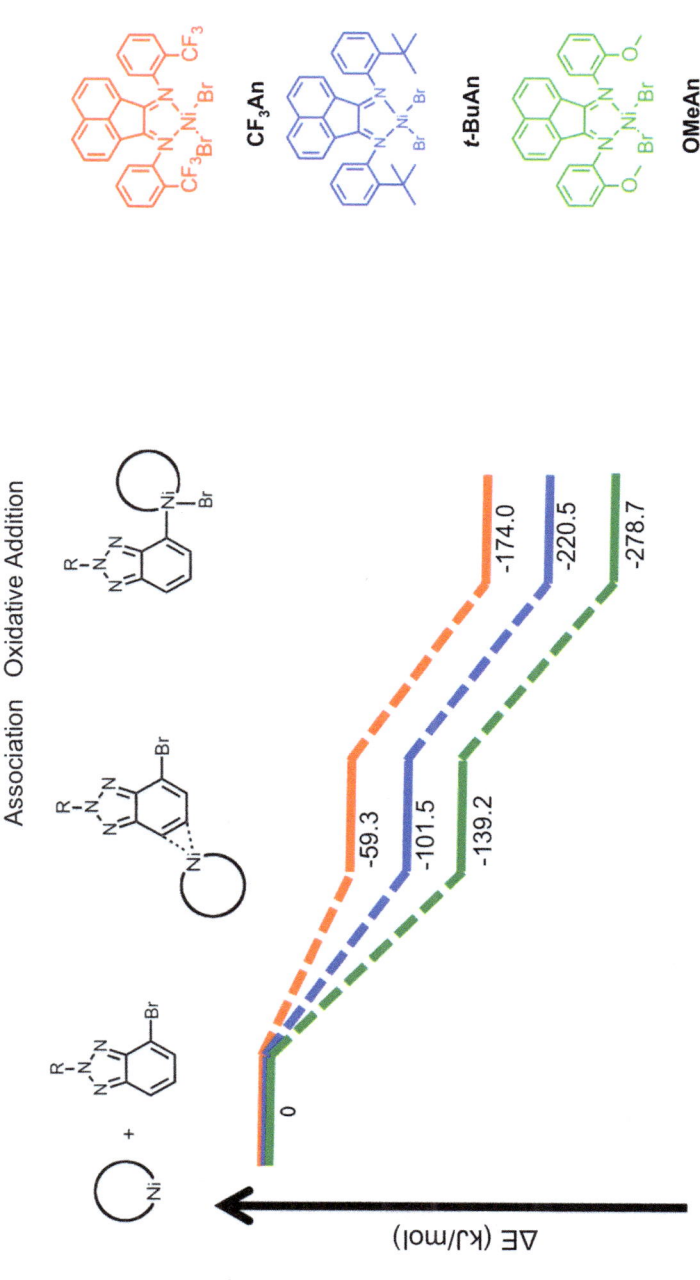

Figure 2.16 Calculated strength of association between each catalyst and BTz. Adapted with permission from C. R. Bridges, T. M. McCormick, G. L. Gibson, J. Hollinger and D. S. Seferos, *J. Am. Chem. Soc.*, 2013, **135**, 13212–13219.[65] Copyright 2013 American Chemical Society.

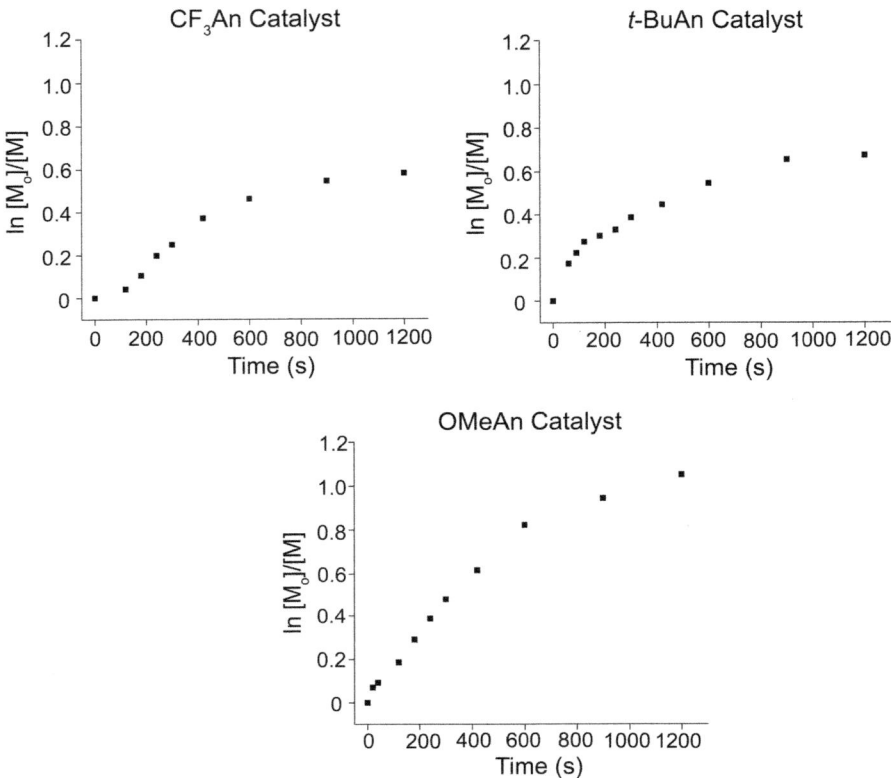

Figure 2.17 Semilogarithmic plots for the polymerization of BTz using each catalyst.
Adapted with permission from C. R. Bridges, T. M. McCormick, G. L. Gibson, J. Hollinger and D. S. Seferos, *J. Am. Chem. Soc.*, 2013, **135**, 13212–13219.[65] Copyright 2013 American Chemical Society.

monomers have a similar electronic structure to thiophene. Taken together, the results in this section highlight the fact that obtaining controlled polymerization with monomers dissimilar to thiophene is certainly possible. However, differences in electronic properties, size, and composition can all disrupt the ring-walking mechanism. In some cases, the effectiveness of CTP can be improved by using external initiators to increase the initiation rate and force unidirectional chain growth; however, understanding the nature of the association complex formed between the catalyst and monomer is the key to improving controlled polymerization behavior. It may be necessary to move away from Ni(II) phosphine catalysts to extend CTP to certain monomers. Using computational chemistry to aid catalyst design is an effective way to predict the strength of the association complex for a particular monomer–catalyst system, which, in turn, can indicate how well the polymerization will be controlled.

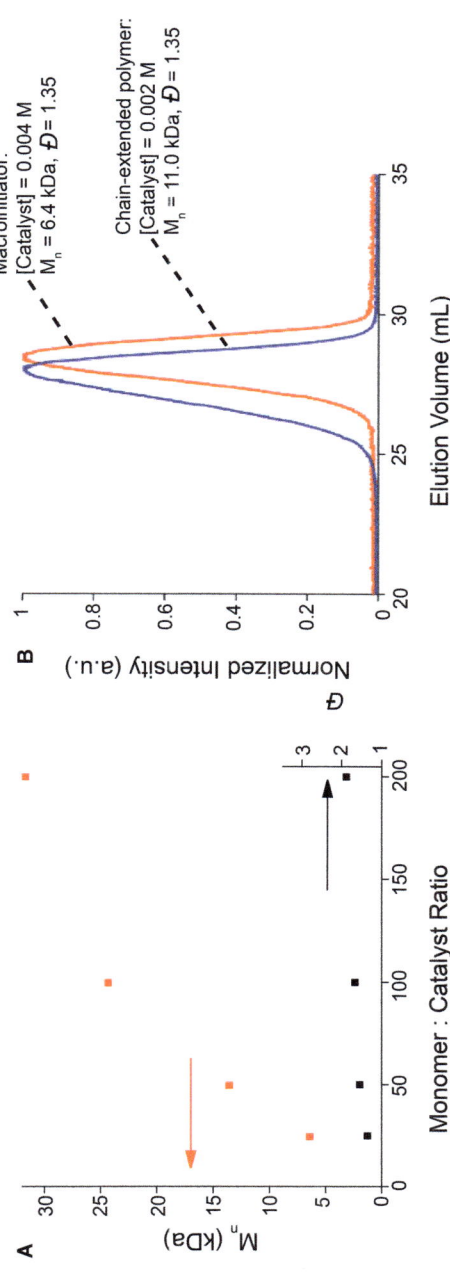

Figure 2.18   (A) Molecular weight (red) and dispersity (black) of PBTz *versus* OMeAn catalyst loading and (B) size-exclusion chromatography elution curves showing chain extension of PBTz using OMeAn. Adapted with permission from C. R. Bridges, T. M. McCormick, G. L. Gibson, J. Hollinger and D. S. Seferos, *J. Am. Chem. Soc.*, 2013, **135**, 13212–13219.[65] Copyright 2013 American Chemical Society.

2.4 Chain-growth Synthesis of Alternating Copolymers

Alternating donor–acceptor copolymers are emerging as the choice materials for high-performance organic electronics, but studies have only recently shown that alternating copolymers can be synthesized using CTP. Ono *et al.*[71] studied the controlled chain-growth synthesis of a thiophene-*p*-phenylene alternating copolymer from an A–B monomer, poly(2-(2,5-diethylhexylphenyl)thiophene) (PTPP). CTP was observed even at high monomer conversions and the polymers reached high molecular weights ($M_n = 20$ kg mol^{-1}) while maintaining narrow dispersities (Figure 2.19).

Traditionally, large monomers make it difficult for the catalyst to ring-walk across the monomer without dissociating prior to oxidative addition. In the case of PTPP, each heterocycle interacts strongly enough with the nickel catalyst for intramolecular oxidative addition to be the predominant pathway, even for a large A–B monomer (Figure 2.20).

Ono *et al.*[71] also prepared PTPP alternating copolymers and used them as macroinitiators for P3HT to form a PTPP-*block*-P3HT block copolymer by chain extension. This is an early example of an alternating copolymer synthesized using CTP; in this instance, all the heterocycles in the polymer were highly amenable to CTP.

Elmalem *et al.*[72] reported the first chain-growth synthesis of a commonly studied high-performance donor–acceptor alternating copolymer, poly((9,9-dialkylfluorene)-2,7-diyl-*alt*-[4,7-bis(3-alkylthien-5-yl)-2,1,3-benzothiadiazole]-2′,2″-diyl) (PF8BT). This copolymer contains an electron-deficient heterocycle, benzothiadiazole, that is not very amenable to CTP. Elmalem *et al.*[72] first used an externally initiating Pd catalyst to prepare PF homopolymers. They observed a single end-group distribution and narrow dispersities, which are typical of controlled polymerization; however, as was the case with most earlier attempts, only low molecular weight PF could be synthesized. When extending this method to the much larger A–B monomer of fluorene and benzothiadiazole (PF8BT), Elmalem *et al.*[72] still observed some chain-growth behavior. The polymers had narrow dispersities and at low monomer conversion a single distribution of polymer chains was observed that steadily increased in molecular weight during the reaction.

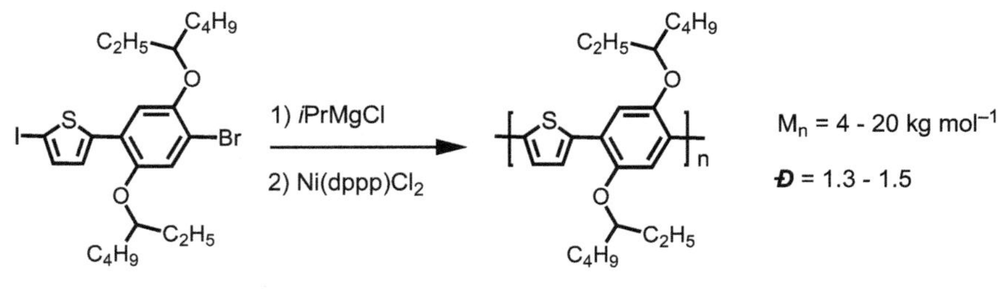

Figure 2.19 Synthesis, molecular weight range and dispersity values for PTPP.

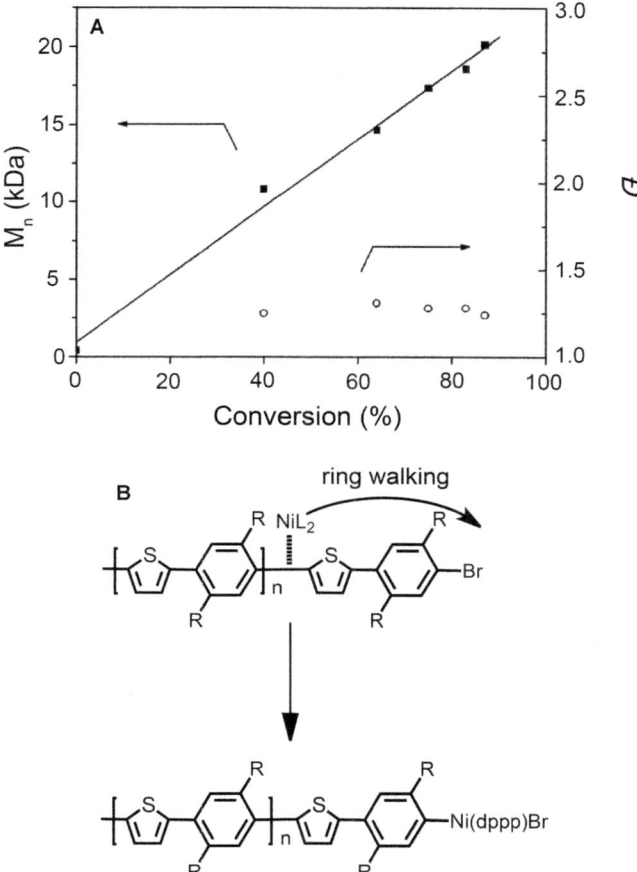

Figure 2.20 (A) Molecular weight as a function of monomer conversion for PTPP polymerization and (B) intramolecular oxidative addition pathway for PTPP.
Adapted with permission from R. J. Ono, S. Kang and C. W. Bielawski, *Macromolecules*, 2012, **45**, 2321–2326.[71] Copyright 2012 American Chemical Society.

These characteristics are not observed with Suzuki polymerization under similar conditions, which typically proceeds *via* a polycondensation mechanism. Most Suzuki condensation polymerizations are carried out with two complementary monomers rather than the single A–B monomer used in this study (Figure 2.21).

In this instance, chain termination began to dominate after PF8BT reached $M_n = 8$ kg mol^{-1} and control was gradually lost. Elmalem *et al.*[72] examined the end-groups of PF8BT using MALDI, which showed two dominant end-groups. The pyrene-H end-groups indicated successful CTP and the pyrene-Br end-groups indicated that intramolecular oxidative addition

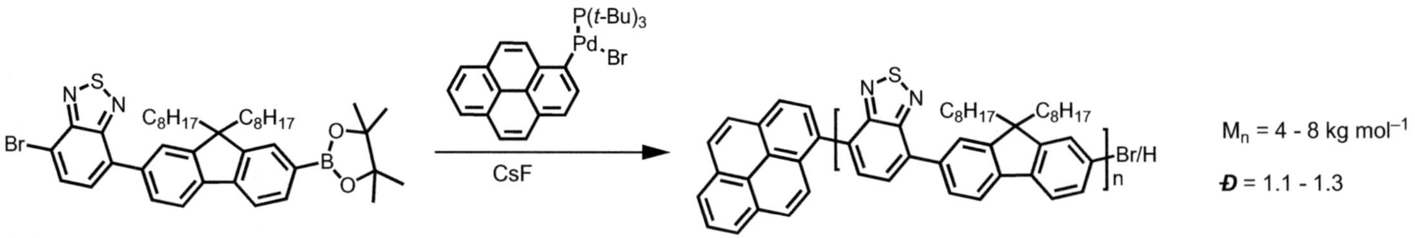

Figure 2.21 Synthesis, molecular weight range and dispersity values for PF8BT.

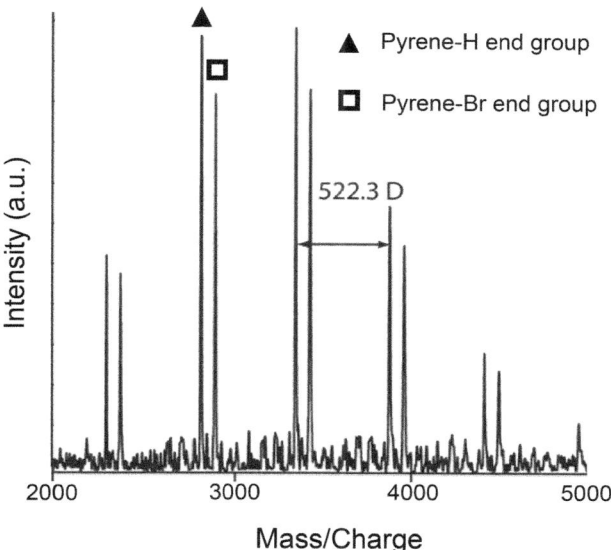

Figure 2.22 MALDI end-group analysis of PF8BT synthesized with a pyrene external initiator.
Adapted with permission from E. Elmalem, A. Kiriy and W. T. S. Huck, *Macromolecules*, 2011, **44**, 9057–9061.[72] Copyright 2011 American Chemical Society.

had not occurred. This is probably the reason for the limited CTP behavior in this example (Figure 2.22).

Not much is known about the nature of any association complex between palladium catalysts and conjugated polymers and, although the molecular weights of the polymers in this study were low, these initial results were promising. Further understanding of how ring-walking occurs in Suzuki-type (or Stille-type) polymerization may extend the utility of controlled polymerization methods.

PTNDIT is a high-performance conjugated polymer that has a high electron mobility in thin-film transistors.[73,74] Although this polymer is commercially available (Activink N2200),[75] it is only available with broad dispersities and molecular weights ranging from 16 to 25 kg mol^{-1}. Developing an efficient CTP synthesis of PTNDIT is desirable as it would afford more control over the chain length and dispersity of the resulting polymer. To this end, Kiriy and co-workers[76] studied the controlled polymerization of PTNDIT and other related polymers. They first reported that the 2,6-bis(2-bromothien-5-yl)naphthalene-1,4,5,8-tetracarboxylic-*N,N'*-bis(2-octyldodecyl) diimide (TNDIT) monomer forms a unique stable radical anion when treated with Zn powder (Figure 2.23).

Treating this radical anion with Ni(dppe)Br$_2$ initiated the polymerization, which had some of the characteristics of CTP. High molecular weight polymers were formed while monomers were still present in the reaction and the

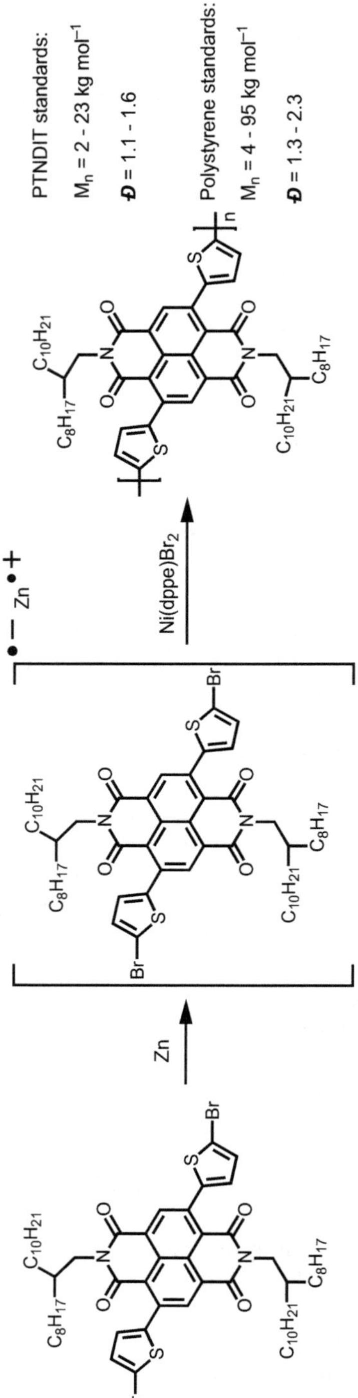

Figure 2.23 Synthesis, molecular weight range and dispersity values for PTNDIT.

molecular weight could be controlled by the initial catalyst loading.[77] Other observations were contrary to CTP, however. For example, the entire population of polymer chains did not continue to grow larger as polymerization progressed, as evidenced by SEC. Early chain termination significantly broadens the dispersity and indicates limited control (Figure 2.24).

DFT calculations were used to probe the strength of interaction between the monomer and catalyst during ring-walking across the TNDIT monomer. They found that Ni(dppe)Br$_2$ exhibits moderately strong binding to TNDIT across the entire conjugated π-system. The DFT calculations indicated that ring-walking can occur with this monomer and the catalyst–monomer association complex was detected by nuclear magnetic resonance. The association complex was identified as a transition state that could lead to intramolecular oxidative addition and, thus, CTP. By studying the molecular weight progression during the polymerization, it can be seen that chain-growth stops after 50% monomer conversion, possibly due to the moderate association between the catalyst and the monomer. A further study[78] reported the polymerization of TNDIT using a Pd catalyst (Pd(P(t-Bu)$_3$)) known to bind more strongly to aromatic systems. This should improve the controlled polymerization character. Similar chain-growth characteristics were observed: high molecular weight polymers formed while the monomers were still present in the polymerization. The dispersity was quite large in these polymerizations, however, and there was little correlation between the catalyst loading and the molecular weight. By decreasing the catalyst loading significantly and increasing the reaction time, extremely high molecular weight polymers were formed ($M_n > 170$ kg mol^{-1}). In these cases, the reaction kinetics resemble typical polycondensation reactions, in which the molecular weight is more greatly affected by the reaction time than by the catalyst loading and the dispersity is large. This Pd catalyst does not seem to be more effective than Ni(dppe)Br$_2$ for controlling the polymerization of TNDIT. When this procedure was applied to the analogous perylene diimide copolymer, 2,6-bis(thien-5-yl)perylene-1,4,5,8-tetracarboxylic-N,N'-bis(2-alkyl)diimide (PTPDIT), the results improved.[79] Although the molecular weights of PTPDIT were much lower than PTNDIT, a linear increase in PTPDIT molecular weight was observed throughout the polymerization, indicating controlled chain growth (Figure 2.25).

Both the PTNDIT and PTPDIT copolymers were subsequently used as n-channel transistors and exhibited a similar performance to polymers synthesized using polycondensation reactions.[73,74,78,79] This is a good example of traditional polycondensation techniques being replaced by CTP techniques to achieve more favorable kinetics and better control over M_n and the chain length distribution. It should be noted that, in these particular cases where the monomer itself has a large hydrodynamic radius, dimers and trimers present in the polymerization may exhibit fairly high molecular weights compared with polystyrene standards. This may cause problems when trying to differentiate between CTP and polycondensation

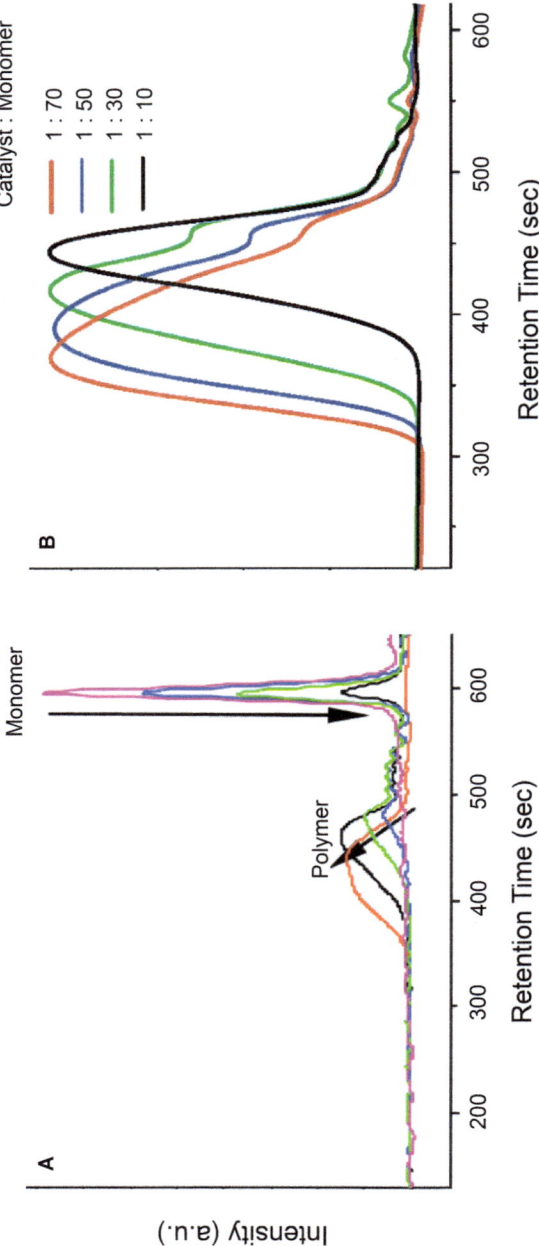

Figure 2.24 (A) Size-exclusion chromatography elution curves showing high molecular weight PTNDIT in the presence of TNDIT monomers during polymerization. Adapted with permission from V. Senkovskyy, R. Tkachov, H. Komber, A. John, J.-U. Sommer and A. Kiriy, *Macromolecules*, 2012, **45**, 7770–7777.[77] Copyright 2012 American Chemical Society. (B) Size-exclusion chromatography elution curves showing the effect of catalyst loading on molecular weight. Adapted with permission from V. Senkovskyy, R. Tkachov, H. Komber, M. Sommer, M. Heuken, B. Voit, W. T. S. Huck, V. Kataev, A. Petr and A. Kiriy, *J. Am. Chem. Soc.*, 2011, **133**, 19966–19970.[76] Copyright 2011 American Chemical Society.

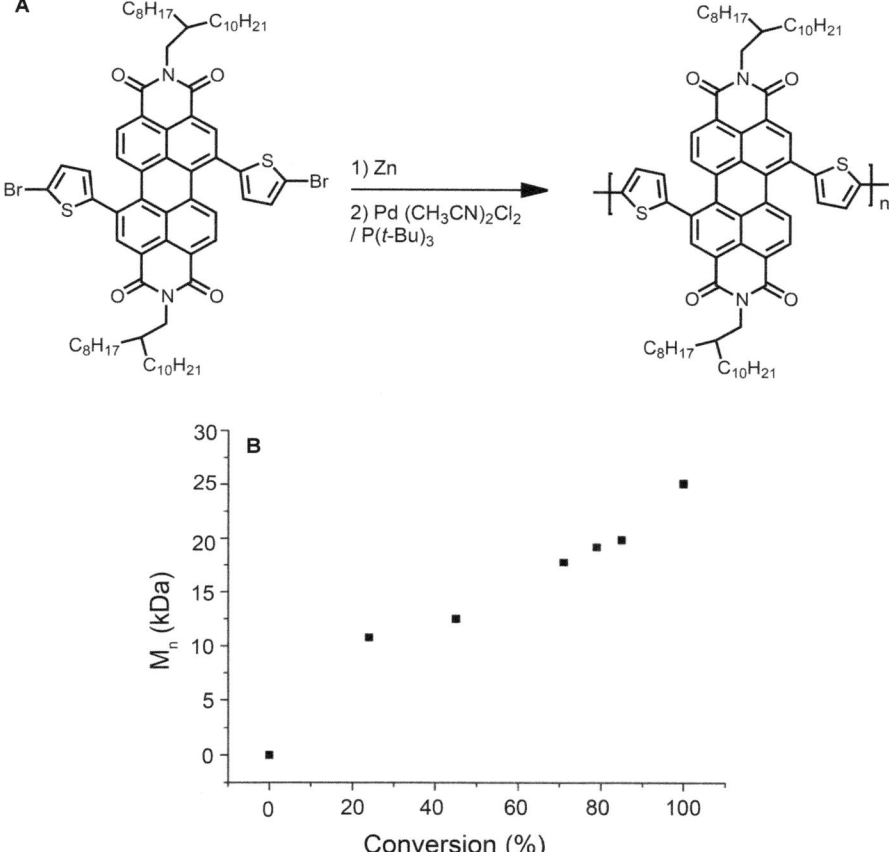

Figure 2.25   (A) Synthesis of PTPDIT and (B) molecular weight as a function of
monomer conversion during polymerization.
Adapted from ref. 79 with permission from the Royal Society of
Chemistry.

mechanisms at low monomer conversions and caution must be used when
interpreting such results. To lend more support to their interpretations,
these researchers calibrated their SEC using oligo- and poly-TNDIT standards, a step which may be necessary when synthesizing similar
polymers.[76,77]

2.5   Mixed-mechanism Chain-growth Synthesis of Block Copolymers

Block copolymers exhibit distinct and composition-dependent material
properties compared with their respective polymer blends and have been
under intense investigation for decades. Extending these methods to a

variety of monomers has allowed researchers to synthesize and study a large number of copolymers. This research was ultimately driven by the desire to obtain greater control over the physical properties of the polymers, such as phase separation, melting transitions, and optical and mechanical properties. After controlled polymerizations had been developed for P3HTs, researchers began to synthesize conjugated block copolymers in an effort to control their optoelectronic and morphological properties. The first block copolymers containing a conjugated block (usually P3HT) always contained a non-conjugated block and were synthesized using an isolated macroinitiator. These block copolymers tended to have high dispersities and were difficult to separate from unreacted macroinitiators. Bielawski and co-workers[80,81] studied an interesting case in which a block copolymer was synthesized using a single bidentate nickel catalyst, but each block proceeded *via* a distinct controlled polymerization mechanism. They showed that the "living" end of a propagating P3HT polymer can be chain-extended using a highly reactive *n*-decyl-4-phenylisocyanobenzoate monomer to form a single block copolymer containing a rigid P3HT block and a non-conjugated poly(*n*-decyl-4-phenylisocyanobenzoate) block. The polymerization proceeded with good control over molecular weight and composition and the resulting polymers had narrow dispersity. Films prepared from these polymers displayed phase-separated nanofibrillar structures reminiscent of P3HT block copolymers.[82–84] Inspired by these results, these workers extended this concept of mixed-mechanism copolymerization to a block copolymer containing P3HT and an isocyanide monomer functionalized with perylene diimide, an effective organic electron-acceptor moiety (Figure 2.26).[85]

Although the backbone of the acceptor-containing copolymer is not fully conjugated, this type of polymer is potentially interesting in organic photovoltaics. As was seen with P3HT-*block*-poly(*n*-decyl-4-phenylisocyanobenzoate), the donor and acceptor morphology can be controlled if the polymers are well defined. Similar to the example of P3HT-*block*-poly(*n*-decyl-4-phenylisocyanobenzoate), the P3HT macroinitiator was chain-extended with the perylene diimide functionalized isocyanate monomer. Successful chain extension was observed up to 14 kg mol^{-1} and the resulting polymers had narrow dispersities. Films of the copolymers once again exhibited a nanofibrillar morphology with distinct phase separation between the donor and acceptor blocks (Figure 2.27).

In this example, the mixed-mechanism nature of these chain-growth polymerizations may limit the monomer scope to at least one non-conjugated block, whereas fully conjugated block copolymers may end up being more useful for electronic applications. Nonetheless, the mixed-mechanism approach to block copolymers represents the first such example of the synthesis of functional conjugated polymers, which could be expanded to a variety of functionalized isocyanide monomers in the future.

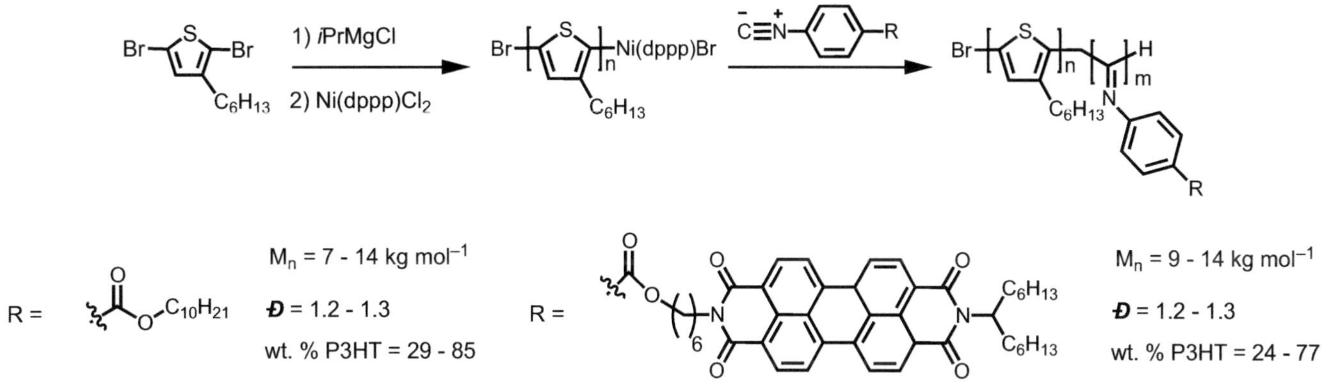

Figure 2.26 Synthesis, molecular weight range and dispersity values for P3HT-*block*-poly(*n*-decyl-4-phenylisocyanobenzoate) and P3HT-*block*-poly(perylene diimideisocyanide).

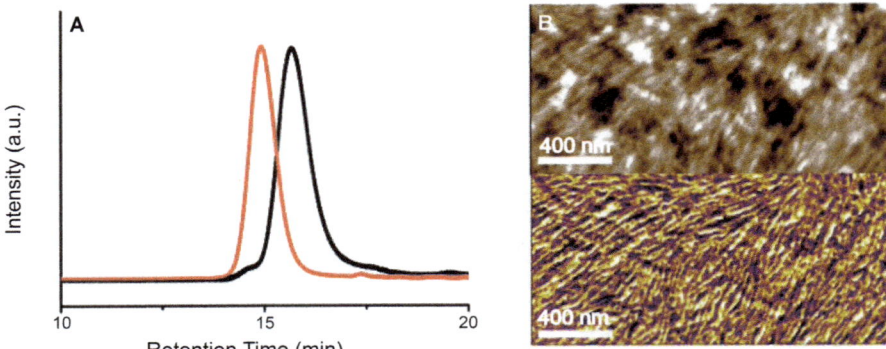

Figure 2.27 (A) Size-exclusion chromatography elution curves for the chain extension of a P3HT macroinitiator (black) to form P3HT-*block*-poly(*n*-decyl-
4-phenylisocyanobenzoate) (red). Adapted with permission from Z.-Q.
Wu, R. J. Ono, Z. Chen and C. W. Bielawski, *J. Am. Chem. Soc.*, 2010, **132**,
14000–14001.[80] Copyright 2010 American Chemical Society. (B) Atomic
force microscopy height and phase images of P3HT-*block*-poly(perylene
diimideisocyanide). Adapted from ref. 81 with permission from John
Wiley and Sons. Copyright © 2014 WILEY-VCH Verlag GmbH & Co.
KGaA, Weinheim.

2.6 Chain-growth Synthesis of Fully Conjugated Block Copolymers

The previous examples have all achieved CTP using a single activated
monomer. Controlled chain-growth syntheses of copolymers are advan-
tageous because they allow for control over monomer incorporation, sequence, and molecular weight.

Hashimoto and co-workers[86] and Jenekhe and co-workers[87] used CTP
methods to synthesize the first examples of fully conjugated block copolymers consisting of poly(3-octylthiophene) and poly(3-butylthiophene)
blocks, or poly(3-(2-ethylhexyl)thiophene) and P3HT blocks (Figure 2.28).
Thiophene copolymers exhibit interesting nanoscale phase separation and
self-assembly and these studies were followed up by others that investigated
related conjugated copolymers with various side-chains, backbone compositions, and sequences. Another system of great interest is copolymers that
contain both thiophene and selenophene because this allows control over
both their morphology and optoelectronic properties.[88–95] Conjugated
block copolymers exhibit distinct morphological or optoelectronic properties
compared with their respective blends and have favorable nanoscale
morphologies, higher thermostabilities, and solvent-switchable properties.

There is interest in extending the CTP mechanism to copolymers consisting of p-type (donor) and n-type (acceptor) units. Willot *et al.*[96] studied an
example of this with the synthesis of a P3HT–PTz block copolymer. Although
the procedure for the controlled synthesis of polythiophene is well

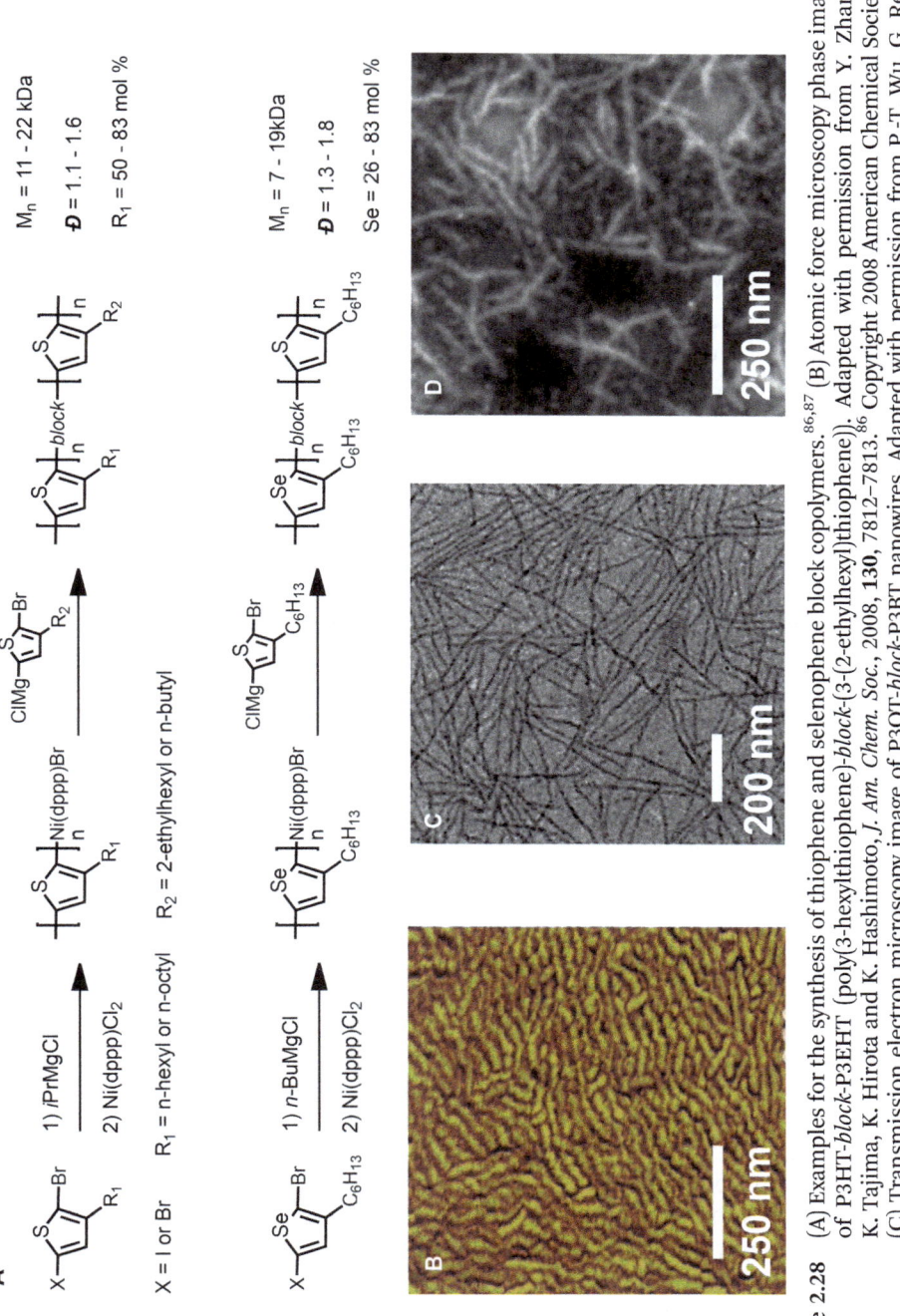

Figure 2.28 (A) Examples for the synthesis of thiophene and selenophene block copolymers.[86,87] (B) Atomic force microscopy phase image of P3HT-*block*-P3EHT (poly(3-hexylthiophene)-*block*-(3-(2-ethylhexyl)thiophene)). Adapted with permission from Y. Zhang, K. Tajima, K. Hirota and K. Hashimoto, *J. Am. Chem. Soc.*, 2008, **130**, 7812–7813.[86] Copyright 2008 American Chemical Society. (C) Transmission electron microscopy image of P3OT-*block*-P3BT nanowires. Adapted with permission from P.-T. Wu, G. Ren, C. Li, R. Mezzenga and S. A. Jenekhe, *Macromolecules*, 2009, **42**, 2317–2320.[87] Copyright 2009 American Chemical Society. (D) Scanning electron microscopy image of P3HS-*block*-P3HT. Adapted with permission from J. Hollinger, A. A. Jahnke, N. Coombs and D. S. Seferos, *J. Am. Chem. Soc.*, 2010, **132**, 8546–8547.[82] Copyright 2010 American Chemical Society.

established, less is known about the polymerization of thienopyrazines. An in-depth study of thienopyrazine polymerization showed that CTP was limited to around eight monomer units, affording little to no ability to control the chain length by varying the catalyst loading. In spite of the very limited control over PTz, a block copolymer can be synthesized by chain extension with PTz when using P3HT as a macroinitiator (Figure 2.29).

This strategy does not allow for control over the molecular weight of the PTz block; however, it shows that it is possible to synthesize fully conjugated block copolymers *via* chain growth as long as the polymerization of the first block is adequately controlled. As PTz is a narrow HOMO–LUMO gap polymer ($E_g = 1.1$ eV), P3HT-*b*-PTz exhibited very broad absorption and films of these block copolymers undergo some degree of phase separation (Figure 2.30). These characteristics make P3HT-*block*-PTz an interesting panchromatic donor–acceptor block copolymer for single-component all-polymer solar cells.

To synthesize donor–acceptor block copolymers using CTP, it is necessary to use a catalyst that can coordinate to both the electron-rich and electron-poor monomers equally strongly. We calculated the association strength of MesAn to thiophene and benzotriazole using DFT and found that it is strong enough to allow for CTP in both instances. When synthesizing P3HT and PBTz homopolymers with MesAn, we observed good control over the molecular weight as a function of catalyst loading and narrow dispersities for both polymers. A more in-depth study of the molecular weight progression throughout the polymerization indicated that the controlled polymerization mechanism remained intact for both monomer types past 80% conversion (Figure 2.31).

After demonstrating that we could control the polymerization of each of these monomers independently, we synthesized the first example of a fully conjugated donor–acceptor block copolymer where controlled CTP occurred for both blocks: P3HT-*block*-PBTz (Figure 2.32).[97]

In this example, a "living" chain of either P3HT or PBTz is extended to form P3HT-*block*-PBTz. After adding the second activated monomer, the SEC trace shows that the polymer distribution completely shifts to higher molecular weight from the macroinitiator. CTP is also evidenced by the fact that, in all cases, the composition of the P3HT-*block*-PBTz closely matched the monomer feed ratio, the molecular weight reflected the catalyst loading, and the resulting block copolymers had narrow dispersities. P3HT-*block*-PBTz exhibited phase separation between the donor and acceptor in thin films (Figure 2.33). This behavior is typical of block copolymers and potentially useful for organic electronic devices.

This block copolymer can be synthesized in either monomer order and was the first example of successful chain extension from a "living" electron-deficient (n-type) polymer chain. This result is important as, until now, all chain extension conjugated block copolymers have relied on a P3HT macroinitiator because it is well controlled during polymerization. This opens the door to the synthesis and study of block copolymers that are free of thiophene blocks.

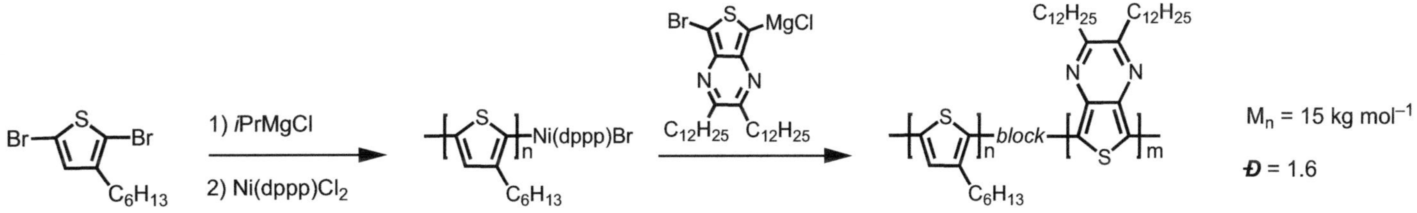

Figure 2.29 Synthesis, molecular weight range and dispersity values for P3HT-*block*-PTz.

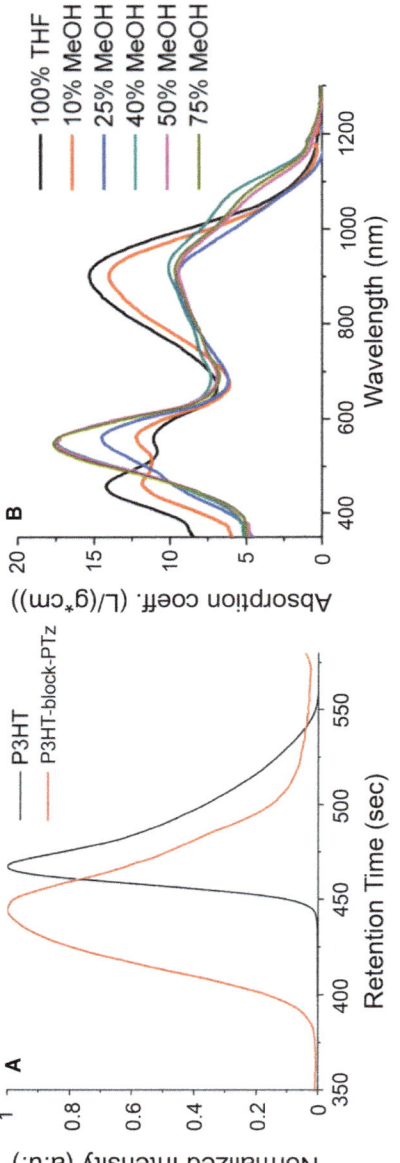

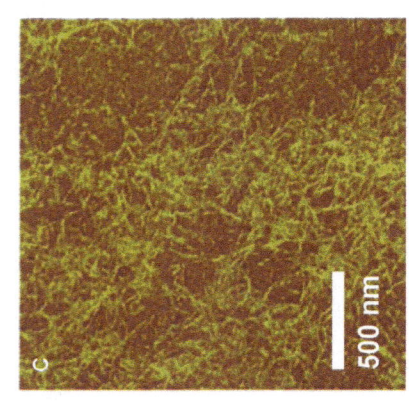

Figure 2.30 (A) Chain extension of the P3HT-*block*-PTz block copolymer. (B) Absorbance spectra of P3HT-*block*-PTz. (C) Atomic force microscopy image of a P3HT-*block*-PTz thin film. Adapted with permission from P. Willot, D. Moerman, P. Leclère, R. Lazzaroni, Y. Baeten, M. Van der Auweraer and G. Koeckelberghs, *Macromolecules*, 2014, *47*, 6671–6678.[96] Copyright 2014 American Chemical Society.

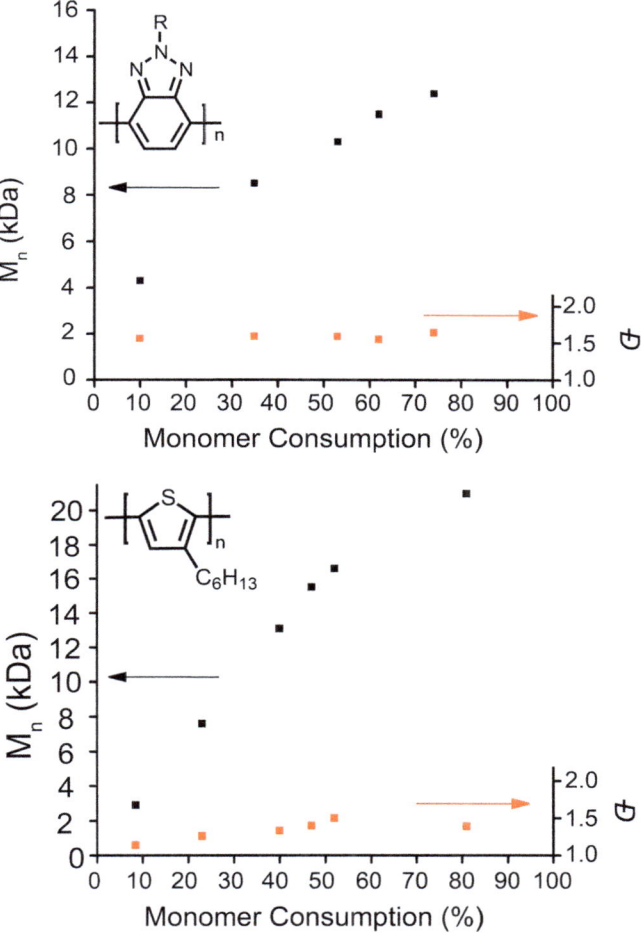

Figure 2.31 Molecular weight (black) and dispersity (red) as a function of monomer consumption for PBTz and P3HT using MesAn as a catalyst. Adapted with permission from C. R. Bridges, H. Yan, A. A. Pollit and D. S. Seferos, *ACS Macro Lett.*, 2014, **3**, 671–674.[97] Copyright 2014 American Chemical Society.

2.7 Chain-growth Synthesis of Other Copolymers

Studies on high-performance donor–acceptor alternating copolymers have shown that the ideal ratio of donor to acceptor in the polymer may not be exactly 50 : 50.[37,38,98–100] From a synthetic standpoint the polycondensation of a 50 : 50 alternating polymer is far easier because it only requires the synthesis of two monomer types. Statistical copolymers synthesized with varying compositions of donor and acceptor monomer require the synthesis of at least three different monomers. CTP is an attractive method to synthesize statistical donor–acceptor copolymers because it only requires the

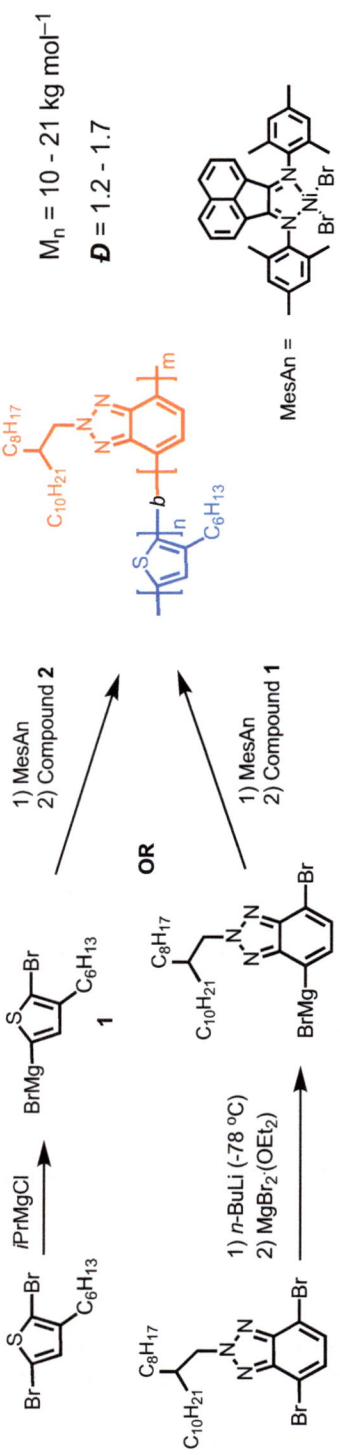

Figure 2.32 Synthesis, molecular weight range and dispersity values for P3HT-*block*-PBTz.

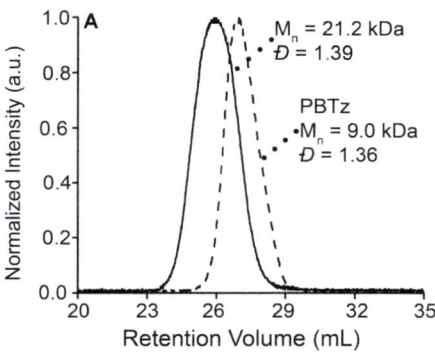
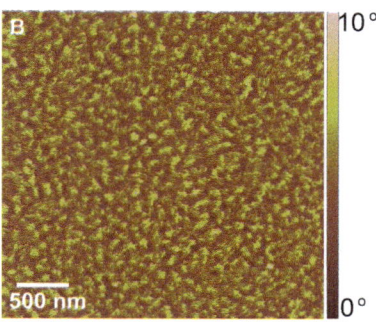

Figure 2.33   (A) Chain extension of a PBTz macroinitiator (dashed line) to form PBTz-*block*-P3HT (solid line). (B) Atomic force microscopy phase image of a PBTz-*block*-P3HT block copolymer thin film.
Adapted with permission from C. R. Bridges, H. Yan, A. A. Pollit and D. S. Seferos, *ACS Macro Lett.*, 2014, 3, 671–674.[97] Copyright 2014 American Chemical Society.

synthesis of one donor and one acceptor monomer with identical reactive groups, the monomer ratios can be varied, and it should control the molecular weight and lead to polymers with narrow dispersities.

Considering that PTNDIT exhibits such favorable n-type electronic properties, copolymers consisting of TNDIT and a p-type monomer may have useful optoelectronic properties. Previous work has shown that both p-type fluorene and n-type TNDIT monomers can be polymerized using a limited CTP mechanism.[78,101] Extending this work, Tkachov *et al.*[102] attempted to synthesize a random copolymer of fluorene and TNDIT by mixing the respective activated monomers with a palladium catalyst (Figure 2.34).

In this example there was a significant difference in reactivity between these two monomers and the fluorene monomer was completely consumed before a significant amount of the TNDIT monomer could react. The resulting polymer was block-like; PF was the first block, followed by a small PF-PTNDIT gradient region, followed by a pure PTNDIT block. The SEC elution curve steadily progressed to higher molecular weights throughout the polymerization; however, it became distinctly bimodal (Figure 2.35).

This indicates that there was a significant amount of PTNDIT homopolymer forming later in the polymerization as the CTP mechanism failed. Thin films of this copolymer–homopolymer mixture formed nanofibres and the domains appeared to undergo macro-scale phase separation.

Wang *et al.*[103] studied a similar system whereby the same TNDIT alternating copolymer was used as a macroinitiator for P3HT. Films composed of this polymer also exhibited some degree of phase separation, presumably between the p and n portions of the block copolymer (Figure 2.36).

Extending controlled chain-growth polymerizations to n-type monomers is beginning to allow the synthesis and study of p–n block copolymers.

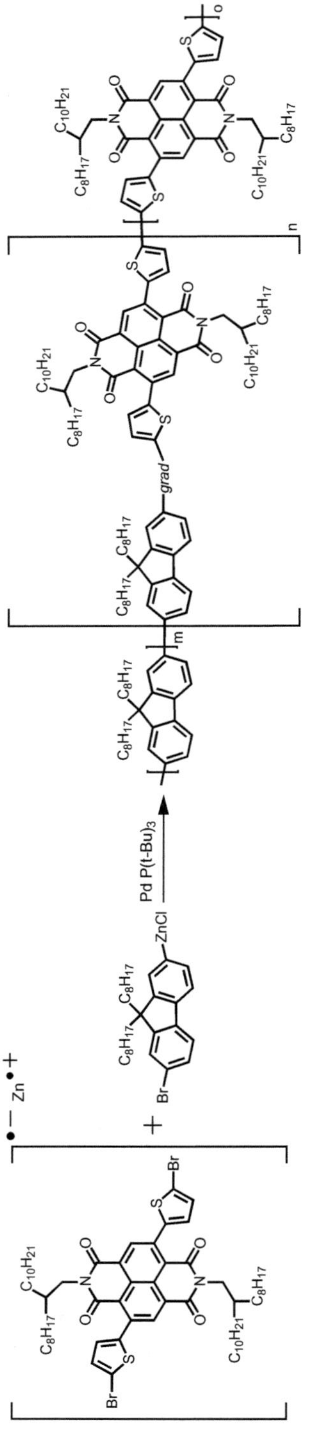

Figure 2.34 Synthesis of PF-*block*-[PF-*grad*-PTNDIT]-*block*-PTNDIT.

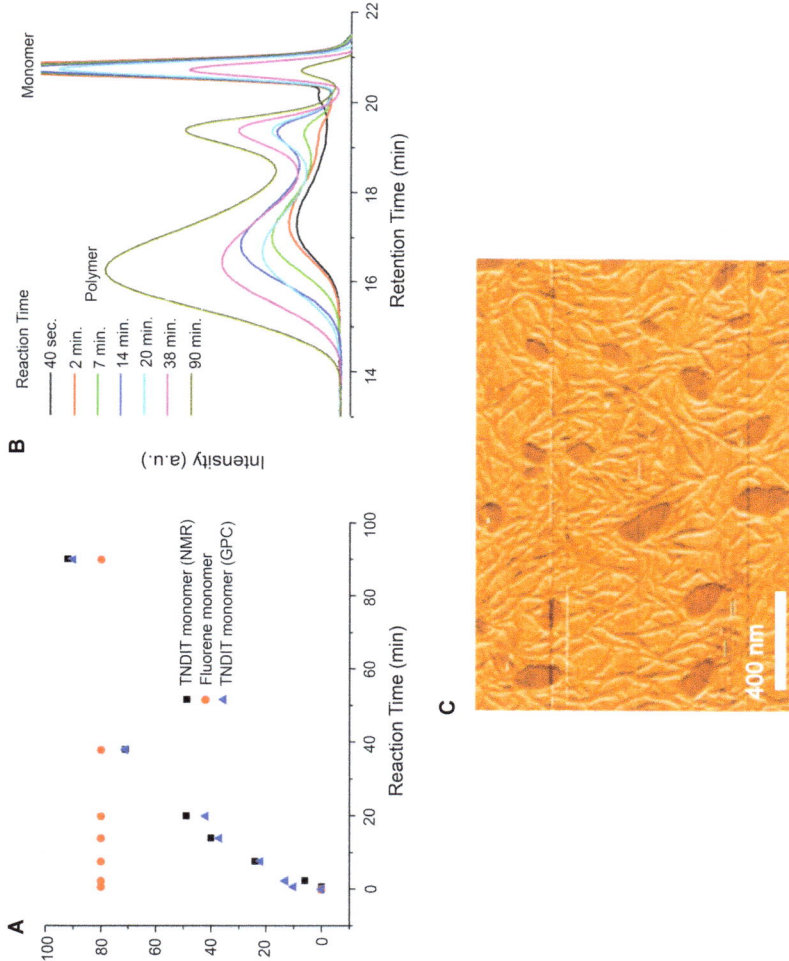

Figure 2.35 (A) Monomer conversion as a function of time for the copolymerization of fluorene and TNDIT. (B) Size-exclusion chromatography elution curves of PF-PTNDIT copolymers. (C) Atomic force microscopy phase image of a PF-PTNDIT copolymer film.

Adapted with permission from R. Tkachov, H. Komber, S. Rauch and A. Lederer, *Macromolecules*, 2014, **15**, 4994–5001.[102] Copyright 2014 American Chemical Society.

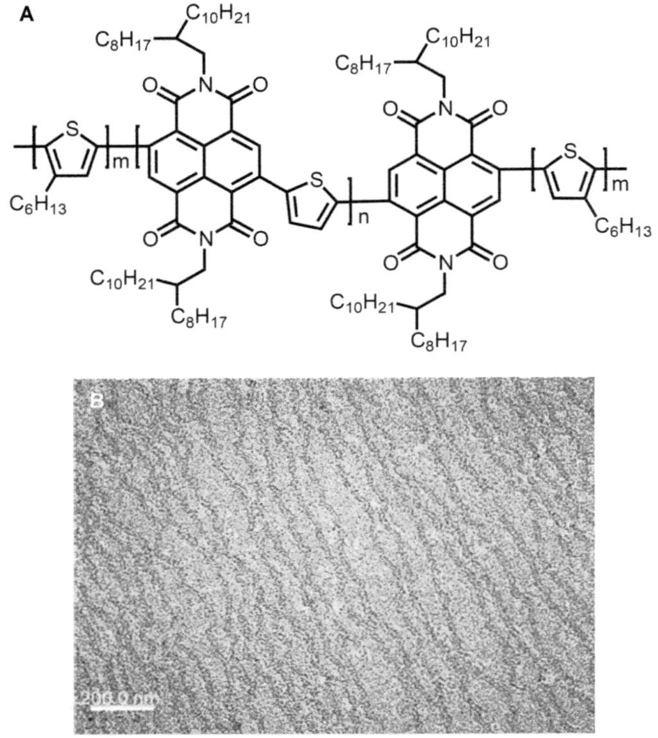

Figure 2.36 (A) Chemical structure and (B) transmission electron microscopy image of a thin film of PNDIT-*block*-P3HT.
Adapted with permission from J. Wang, M. Ueda and T. Higashihara, *ACS Macro Lett.*, 2013, **2**, 506–510.[103] Copyright 2014 American Chemical Society.

The nanoscale phase separation observed between donor (p-type) and acceptor (n-type) regions in the previously presented block copolymers has long been sought in organic solar cells. Using block copolymers in the active layer of solar cells might also afford some control over the nanoscale morphology, which is important for charge separation and transport. A polymer consisting of a wide bandgap donor and a narrow bandgap acceptor would also allow panchromatic absorption of the solar spectrum. Greater thermal stability for single polymer devices is also expected.

We have also studied the statistical copolymerization of a 4,4-bis(2-ethylhexyl)-4*H*-silolo[3,2-*b*:4,5-*b*′]dithiophene (DTS) donor and a BTz acceptor using the previously developed MesAn catalyst.[104] As each individual monomer exhibited controlled chain-growth behavior under similar conditions, donor–acceptor statistical copolymers may be synthesized simply by adding MesAn to a solution containing variable amounts of each activated monomer (Figure 2.37).

Figure 2.37 Synthesis, molecular weight range and dispersity values for BTz-*stat*-DTS copolymers.

If the reaction proceeds by the CTP mechanism, it would also afford a way to concurrently control the composition and molecular weight of the copolymer. These polymers exhibit the broad, red-shifted absorbance typical of donor–acceptor copolymers and composition-dependent optical properties. Unfortunately, the resulting BTz-*stat*-DTS copolymers could only be synthesized with low molecular weights and relatively broad dispersities. This is unexpected because high molecular weight, narrow dispersity homopolymers could be produced using each monomer under identical conditions. For statistical copolymerizations of this type, problems can arise for several reasons. As both monomers must be activated in the same solution, Grignard exchange between the monomers is likely and will reach an equilibrium dictated by the basicity of each Grignard species. These exchange reactions can form chain-terminating or chain-coupling species and limit controlled chain growth. The reactivity of the catalyst toward both monomers must also be very similar to prevent block copolymer formation as opposed to the desired statistical copolymers. With these monomers, cross metathesis resulted in early chain termination, making it difficult to study any chain-growth character of the polymerization. To show that CTP could occur with this monomer combination, we conducted a chain extension experiment whereby a PBTz macroinitiator was extended with a mixture of activated DTS and BTz monomers, forming a PBTz-*block*-(DTS-*stat*-BTz) copolymer (Figure 2.38).

Because termination reactions are not able to occur until later in the polymerization, this copolymer can be synthesized with much higher molecular weights (16 kg mol^{-1}). This result shows that CTP may be possible for statistical copolymers, but having multiple activated monomers present in one solution complicates the polymerization and potentially disrupts the CTP mechanism. This work represents the first copolymerization using chain-growth methods where neither monomer is thiophene and the polymer itself is structurally very similar to many of the high-performance donor–acceptor alternating copolymers. Further optimization of the reaction conditions and monomer combinations should lead to the controlled synthesis of fully conjugated statistical copolymers.

2.8   Future Outlook

From thiophenes to perylene diimides, great progress has been made in extending controlled chain-growth polymerization to a variety of monomers and polymerization methods. The monomer scope for CTP has been extended to many aromatic species, including a few very large and/or electron-deficient monomers. Improving control over the polymerization using judicious monomer selection and catalyst design has been a key factor in these developments and was made possible by previous fundamental studies on the underlying mechanism for these controlled polymerizations. However, what is most exciting is what is just around the corner. Although controlled chain-growth polymerization methods for conjugated polymers have traditionally been Kumada-type CTP, Suzuki coupling polymerizations

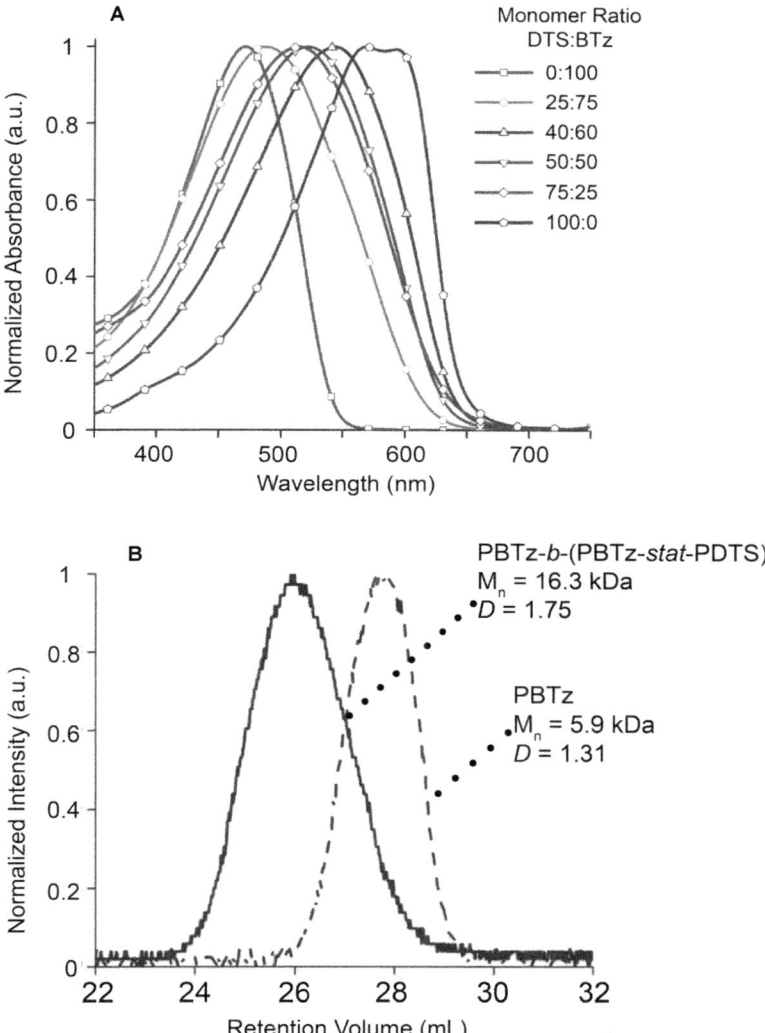

Figure 2.38 (A) Composition dependence on the optical properties of PBTz-*stat*-PDTS. (B) Size-exclusion chromatography elution curves showing the chain extension of PBTz-*block*-(PBTz-*stat*-PDTS) from PBTz.
Adapted from ref. 104 with permission from John Wiley and Sons. Copyright © 2014 WILEY-VCH Verlag GmbH & Co. KGaA, Weinheim.

and, more recently, C–H activation reactions have been shown to exhibit some characteristics of CTP.[36–38,72,105] Advancing all these methods will allow for a wider range of monomer types to become applicable to controlled polymerization. Important work remains, however. For instance, we need to extend CTP to be more generally applicable to monomer units commonly found in high-performance organic electronics. Doing so would allow for the

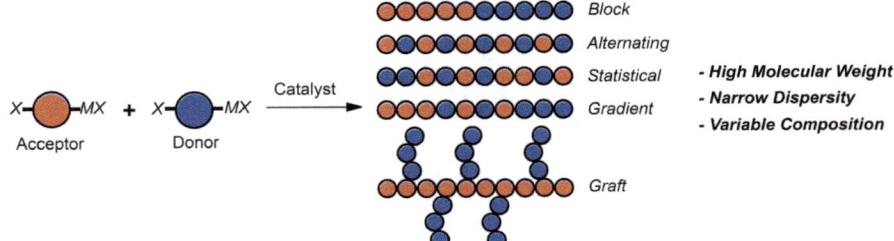

Figure 2.39 Synthesis of arbitrary donor–acceptor fully conjugated copolymers of varying architecture and composition.

synthesis of narrow dispersity, high molecular weight donor–acceptor statistical, block, gradient, and alternating copolymers (Figure 2.39).

There is also a need to move beyond linear chains to form branched systems. This will not only allow researchers more control over polymer composition, molecular weight and end-groups, but also crystallinity, phase structure, and thermal properties. These advances seem imminent and will undoubtedly generate a wealth of new knowledge in the field of polymer science. They also may boost the effectiveness of polymeric materials in devices into the commercial realm.

References

1. A. Yokoyama, R. Miyakoshi and T. Yokozawa, *Macromolecules*, 2004, **37**, 1169–1171.
2. R. Miyakoshi, A. Yokoyama and T. Yokozawa, *J. Am. Chem. Soc.*, 2005, **127**, 17542–17547.
3. T. Yokozawa and Y. Ohta, *Chem. Rev.*, 2016, **116**, 1950–1968.
4. E. E. Sheina, J. Liu, M. C. Iovu, D. W. Laird and R. D. McCullough, *Macromolecules*, 2004, **37**, 3526–3528.
5. M. C. Iovu, E. E. Sheina, R. R. Gil and R. D. McCullough, *Macromolecules*, 2005, **38**, 8649–8656.
6. R. S. Loewe, P. C. Ewbank, J. Liu, L. Zhai and R. D. McCullough, *Macromolecules*, 2001, **34**, 4324–4333.
7. G. Odian, *Principles of Polymerization*, Wiley, New York, 4th edn, 2004.
8. K. Matyjaszewski and J. Xia, *Chem. Rev.*, 2001, **101**, 2921–2990.
9. J. Liu, R. S. Loewe and R. D. McCullough, *Macromolecules*, 1999, **32**, 5777–5785.
10. F. Monnaie, W. Brullot, T. Verbiest, J. De Winter, P. Gerbaux, A. Smeets and G. Koeckelberghs, *Macromolecules*, 2013, **46**, 8500–8508.
11. L. M. Kozycz, D. Gao and D. S. Seferos, *Macromolecules*, 2013, **46**, 613–621.
12. M. Verswyvel, F. Monnaie and G. Koeckelberghs, *Macromolecules*, 2011, **44**, 9489–9498.
13. Y. Zhang, K. Tajima, K. Hirota and K. Hashimoto, *J. Am. Chem. Soc.*, 2008, **130**, 7812–7813.

14. U. Scherf, S. Adamczyk, A. Gutacker and N. Koenen, *Macromol. Rapid Commun.*, 2009, **30**, 1059–1065.
15. C.-A. Dai, W.-C. Yen, Y.-H. Lee, C.-C. Ho and W.-F. Su, *J. Am. Chem. Soc.*, 2007, **129**, 11036–11038.
16. Z. Chen, H. Lemke, S. Albert-Seifried, M. Caironi, M. M. Nielsen, M. Heeney, W. Zhang, I. McCulloch and H. Sirringhaus, *Adv. Mater.*, 2010, **22**, 2371–2375.
17. Y. Chujo, *Conjugated Polymer Synthesis*, Wiley-VCH, Weinheim, 2010.
18. M. Leclerc and J.-F. Morin, *Design and Synthesis of Conjugated Polymers*, Wiley-VCH, Weinheim, 2010.
19. S. Himmelberger, K. Vandewal, Z. Fei, M. Heeney and A. Salleo, *Macromolecules*, 2014, **47**, 7151–7157.
20. D. M. DeLongchamp, R. J. Kline, E. K. Lin, D. A. Fischer, L. J. Richter, L. A. Lucas, M. Heeney, I. McCulloch and J. E. Northrup, *Adv. Mater.*, 2007, **19**, 833–837.
21. R. J. Kline, M. D. McGehee, E. N. Kadnikova, J. Liu, J. M. J. Fréchet and M. F. Toney, *Macromolecules*, 2005, **38**, 3312–3319.
22. R. J. Kline, M. D. McGehee, E. N. Kadnikova, J. Liu and J. M. J. Fréchet, *Adv. Mater.*, 2003, **15**, 1519–1522.
23. L. M. Kozycz, D. Gao, J. Hollinger and D. S. Seferos, *Macromolecules*, 2012, **45**, 5823–5832.
24. A. Yokoyama, A. Kato, R. Miyakoshi and T. Yokozawa, *Macromolecules*, 2008, **41**, 7271–7273.
25. E. F. Palermo and A. J. McNeil, *Macromolecules*, 2012, **45**, 5948–5955.
26. F. Boon, N. Hergué, G. Deshayes, D. Moerman, S. Desbief, J. De Winter, P. Gerbaux, Y. H. Geerts, R. Lazzaroni and P. Dubois, *Polym. Chem.*, 2013, **4**, 4303–4307.
27. F. Monnaie, M.-P. Van Den Eede and G. Koeckelberghs, *Macromolecules*, 2015, **48**, 8121.
28. R. Miyakoshi, K. Shimono, A. Yokoyama and T. Yokozawa, *J. Am. Chem. Soc.*, 2006, **128**, 16012–16013.
29. S. Wu, L. Bu, L. Huang, X. Yu, Y. Han, Y. Geng and F. Wang, *Polymer*, 2009, **50**, 6245–6251.
30. M. Heeney, W. Zhang, D. J. Crouch, M. L. Chabinyc, S. Gordeyev, R. Hamilton, S. J. Higgins, I. McCulloch, P. J. Skabara, D. Sparrowe and S. Tierney, *Chem. Commun.*, 2007, **47**, 5061–5063.
31. J. Kettle, M. Horie, L. A. Majewski, B. R. Saunders, S. Tuladhar, J. Nelson and M. L. Turner, *Sol. Energy Mater. Sol. Cells*, 2011, **95**, 2186–2193.
32. P. M. Beaujuge, C. M. Amb and J. R. Reynolds, *Acc. Chem. Res.*, 2010, **43**, 1396–1407.
33. H. Zhou, L. Yang, S. Stoneking and W. You, *ACS Appl. Mater. Interfaces*, 2010, **2**, 1377–1383.
34. Z. G. Zhang and J. Wang, *J. Mater. Chem.*, 2012, **22**, 4178–4187.
35. M. Schubert, D. Dolfen, J. Frisch, S. Roland, R. Steyrleuthner, B. Stiller, Z. H. Chen, U. Scherf, N. Koch, A. Facchetti and D. Neher, *Adv. Energy Mater.*, 2012, **2**, 369–380.

36. T. Yokozawa, R. Suzuki, M. Nojima, Y. Ohta and A. Yokoyama, *Macromol. Rapid Commun.*, 2011, **32**, 801–806.
37. W. A. Braunecker, S. D. Oosterhout, Z. R. Owczarczyk, N. Kopidakis, E. L. Ratcliff, D. S. Ginley and D. C. Olson, *ACS Macro Lett.*, 2014, **3**, 622–627.
38. M. J. Robb, S.-Y. Ku and C. J. Hawker, *Adv. Mater.*, 2013, **25**, 5686–5700.
39. A. Kiriy, V. Senkovskyy and M. Sommer, *Macromol. Rapid Commun.*, 2011, **32**, 1503–1517.
40. R. Tkachov, V. Senkovskyy, H. Komber, J.-U. Sommer and A. Kiriy, *J. Am. Chem. Soc.*, 2010, **132**, 7803–7810.
41. Z. J. Bryan and A. J. McNeil, *Macromolecules*, 2013, **46**, 8395–8405.
42. S. R. Lee, J. W. G. Bloom, S. E. Wheeler and A. J. McNeil, *Dalton Trans.*, 2013, **42**, 4218–4222.
43. S. R. Lee, Z. J. Bryan, A. M. Wagner and A. J. McNeil, *Chem. Sci.*, 2012, **3**, 1562–1566.
44. E. L. Lanni, J. R. Locke, C. M. Gleave and A. J. McNeil, *Macromolecules*, 2011, **44**, 5136–5145.
45. E. L. Lanni and A. J. McNeil, *Macromolecules*, 2010, **43**, 8039–8044.
46. Z. J. Bryan and A. J. McNeil, *Chem. Sci.*, 2013, **4**, 1620–1624.
47. T. Beryozkina, V. Senkovskyy, E. Kaul and A. Kiriy, *Macromolecules*, 2008, **41**, 7817–7823.
48. T. Yokozawa, Y. Nanashima and Y. Ohta, *ACS Macro Lett.*, 2012, **1**, 862–866.
49. L. Wen, B. C. Duck, P. C. Dastoor and S. C. Rasmussen, *Macromolecules*, 2008, **41**, 4576–4578.
50. D. D. Kenning and S. C. Rasmussen, *Macromolecules*, 2003, **36**, 6298–6299.
51. D. D. Kenning, K. A. Mitchell, T. R. Calhoun, M. R. Funfar, D. J. Sattler and S. C. Rasmussen, *J. Org. Chem.*, 2002, **67**, 9073–9076.
52. L. Huang, S. Wu, Y. Qu, Y. Geng and F. Wang, *Macromolecules*, 2008, **41**, 8944–8947.
53. M. C. Stefan, A. E. Javier, I. Osaka and R. D. McCullough, *Macromolecules*, 2009, **42**, 30–32.
54. A. E. Javier, S. R. Varshney and R. D. McCullough, *Macromolecules*, 2010, **43**, 3233–3237.
55. A. Sui, X. Shi, S. Wu, H. Tian and Y. Geng, *Macromolecules*, 2012, **45**, 5436–5443.
56. Z. J. Bryan, M. L. Smith and A. J. McNeil, *Macromol. Rapid Commun.*, 2012, **33**, 842–847.
57. C. A. Traina, R. C. Bakus II and G. C. Bazan, *J. Am. Chem. Soc.*, 2011, **133**, 12600–12607.
58. E. Elmalem, F. Biedermann, K. Johnson, R. H. Friend and W. T. S. Huck, *J. Am. Chem. Soc.*, 2012, **134**, 17769–17777.
59. H. Bronstein, M. Hurhangee, E. C. Fregoso, D. Beatrup, Y. W. Soon, Z. Huang, A. Hadipour, P. S. Tuladhar, S. Rossbauer and E.-H. Sohn, *Chem. Mater.*, 2013, **25**, 4239–4249.
60. F. Pammer and U. Passlack, *ACS Macro Lett.*, 2014, **3**, 170–174.

61. F. Pammer, J. Jäger, B. Rudolf and Y. Sun, *Macromolecules*, 2014, **17**, 5904–5912.
62. Y. Nanashima, R. Shibata, R. Miyakoshi, A. Yokoyama and T. Yokozawa, *J. Polym. Sci., Part A: Polym. Chem.*, 2012, **50**, 3628–3640.
63. Y. Nanashima, A. Yokoyama and T. Yokozawa, *Macromolecules*, 2012, **45**, 2609–2613.
64. Y. Nanashima, A. Yokoyama and T. Yokozawa, *J. Polym. Sci., Part A: Polym. Chem.*, 2012, **50**, 1054–1061.
65. C. R. Bridges, T. M. McCormick, G. L. Gibson, J. Hollinger and D. S. Seferos, *J. Am. Chem. Soc.*, 2013, **135**, 13212–13219.
66. M. Verswyvel, P. Verstappen, L. De Cremer, T. Verbiest and G. Koeckelberghs, *J. Polym., Sci. Part A: Polym. Chem.*, 2011, **49**, 5339–5349.
67. M. Verswyvel, J. Steverlynck, S. Hadj Mohamed, M. Trabelsi, B. Champagne and G. Koeckelberghs, *Macromolecules*, 2014, **47**, 4668–4675.
68. P. Willot and G. Koeckelberghs, *Macromolecules*, 2014, **47**, 8548–8555.
69. P. Willot, S. Govaerts and G. Koeckelberghs, *Macromolecules*, 2013, **46**, 8888–8895.
70. M. Verswyvel, C. Hoebers, J. De Winter, P. Gerbaux and G. Koeckelberghs, *J. Polym. Sci., Part A: Polym. Chem.*, 2013, **51**, 5067–5074.
71. R. J. Ono, S. Kang and C. W. Bielawski, *Macromolecules*, 2012, **45**, 2321–2326.
72. E. Elmalem, A. Kiriy and W. T. S. Huck, *Macromolecules*, 2011, **44**, 9057–9061.
73. H. Yan, Z. Chen, Y. Zheng, C. Newman, J. R. Quinn, F. Dötz, M. Kastler and A. Facchetti, *Nature*, 2008, **457**, 679–686.
74. Z. Chen, Y. Zheng, H. Yan and A. Facchetti, *J. Am. Chem. Soc.*, 2009, **131**, 8–9.
75. ActiveInk™ N2200 www.polyera.com.
76. V. Senkovskyy, R. Tkachov, H. Komber, M. Sommer, M. Heuken, B. Voit, W. T. S. Huck, V. Kataev, A. Petr and A. Kiriy, *J. Am. Chem. Soc.*, 2011, **133**, 19966–19970.
77. V. Senkovskyy, R. Tkachov, H. Komber, A. John, J.-U. Sommer and A. Kiriy, *Macromolecules*, 2012, **45**, 7770–7777.
78. R. Tkachov, Y. Karpov, V. Senkovskyy, I. Raguzin, J. Zessin, A. Lederer, M. Stamm, B. Voit, T. Beryozkina, V. Bakulev, W. Zhao, A. Facchetti and A. Kiriy, *Macromolecules*, 2014, **47**, 3845–3851.
79. W. Liu, R. Tkachov, H. Komber, V. Senkovskyy, M. Schubert, Z. Wei, A. Facchetti, D. Neher and A. Kiriy, *Polym. Chem.*, 2014, **5**, 3404–3411.
80. Z.-Q. Wu, R. J. Ono, Z. Chen and C. W. Bielawski, *J. Am. Chem. Soc.*, 2010, **132**, 14000–14001.
81. R. J. Ono, A. D. Todd, Z. Hu, D. A. Vanden Bout and C. W. Bielawski, *Macromol. Rapid Commun.*, 2013, **35**, 204–209.
82. J. Hollinger, A. A. Jahnke, N. Coombs and D. S. Seferos, *J. Am. Chem. Soc.*, 2010, **132**, 8546–8547.
83. J. Hollinger and D. S. Seferos, *Macromolecules*, 2014, **15**, 5002–5009.

84. J. Hollinger, P. M. DiCarmine, D. Karl and D. S. Seferos, *Macromolecules*, 2012, **45**, 3772–3778.

85. A. Facchetti, *Mater. Today*, 2013, **16**, 123–132.

86. Y. Zhang, K. Tajima, K. Hirota and K. Hashimoto, *J. Am. Chem. Soc.*, 2008, **130**, 7812–7813.

87. P.-T. Wu, G. Ren, C. Li, R. Mezzenga and S. A. Jenekhe, *Macromolecules*, 2009, **42**, 2317–2320.

88. D. Gao, J. Hollinger and D. S. Seferos, *ACS Nano*, 2012, **6**, 7114–7121.

89. J. Hollinger, J. Sun, D. Gao, D. Karl and D. S. Seferos, *Macromol. Rapid Commun.*, 2013, **34**, 437–441.

90. M. Baghgar, A. M. Barnes, E. Pentzer, A. J. Wise, B. A. G. Hammer, T. Emrick, A. D. Dinsmore and M. D. Barnes, *ACS Nano*, 2014, **8**, 8344–8349.

91. U. Scherf, S. Adamczyk, A. Gutacker and N. Koenen, *Macromol. Rapid Commun.*, 2009, **30**, 1059–1065.

92. A. Yokoyama, A. Kato, R. Miyakoshi and T. Yokozawa, *Macromolecules*, 2008, **41**, 7271–7273.

93. R. Verduzco, I. Botiz, D. L. Pickel, S. M. Kilbey II, K. Hong, E. Dimasi and S. B. Darling, *Macromolecules*, 2011, **44**, 530–539.

94. S. Wu, L. Bu, L. Huang, X. Yu, Y. Han, Y. Geng and F. Wang, *Polymer*, 2009, **50**, 6245–6251.

95. C. Guo, Y.-H. Lin, M. D. Witman, K. A. Smith, C. Wang, A. Hexemer, J. Strzalka, E. D. Gomez and R. Verduzco, *Nano Lett.*, 2013, **13**, 2957–2963.

96. P. Willot, D. Moerman, P. Leclère, R. Lazzaroni, Y. Baeten, M. Van der Auweraer and G. Koeckelberghs, *Macromolecules*, 2014, **47**, 6671–6678.

97. C. R. Bridges, H. Yan, A. A. Pollit and D. S. Seferos, *ACS Macro Lett.*, 2014, **3**, 671–674.

98. A. E. Rudenko, P. P. Khlyabich and B. C. Thompson, *ACS Macro Lett.*, 2014, **3**, 387–392.

99. P. P. Khlyabich, B. Burkhart, C. F. Ng and B. C. Thompson, *Macromolecules*, 2011, **44**, 5079–5084.

100. B. Burkhart, P. P. Khlyabich and B. C. Thompson, *Macromol. Chem. Phys.*, 2012, **214**, 681–690.

101. R. Tkachov, V. Senkovskyy, T. Beryozkina, K. Boyko, V. Bakulev, A. Lederer, K. Sahre, B. Voit and A. Kiriy, *Angew. Chem., Int. Ed.*, 2014, **53**, 2402–2407.

102. R. Tkachov, H. Komber, S. Rauch and A. Lederer, *Macromolecules*, 2014, **47**, 4994–5001.

103. J. Wang, M. Ueda and T. Higashihara, *ACS Macro Lett.*, 2013, **2**, 506–510.

104. A. A. Pollit, C. R. Bridges and D. S. Seferos, *Macromol. Rapid Commun.*, 2015, **36**, 65–70.

105. J. Zhang, W. Chen, A. J. Rojas, E. V. Jucov, T. V. Timofeeva, T. C. Parker, S. Barlow and S. R. Marder, *J. Am. Chem. Soc.*, 2013, **135**, 16376–16379.

CHAPTER 3

Application of Catalyst Transfer Polymerizations: From Conjugated Copolymers to Polymer Brushes

YANHOU GENG* AND AIGUO SUI

State Key Laboratory of Polymer Physics and Chemistry, Changchun Institute of Applied Chemistry, Chinese Academy of Sciences, Changchun 130022, China
*Email: yhgeng@ciac.ac.cn

3.1 Introduction

Compared with polycondensation reactions catalyzed by transition metals that are generally used for the synthesis of conjugated polymers, catalyst transfer polymerization (CTP) reactions are characterized by a chain-growth mechanism based on the repeated intramolecular transfer of the catalyst species toward the chain ends.[1,2] CTP reactions based on Kumada, Negeshi, Suzuki–Miyaura and Stille coupling reactions have been demonstrated.[3-8] The CTP of some aromatic monomers exhibit a "living" or "quasi-living" nature, allowing the controlled synthesis of novel conjugated polymers such as conjugated block copolymers, gradient-conjugated copolymers, and conjugated polymer brushes,[2,9-12] which are not accessible *via* conventional polycondensation reactions.

RSC Polymer Chemistry Series No. 21
Semiconducting Polymers: Controlled Synthesis and Microstructure
Edited by Christine Luscombe

CTP is very important as a new protocol for at least four reasons. First, the properties of conjugated polymers are often influenced by structural parameters, such as the molecular weight, dispersity, and end-groups. The ability to control CTP reactions allows these types of materials to be produced with defined structures in a highly reproducible way. Second, both the optoelectronic properties and chemical structures of conjugated polymers are strongly influenced by their micro- and nanostructure in solid or aggregated states and morphology control has been shown to be a key factor in the fabrication of high-performance electronic devices using conjugated polymers. For instance, nanostructured films consisting of interpenetrated bi-continuous phases with nanometer-scale interspacings formed by donor and acceptor materials are highly desirable for the preparation of high-performance organic photovoltaic (OPV) solar cells. Block copolymers are capable of self-organizing into microphase-separated nanostructures driven by immiscibility or the difference in crystallinity between the two blocks. However, it is challenging to synthesize conjugated block copolymers with well-defined structures *via* conventional polycondensation. The chain-growth and "living" characteristics of CTPs make this synthesis feasible. Third, using this new protocol, conjugated polymers with defined chain ends can be produced and other functional groups or polymer chains can then be introduced *via* post-macromolecular reactions or polymerizations. These new conjugated polymers are intriguing in a number of interdisciplinary fields, such as supramolecular chemistry, biosensing and bio-targeting. Fourth, CTP also allows the preparation of conjugated polymers with special topologies – such as conjugated polymer brushes, star-like conjugated polymers, and conjugated macro-rings – promoting the exploration of new applications of conjugated polymers. This chapter highlights the applications of CTPs in the synthesis of conjugated block copolymers, gradient-conjugated copolymers, conjugated polymer brushes, and polymers with special topologies and the properties of the resulting polymers are discussed.

3.2 Fully Conjugated Block Copolymers

Fully conjugated block copolymers are very important as a result of their potential in morphology control and the development of novel polymeric semiconductors with the integrated properties of different conjugated polymers. These types of block copolymers can be divided into two classes: donor–donor (D–D) conjugated block copolymers and donor–acceptor (D–A) conjugated block copolymers, depending on the properties of the blocks used.

3.2.1 Donor–Donor Type Conjugated Block Copolymers

In principle, if the "living" polymerization of two different monomers is achieved under the same conditions, block copolymers can be synthesized in a one-pot reaction by the sequential addition of monomers. The CTP of some

A–B monomers based on electron-rich aromatic units, such as phenylene, thiophene, pyrrole and fluorene, exhibit "living" characteristics under almost the same conditions (*e.g.* catalyst, solvent).[9–12] The resulting polymers are generally recognized as D- or p-type semiconductors. D–D type block copolymers based on these monomers, including diblock and triblock, can be synthesized in a one-pot reaction by sequential monomer addition (Figure 3.1). It should be noted that the order of addition is important for successful block polymerization in some examples, especially when the fully conjugated block copolymers consist of blocks with different conjugated backbones. For example, to synthesize poly(2,5-dihexyloxy-1,4-phenylene)-*b*-poly(3-hexylthiophene) (PP-*b*-P3HT) *via* a Kumada CTP, the phenyl monomer has to be polymerized first. The copolymerization in an inverted addition order, *i.e.* thienyl and then phenyl monomers, gives a mixture containing both block copolymers and end-capped P3HT (Scheme 3.1).[13] This can be attributed to the higher π-donor ability of the P3HT block than its PP counterpart. Consequently, the catalyst species prefer "ring-walking" on the P3HT block,[14] resulting in inefficient catalyst transfer from the P3HT block to the phenylene terminus.

A large number of D–D type fully conjugated block copolymers, including thiophene–thiophene, phenylene–phenylene, fluorene–fluorene, thiophene–selenophene, phenylene–thiophene, phenylene–pyrrole, and

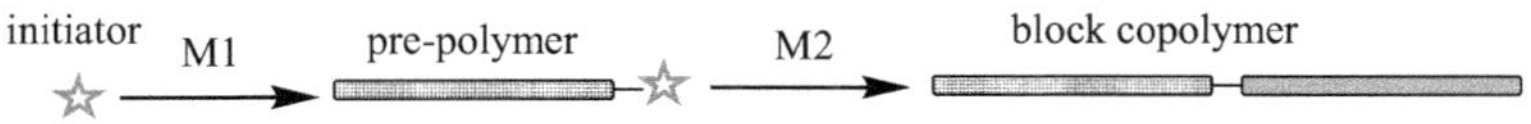

Figure 3.1 One-pot synthesis of donor–donor (D–D) type fully conjugated block copolymers *via* sequential monomer addition.

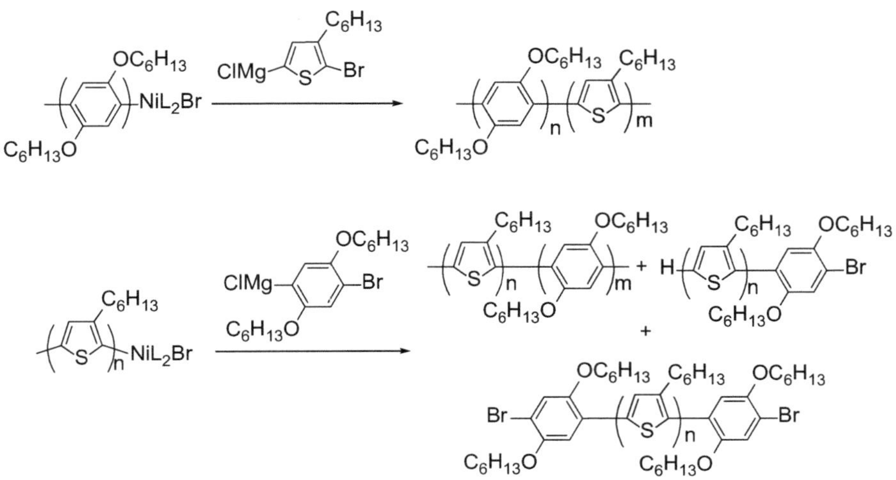

Scheme 3.1 Synthetic route for phenylene–thiophene diblock copolymers.

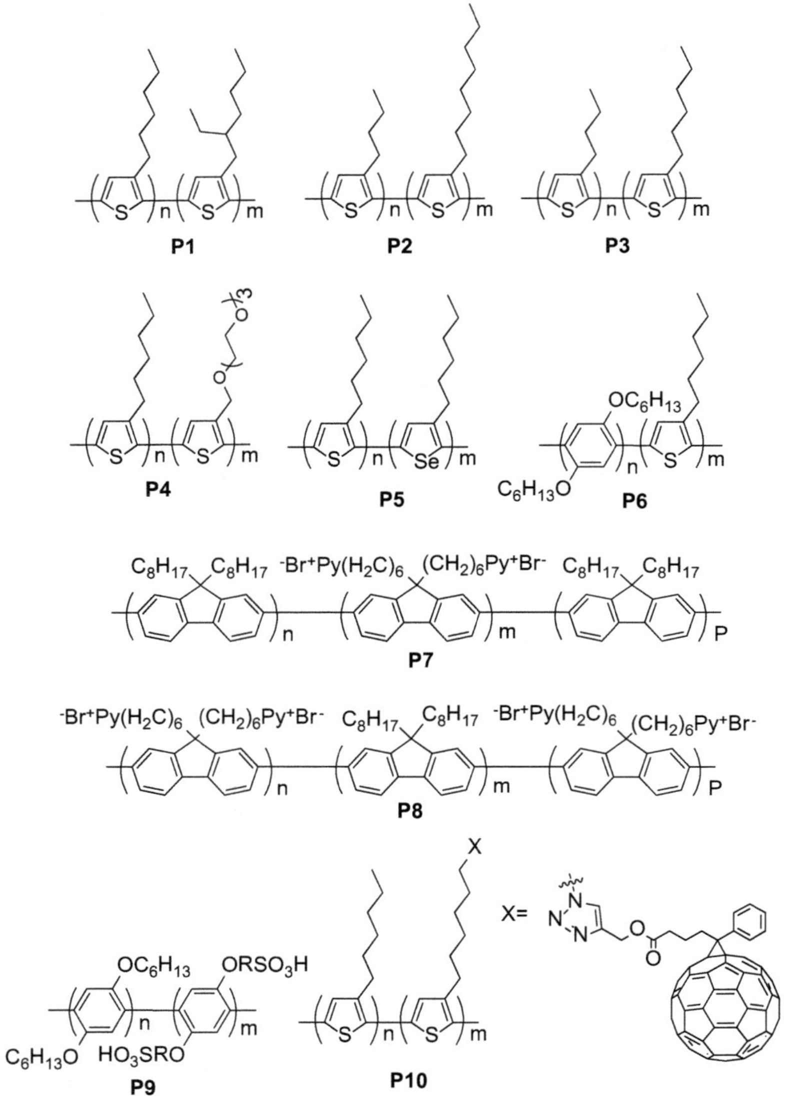

Scheme 3.2   Chemical structures of representative D–D type fully conjugated block copolymers **P1**–**P10**.

fluorene–thiophene, have been prepared by one-pot CTP.[6,13,15–47] Representatives structures are outlined in Scheme 3.2. These block copolymers can be divided into three categories: (1) polymers with conjugated backbones based on the same aromatic units, but with different side-chains; (2) polymers with blocks based on different aromatic units; and (3) polymers in which one or two blocks carry functional groups introduced by post-functionalization.

Fully conjugated block polymers with a thiophene-based conjugated backbone are the most studied D–D type block copolymers.[22-32,35-37,39-42,45-47] This type of block copolymer is usually synthesized by Kumada or Negishi CTP. When functional groups, such as esters, are involved in the side-chains, the later synthesis is more appropriate because it is more tolerant of functional groups.

Microphase separation driven by crystallization or the immiscibility of the different blocks is an important property of block copolymers, which can also take place in copolymers **P1** and **P2** (see Scheme 3.2) when the only difference between the two blocks is the alkyl side-chains. The crystallinity of poly(3-alkylthiophene) was drastically decreased when a branched alkyl group was used instead of a linear alkyl group. The block based on 3-(2-ethylhexyl)thiophene in **P1** exhibited a lower crystallinity, while the block based on 3-hexylthiophene was highly crystalline. Therefore the polymer spontaneously formed microphase-separated nanopatterns with a size dependent on the block ratio.[22,23] Both two blocks in **P2** were highly crystalline and differed only in the length of the linear alkyl groups (butyl *versus* octyl). In films, the two blocks of the polymer microphase separated into two distinct crystalline domains with a lamellar nanostructure.[24,27] Alkyl side-chains differing by more than two carbon atoms were necessary for this type of block copolymer to form microphase-separated nanostructures.[24] Two blocks in **P3**, in which butyl and hexyl groups were used, tended to co-crystallize into a uniform crystal domain.[24] By utilizing the distinct self-assembly properties of **P2** and **P3** at the exciton diffusion length scale, OPV devices with improved power conversion efficiency (PCE) over their homopolymer analogs have been demonstrated.[26,28]

When the flexible side-chains are alkyl groups in one block and ethylene glycol derivatives in the second block, the resulting amphiphilic diblock copolymers can self-assemble into nanostructures depending on the composition of the polymers and the solvents used. Hayward and coworkers[40] synthesized **P4** (see Scheme 3.2) *via* Kumada CTP, in which P3HT and poly(3-triethylene glycol thiophene) (P3(TEG)T) served as the hydrophobic and hydrophilic blocks, respectively. They found that the polymer self-assembled into crystalline fibrils when methanol was added into a chloroform solution. A polymer with a P3HT : P3(TEG)T ratio of about 2 : 1 formed helical nanowires when potassium ions in methanol were added. Park and coworkers[42] found that the same polymer could self-assemble into vesicles or micelles when tetrahydrofuran was used as a good solvent for the P3HT block. They also found that the vesicles formed by this type of block copolymer could encapsulate hydrophobic guests, leading to potential applications in biosensors, nanoreactors, photonic devices, or templates for building micro- to nano-sized structures.[41]

Various D–D type diblock copolymers consisting of different aromatic units have been prepared by one-pot CTP reactions.[6,13,15-21,33,34,38] In principle, this type of conjugated polymer can be prepared in a one-pot reaction once the controlled CTP of two monomers has been achieved with the same catalyst. The order of monomer addition is crucial in some systems, such as phenylene–thiophene and fluorene–thiophene type copolymers.[13,20]

P5 and **P6** (Scheme 3.2) are two examples. The two blocks in **P5** only differ in the heteroatoms (S *versus* Se) in the monomeric units. However, the polymer can still self-assemble into nanostructured films *via* the microphase separation of the two blocks.[33,34] This type of phase separation can be used to align nanoparticles along phase boundaries.[48] Microphase separation and crystallization in thin films of **P6** have been studied in detail and the composition of the copolymer plays a key part in determining the crystallization and final film morphology.[15] Dai and coworkers[16] found that **P6** containing 29 phenyl rings and 163 thienyl rings was characterized by an enhanced crystallinity, enabling the P3HT component to self-assemble into thin films of interpenetrated crystalline fibrils ordered over a long range. The resulting films showed a higher hole mobility than those made from P3HT homopolymers. The phenyl-based block facilitated intimate contact between the copolymer and dye molecules absorbed on the nanoporous TiO_2 layer. Thus the block copolymer was an effective hole conductor for dye-sensitized OPV cells.

Conjugated polyelectrolytes are a class of very important conjugated polymers used in fluorescence-based sensors and multilayer optoelectronic devices.[49] Fully conjugated block copolymers with one block carrying ionic pendant groups exhibit a preferred tendency to self-assemble into structured aggregates, both in solution and in the solid state. Fully conjugated D–D type block copolymers containing polyelectrolyte blocks with well-defined block ratios can be synthesized in a two-step approach, *i.e.* one-pot CTP reactions followed by post-functionalization. Their unique self-assembly properties endow them with new applications in addition to supramolecular chemistry. For instance, fluorene-based conjugated polyelectrolytes have been widely used as interfacial layers in polymer light-emitting diodes, OPVs, and organic thin film transistors (OTFTs). However, dewetting problems may be encountered as a result of interfacial instability between the ionic conjugated polyelectrolytes and the typically hydrophobic organic semiconductors. Bazan and coworkers[43] synthesized the fluorene-based triblock copolymers **P7** and **P8** (Scheme 3.2) *via* Suzuki–Miyaura CTP. Both these polymers are effective interfacial materials for polymer light-emitting diodes, providing a potential way to solve the dewetting problems in conventional conjugated polyelectrolytes. Sulfonated polyphenylenes have been considered as potential membrane materials for polymer electrolyte fuel cells. However, it was found that sulfonated aromatic polymers generally require a much higher ion-exchange capacity to obtain a suitable conductivity, resulting in an excess uptake of water and a drastic loss of mechanical properties. The block copolymerization of sulfonated aromatic units with hydrophobic aromatic units has proved to be an effective way of overcoming these drawbacks. Rikukawa and coworkers[44] synthesized a phenylene-based diblock conjugated polymer–ionomer **P9** in which the hydrophobic and hydrophilic blocks consisted of 164 and 74 phenyl rings, respectively. The polymer exhibited a clear phase separation in films. The well-developed microphase separation of the polymer provided controlled water uptake and sufficiently high proton conductivity. At low relative humidity, the diblock copolymer displayed proton conductivity one magnitude higher than its random copolymer counterpart, making the polymer a high-performance polyelectrolyte for fuel cell applications.

Phase-separated nanostructures characterized by interpenetrated bi-continuous phases at the nanometer scale are highly desirable for bulk heterojunction OPV devices. However, it is generally difficult to obtain suitable nanostructures by the simple blending of donor and acceptor materials. Moreover, the morphology of films spin-cast from the blends is kinetically controlled and is not usually stable. The microphase separation of block copolymers provides an ideal way to construct morphology-stable nanostructured films. Based on this idea, several all-thiophene conjugated diblock copolymers have been developed with one block having fullerene pendant groups attached.[46,47] **P10** (Scheme 3.2) is an example in which the fullerene derivative is attached as pendant groups in one of the two blocks. The polymer formed clear nanostructures, with sizes of about 20 nm, driven by the crystallization of the neat P3HT block and the aggregation of the fullerene groups. OPV devices based on this diblock copolymer exhibited a PCE of 2.5% and a fill factor as high as 0.63.[47] The device was very stable under high-temperature thermal annealing. In contrast, the performance of the device based on blends degraded dramatically after thermal annealing under the same conditions.[46]

3.2.2 Donor–Acceptor Type Fully Conjugated Block Copolymers

D–A type fully conjugated block copolymers are polymers in which the electron-rich block(s) and electron-deficient block(s) are covalently bonded. This type of block copolymer has attracted strong interest because of its potential application in bulk heterojunction OPVs.[50–52] Three methods have been developed to synthesize these polymers (Figure 3.2): (1) one-pot CTP reactions; (2) macro-initiator polymerization; and (3) end-capping of the polycondensation reactions with a mono-functionalized macro-reactant.

Unlike their D–D counterparts, fewer D–A fully conjugated block copolymers have been synthesized *via* one-pot CTPs because only a few conjugated polymers with electron-accepting characteristics can be synthesized in a controlled manner by CTP.[53–56] The most successful example is the diblock copolymer **P11**, with P3HT and polybenzotriazole as the electron-rich and electron-deficient blocks, respectively. Seferos and coworkers[57] found that a Ni^{II} diimine-type catalyst was able to control the polymerization of both the electron-rich thiophene and electron-deficient benzotriazole monomers as a result of the strong association of Ni^0 to both monomer types. Surprisingly, the copolymerization was independent of the

(a) One-pot polymerization approach

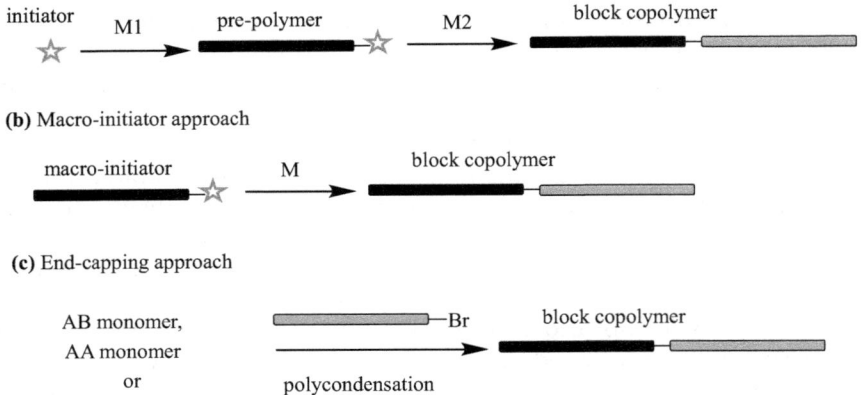

(b) Macro-initiator approach

(c) End-capping approach

Figure 3.2   General approaches for the synthesis of donor–acceptor (D–A) type fully conjugated block copolymers.

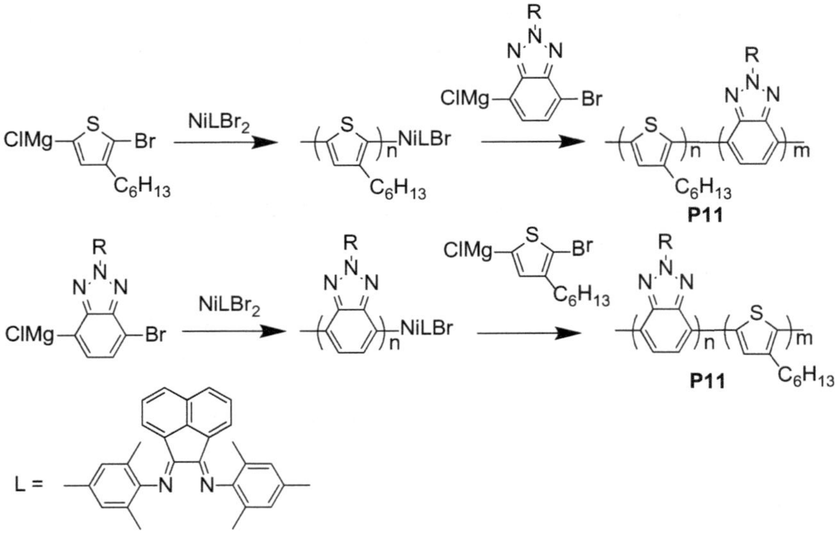

Scheme 3.3   Synthetic route of D–A type fully conjugated diblock copolymer **P11**.

order of addition of the monomer. As shown in Scheme 3.3, either P3HT or polybenzotriazole can serve as the macro-initiator or pre-polymer in the block copolymerization.

As most of the conjugated polymers used as electron-acceptor materials or n-type semiconductors cannot be synthesized in a controlled way, *i.e. via* CTP, an alternative approach to synthesizing D–A block copolymers is the CTP of an electron-rich monomer with an electron-deficient conjugated

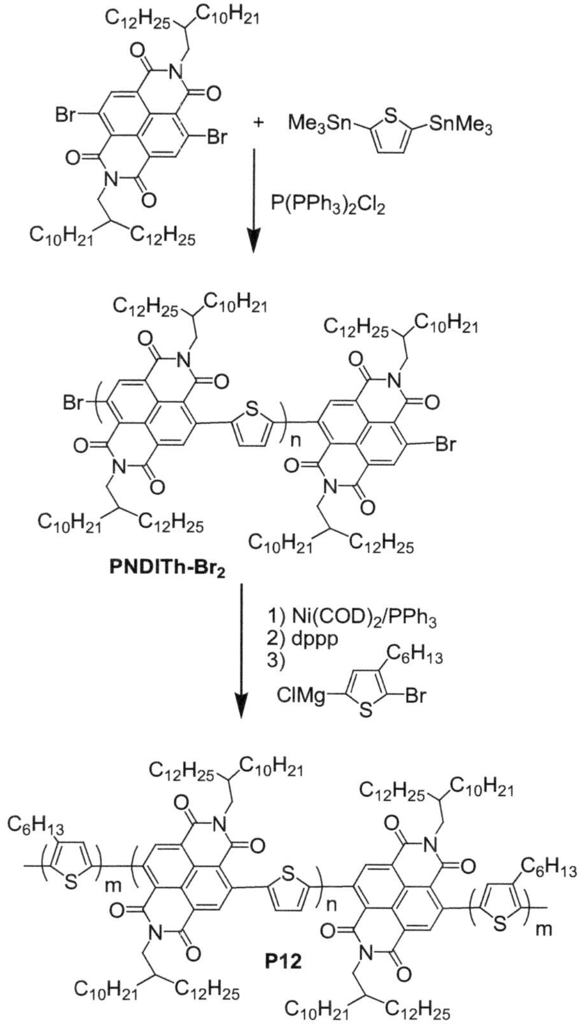

Scheme 3.4 Synthetic route of D–A type fully conjugated triblock copolymer **P12**.

polymer as the macro-initiator (Figure 3.2b). Scheme 3.4 shows the route developed by Higashihara and coworkers[58] for the synthesis of D–A–D triblock copolymers. They first synthesized PNDITh-Br$_2$ *via* Stille coupling by controlling the molar feed ratio of the monomers. The macro-initiator was prepared by treating PNDITh-Br$_2$ with Ni(COD)$_2$/PPh$_3$, followed by ligand exchange with dppp (dppp = 1,3-bis(diphenylphosphino)propane). Then the triblock copolymers (**P12**) with different P3HT contents and low dispersity ($Đ < 1.3$) were obtained by adding different amounts of the thiophene monomer. In the solid state, the resulting copolymer could microphase-separate to a lamellar structure in the range 10–20 nm, which was stable even under high-temperature thermal annealing. Diblock D–A conjugated copolymers were also synthesized in a similar way, but using mono-brominated PNDITh (*i.e.* PNDITh-Br).[59]

P3HT with one end terminated with bromine (P3HT-Br) can be easily prepared by Kumada CTP. By using P3HT-Br as an end-capper in poly-condensation reactions, it should, in principle, be possible to obtain fully conjugated block copolymers, including diblock and triblock copolymers, following the approach shown in Figure 3.2c. Most of the D–A type diblock and triblock conjugated copolymers studied so far have been synthesized by this approach, which was first demonstrated by Scherf and coworkers[60,61] Suzuki–Miyaura, Stille, and Yamamoto couplings have all been used and AB, AA and AA + BB monomers can be used depending on the synthetic route and the structures of the targeted polymers. The fully conjugated block co-polymers prepared by this approach are often contaminated with homo-polymer impurities.[62,63]

Poly(2,7-(9,9-dioctyl-fluorene)-*alt*-[4,7-bis(thien-5-yl)-2,1,3-benzothiadiazole] (PFTBT) is a well-studied electron-acceptor material for all-polymer OPVs. Scherf and coworkers[64] synthesized P3HT-*b*-PFTBT (**P13**) *via* Stille coupling of the AB monomer followed by end-capping with P3HT-Br (Scheme 3.5a). Although the product is a mixture of P3HT-*b*-PFTBT (53%), PFTBT (42%), and P3HT (5%), it could still self-assemble into lamellar nanostructures with a period of about 25 nm under thermal annealing. The polymer also showed potential as a compatibilizer to stabilize OPV device performance against exposure to high temperatures. The devices based on the ternary blend dis-played remarkably high stability and no degradation was observed when they were annealed at 200 °C. In contrast, the device performance of the homo-polymer blend decreased by 30% when treated under the same conditions.

Verduzco and coworkers[65,66] synthesized the diblock copolymer P3HT-*b*-PFTBT (**P14**) *via* Suzuki–Miyaura coupling of AA + BB monomers (Scheme 3.5b). The difference between **P13** and **P14** is that the P3HT block is covalently bonded with fluorene in **P14**, whereas it is covalently bonded with thiophene in **P13**. The polymer was also contaminated with some of the homopolymer P3HT. After the film had been thermally annealed at 165 °C, the block copolymer self-assembled into lamellar microdomains with a do-main-spacing of about 18 nm, corresponding to individual domain sizes of about 9 nm, as evaluated by resonant soft X-ray scattering. This value is comparable with the exciton diffusion length in organic semiconductors. OPV devices based on this block copolymer exhibited a PCE of 3.1%. This efficiency is three times that obtained with a blend of homopolymers and is the highest reported to date for OPV devices based on fully conjugated block copolymers. The devices surprisingly displayed a very high open circuit voltage. These results suggest that the covalent control of D–A interfaces is a route to controlling the interfacial molecular order, exciton dissociation and charge recombination to enhance excitonic solar cell performance.[66] A similar phenomenon was also observed in the D–A conjugated co-oligomer counterparts.[67–69]

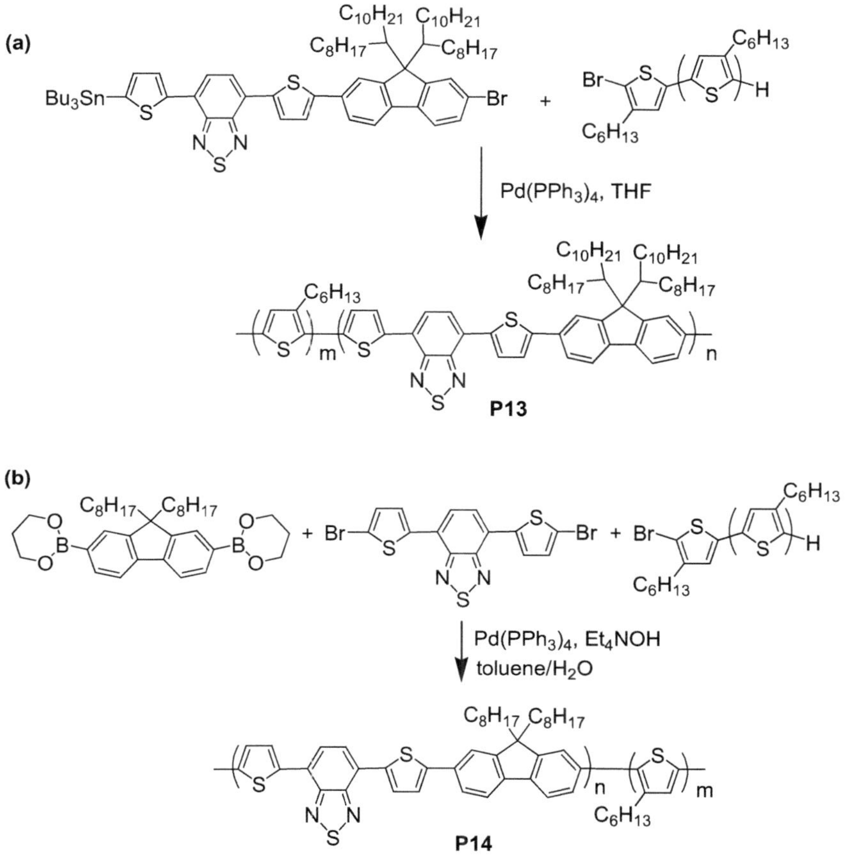

Scheme 3.5 Synthetic route of D–A type fully conjugated diblock copolymers (a) **P13** and (b) **P14**.

Cyano-substituted poly(phenylenevinylene) (CNPPV) is another important acceptor polymer for all-polymer OPVs. Scherf and coworkers[70] synthesized the D–A–D triblock copolymer P3HT-*b*-CNPPV-*b*-P3HT (**P15**) by Yamamoto polycondensation using P3HT-Br as an end-capper (Scheme 3.6a). The triblock copolymer showed a distinctly different morphology from the corresponding polymer blend.

Conjugated polymers based on naphthalene bisimide are important n-type polymeric semiconductors as a result of their low-lying lowest unoccupied molecular orbital levels and high electron mobility. The D–A–D conjugated triblock copolymers (P3HT-PNBI-P3HT, **P16**) shown in Scheme 3.6b, were synthesized by Yamamoto coupling of the dithienyl naphthalene bisimide monomer in the presence of a P3HT-Br end-capper.[71] The resulting triblock copolymers were contaminated with homo-coupled P3HT-Br and the diblock copolymer P3HT-PNBI. The copolymers exhibited two melting transitions at

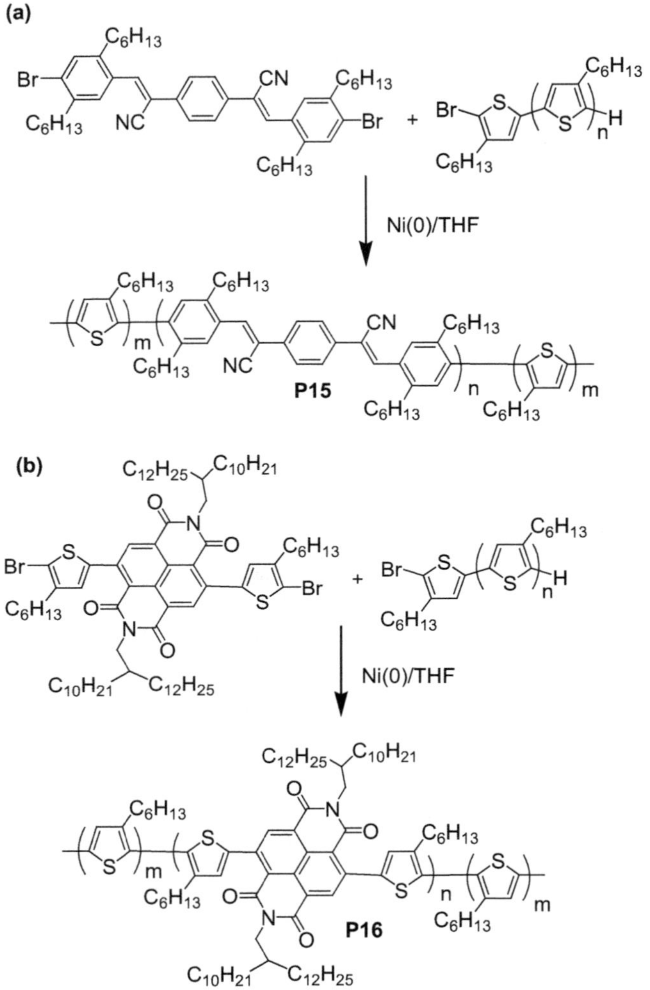

Scheme 3.6   Synthetic route of D–A type fully conjugated triblock copolymers (a) **P15** and (b) **P16**.

115 and 199 °C, which is consistent with those of PNBI and P3HT and indicates the presence of microphase separation. All-polymer solar cells were prepared using the P3HT:P3HT-PNBI-P3HT blend films and a PCE of 1.28% was achieved after thermal annealing at 200 °C.

The block copolymers (P3HT-*b*-PDPP, **P17**) with poly(diketopyrrolopyrrole-terthiophene) (PDPP) and P3HT as the A and D blocks were synthesized in a single reaction step using the Stille coupling polymerization of AA + BB monomers and using P3HT-Br as the end-capper (Scheme 3.7a).[72] Homopolymer contaminants could be completely removed by Soxhlet extraction, as evidenced by gel-permeation chromatography characterization using both

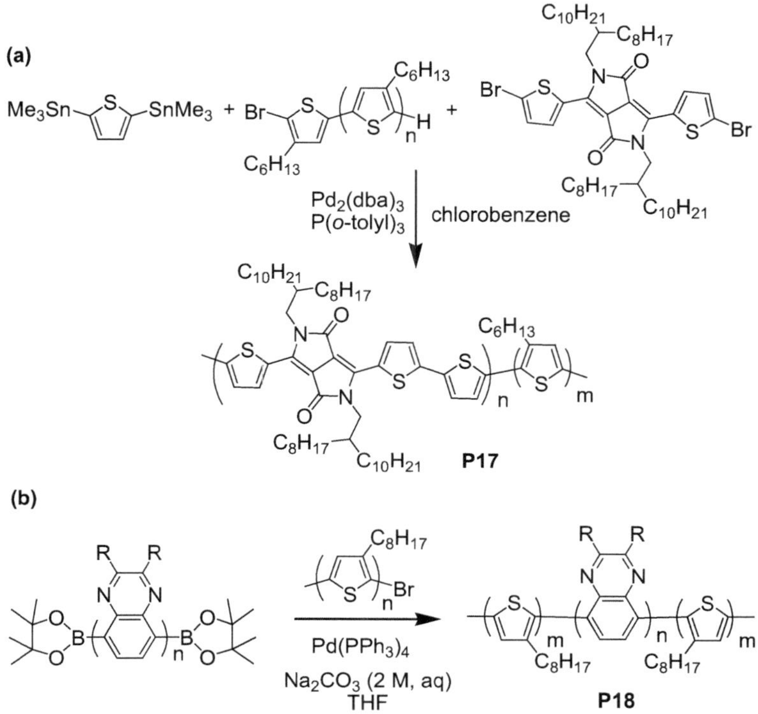

Scheme 3.7 Synthetic route of D–A type fully conjugated block copolymers (a) **P17** and (b) **P18**.

refractive index and ultraviolet detectors. Using this approach, the block copolymers with a number-average molecular weight $(M_n) > 100\,000$ g mol^{-1} can be obtained. These block copolymers showed a rich self-assembly behavior and could self-assemble into two separate crystalline domains, allowing the construction of an ordered microstructure on multiple length scales.

The D–A–D type triblock copolymers P3OT-*b*-PQ-*b*-P3OT (**P18**) with poly(quinoxaline) (PQ) and poly(3-octylthiophene) (P3OT) as the D and A blocks, respectively, were synthesized by macromolecular Suzuki–Miyaura coupling (Scheme 3.7b).[73] The authors first synthesized PQ with Br atoms in both termini and then subsequently used the boronic ester macro-reactant, which was reacted with P3OT-Br in a Suzuki–Miyaura coupling reaction to yield the targeted triblock copolymers. This approach may be limited in the preparation of high molecular weight block copolymers.

3.3 Conjugated Gradient Copolymers

Gradient copolymers, which exhibit a gradual change in composition along the normalized chain length, are more effective compatibilizers of homopolymer physical blends than block copolymers.[74] Therefore it is rational to postulate that conjugated gradient copolymers are ideal compatibilizers of conjugated polymer blends to solve a long-standing problem in polymer-based OPV devices by providing access to stable, nanostructured polymer blends. McNeil and coworkers[75–78] systematically studied the synthesis of thiophene-based conjugated gradient copolymers *via* Kumada CTP. In 2010, they studied the copolymerization of two thiophene monomers (3-hexylthiophene and 3-((hexyloxy)methyl)thiophene), which carry different side-chains, and found that these two monomers exhibited reactivity ratios of $r_1 = 1.12 \pm 0.04$ and $r_2 = 1.09 \pm 0.02$. This means that copolymers synthesized using the batch method have a random sequence. They then synthesized the conjugated gradient copolymer **P19** (Scheme 3.8) by the semi-batch method with Ni(dppp)Cl$_2$ as the catalyst.[75] Later, they synthesized a conjugated gradient copolymer based on 3-hexylthiophene and 3-hexylselenophene (**P20**, Scheme 3.8). An external initiator was used to avoid propagation from both ends of the polymer and to enable precise control over the polymer sequence.[76] The thin film morphology and phase separation behavior of **P20** were compared with diblock and random copolymers. Interestingly, the properties of the gradient copolymer were intermediate between those of the random and block analogs.

To explore the potential of conjugated gradient copolymers as compatibilizers of physical blends, gradient copolymer **P21** (Scheme 3.8), along with its diblock and random counterparts of similar molecular weights, were synthesized using 3-hexylthiophene and 3-(6-bromohexyl)-thiophene monomers.[77] The physical blends of the corresponding homo-polymers phase-separated into micrometer-sized P3HT-rich domains within a poly(3-(6-bromohexyl)thiophene) (P3BrHT) matrix. After adding 20 wt% **P21** to the blend, the domain size of P3HT was decreased to one third of that without **P21**, suggesting that **P21** is compatibilizing the polymer blend.

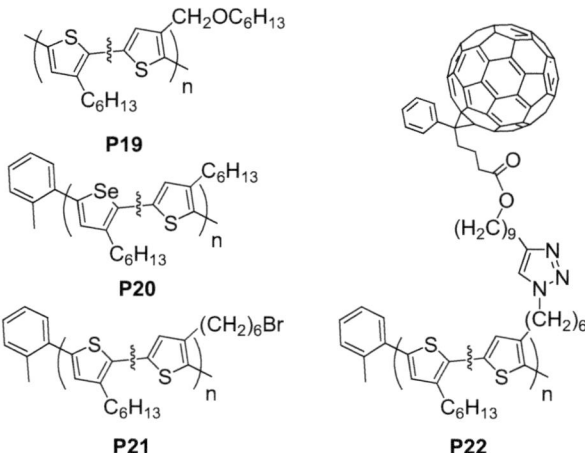

Scheme 3.8 Chemical structures of conjugated gradient copolymers **P19–P22**.

By comparison, the random and block copolymers were less effective as compatibilizers. This study confirmed that the gradient sequence copolymer is well suited to tailor the morphology of immiscible polymer blends.

The conjugated gradient copolymer **P22** (Scheme 3.8) was synthesized *via* the post-functionalization of **P21** with fullerene derivative.[78] The film morphology of the blends of P3HT and phenyl-C61-butyric acid methyl ester (PCBM) with and without the additive **P22** was studied. The 5–30 μm length and ~1 μm width PCBM aggregates were observed in the P3HT/PCBM blend after thermal annealing at 150 °C for 1 h. By contrast, no obvious microscale phase separation was observed after thermal annealing when 10 wt% **P22** was added as a compatibilizer. Although the efficiency of OPV devices was decreased when **P22** was added, the results clearly proved that the gradient copolymer could be used to stabilize the morphology of conjugated polymer/PCBM blends, providing an effective protocol for improving the stability of bulk heterojunction OPV devices.

3.4 Block Copolymers Consisting of Conjugated and Non-conjugated Blocks

Block copolymers consisting of conjugated and non-conjugated blocks can self-assemble into various nanostructures and thus are very attractive for optoelectronic applications. CTP reactions with "living" characteristics enable the deterministic incorporation of various functional groups at one or two ends of the conjugated polymers by either using appropriate external initiators or by end-capping with functional aromatics.[79–81] The block copolymers can then be synthesized by coupling reactions between the conjugated polymers with active end-groups and other active polymer chains (Figure 3.3a). It also allows the preparation of conjugated polymer-based macro-initiators for various controlled polymerizations, such as atom transfer radical polymerization (ATRP), nitroxide-mediated free radical polymerization (NMP), reverse addition fragmentation chain transfer polymerization (RAFT), ring-opening polymerization (ROP), and anionic living polymerization (Figure 3.3b). Recent studies have also revealed that the controlled polymerization of some non-aromatic monomers could be realized by using conjugated polymers carrying active chain ends as the initiator, allowing the one-pot synthesis of block copolymers consisting of conjugated and non-conjugated blocks (Figure 3.3c).

3.4.1 Synthesis *via* Macromolecular Coupling Reactions

Reactive groups can be introduced into the chain ends of conjugated polymers prepared *via* CTP. This allows coupling with other polymer chains to form block copolymers.

Click chemistry is well-recognized as an efficient protocol for polymer post-functionalization.[82] An ethynyl-terminated conjugated polymer can be

(a) Coupling approach

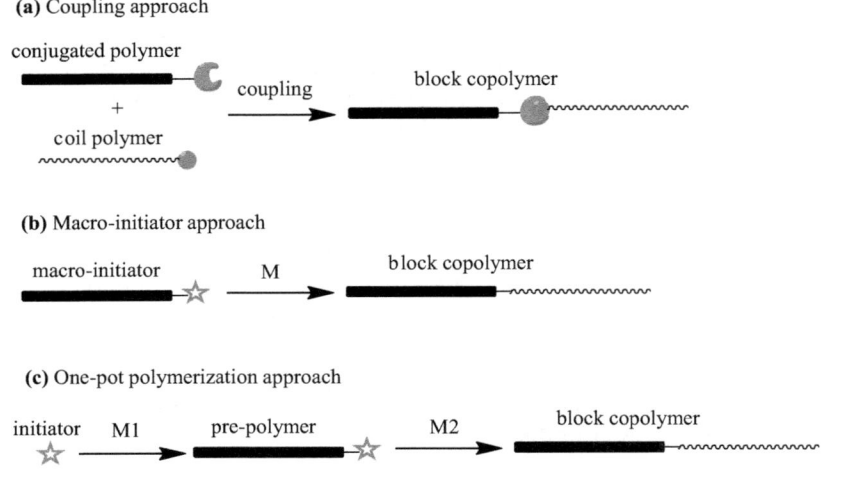

(b) Macro-initiator approach

(c) One-pot polymerization approach

Figure 3.3   General approaches for the synthesis of block copolymers consisting of conjugated and non-conjugated blocks.

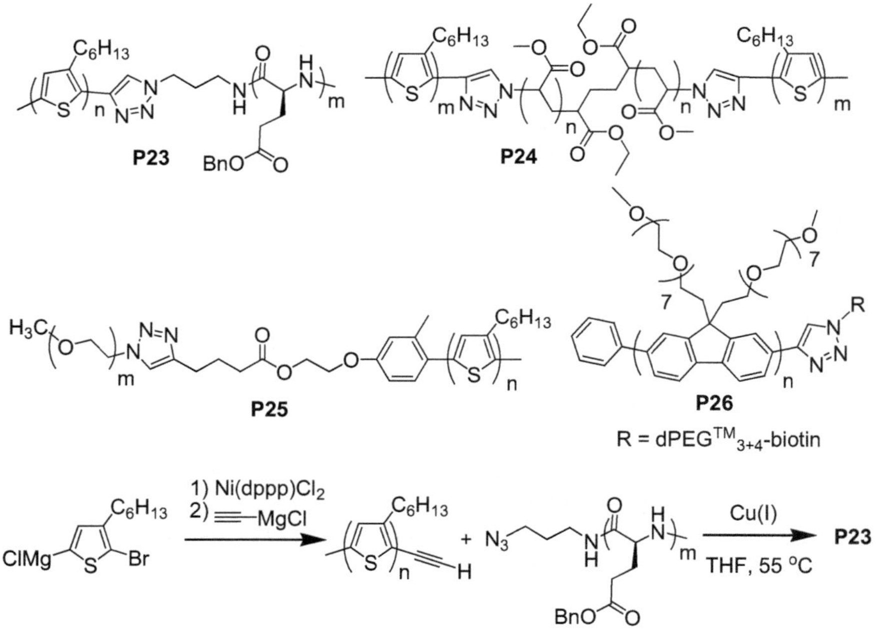

Scheme 3.9   Chemical structures of block copolymers **P23**–**P26** and synthetic route for **P23**.

obtained by quenching Kumada CTP.[80] The other polymer chain can then be covalently bonded to this chain end *via* click chemistry. Scheme 3.9 shows some block copolymers synthesized by this method[83–89] and the synthetic route for **P23** is included as an example. Ethynyl-terminated P3HT was first synthesized *via* Kumada CTP and was then coupled with azide end-functionalized poly(*g*-benzyl ʟ-glutamate) *via* Cu^I-catalyzed click chemistry.[83] Using this route, **P23** can be prepared with different molecular weights and block ratios. The block copolymer is capable of undergoing phase separation in the solid state and its spin-cast film is characterized by a nanofibrillar morphology. Qiu and coworkers[84] synthesized a rod–coil–rod triblock co-polymer **P24** in which P3HT and poly(methyl acrylate) (PMA) served as the rod and coil blocks, respectively. The polymer behaves as a thermoplastic elastomer and can be used for the fabrication of stretchable OTFTs. Verduzco and coworkers[85] synthesized the amphiphilic rod–coil block copolymer **P25** with P3HT as the rod and poly(ethylene glycol) as the coil *via* click-coupling. Alkynyl-terminated P3HT was prepared *via* Kumada CTP with a modified nickel initiator followed by end-group modification. The co-polymer was capable of forming micelles with crystalline and hydrophobic cores. PS-*b*-P3HT-*b*-PS (PS = polystyrene) triblock copolymers have been synthesized *via* this approach. Rod–rod conjugated block copolymers can also be constructed using this approach.[90,91]

Fluorescent sensing is one of the most important applications of conjugated polymers. As a result of π-electron delocalization along the conjugated backbone, conjugated polymer sensors usually exhibit a much higher sensitivity to analytes than their small molecule counterparts. An ideal conjugated polymer for biosensor applications should have several properties: a high photoluminescence quantum yield; water solubility; uniformity in chain length; resistance to nonspecific interactions; and a single site for biorecognition. Conjugated polymers with a combination of all these properties cannot be synthesized by polycondensation. Bazan and coworkers[86] synthesized a water-soluble ethynyl-terminated polyfluorene *via* Kumada CTP. At the chain end, each chain was functionalized with a biotin group *via* click chemistry (**P26**) (Scheme 3.9). The polymer proved to be a well-defined system that was reactive for binding in solution and on surfaces; it holds promise as a luminescent reporter for a number of biological recognition applications.

"Living" anions produced by "living" anionic polymerization can react with ethenyl or carbonyl groups. Several rod–coil block copolymers, such as **P27–P30**,[92–95] have been synthesized by the reaction of this type of "living" anion with terminal functionalized P3HT (Scheme 3.10). Some of these polymers exhibit unique properties. For example, **P30** is an elastomeric semiconducting material and can be used to fabricate stretchable OTFTs.[95]

The other reactive groups can also be introduced by a functional external initiator to attach different polymer chains. Scheme 3.11 gives an example. Fluorescent nanoparticles in water are of strong interest for their applications in optoelectronics, live cell imaging, and biosensing. In general, the stability of these nanoparticles against aggregation is guaranteed by physisorbed surfactants. Amphiphilic block copolymers in water can form stable nanoparticles with a controlled particle size *via* self-assembly and therefore

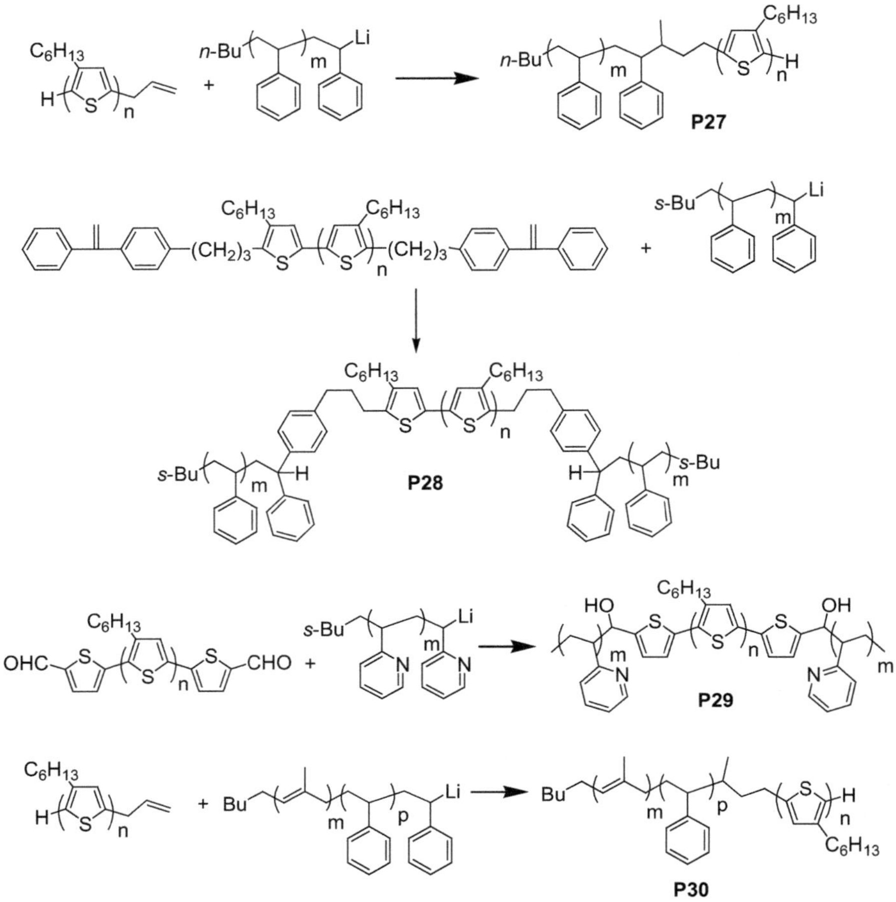

Scheme 3.10 Chemical structures and synthetic routes of block copolymers **P27**–
         P30.

this is an elegant alternative approach to prepare nanoparticles dispersed in
water. Using Suzuki–Miyaura CTP, Mecking and coworkers[96] synthesized
polyfluorene bearing a reactive group at one end and a perylene monoimide
unit at the other end. The amphiphilic block copolymer **P31** (Scheme 3.11)
was obtained by covalently attaching polyethylene glycol to the reactive end-
group. This block copolymer self-assembled into stable nanoparticles with
diameters in the range 25–50 nm when dispersed in water; the fluorescence
quantum yield of these particles was as high as 84%.

3.4.2   Synthesis with Macromolecular Initiators

Macro-initiators can be prepared either by using a functional external initiator for CTP or *via* post-functionalization of the polymer chain ends.

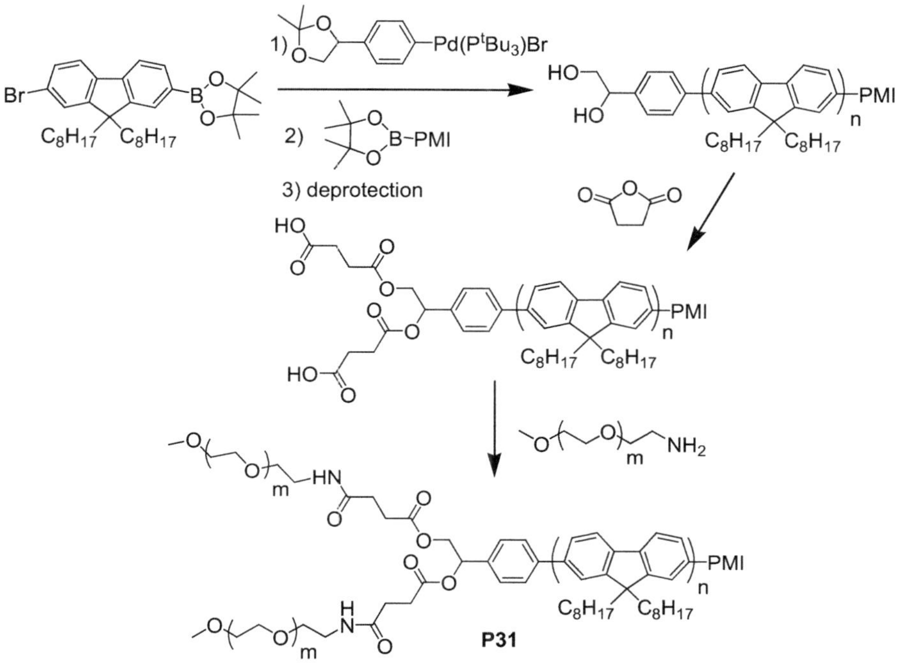

Scheme 3.11 Chemical structure and synthetic route of block copolymer **P31**.

Scheme 3.12 shows several approaches for making macro-initiators, which can be used to initiate controlled radical polymerization.

McCullough and coworkers[97–99] reported the preparation of P3HT-*b*-PMA with PMA up to 57% by ATRP with a P3HT macro-initiator **MI-1**, which was prepared from vinyl- or allyl-terminated P3HT (Scheme 3.12a). These diblock copolymers showed good processability and could form films with a nanofibrillar morphology. The dependence of both the conductivity (doping with I_2) and field-effect mobility on the PMA content was studied. It was found there was no obvious decrease in conductivity with 48% PMA[97] and that the field-effect mobility was still as high as 0.048 cm^2 V^{-1} s^{-1} with 57% PMA.[98] High purity P3HT-*b*-PMA was also synthesized by Kim and coworkers[100] in a similar way, but with a modified route for the synthesis of the macro-initiator. The amphiphilic bock copolymer P3HT-*b*-poly(*N,N*-dimethylamino-2-ethyl methacrylate) has also been prepared *via* this approach.[101] With **MI-1** as the initiator, Chen and coworkers[102] synthesized a P3HT-based rod–coil diblock copolymer with 2,5-diphenyl-1,3,4-oxadiazole pedant groups in the coil block. The polymer was used to fabricate polymer memory devices. The macro-initiators **MI-2** and **MI-3** were prepared according to Scheme 3.12b and 3.12c and the diblock copolymers P3HT-*b*-polyisoprene and P3HT-*b*-PS were successfully synthesized by NMP and RAFT polymerization, respectively.[103] **MI-2** was also used for the successful synthesis of P3HT-*b*-poly(4-vinylpyridine) (P3HT-*b*-P4VP). A better dispersion of CdSe

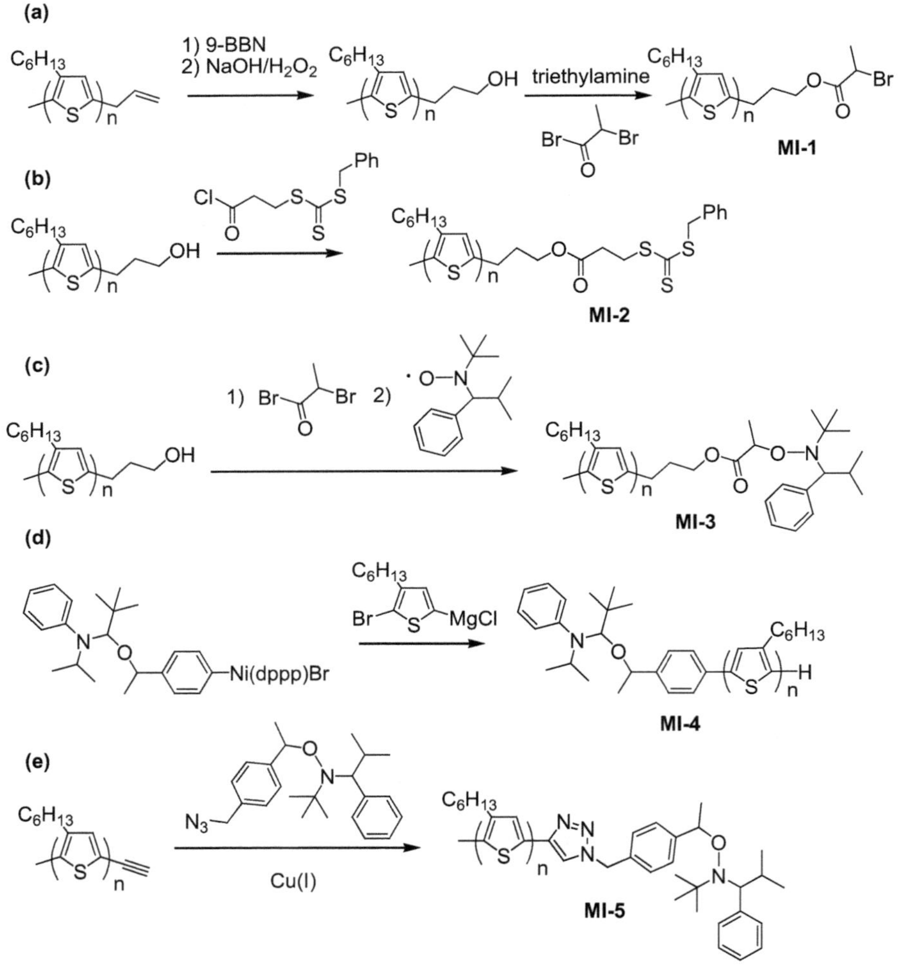

(a)

(b)

(c)

(d)

(e)

Scheme 3.12   Synthetic routes of macro-initiators **MI-1**, **MI-2**, **MI-3**, **MI-4**, and **MI-5**.

quantum dots within the polymer matrix was found as a result of the presence of the P4VP block.[104]

Kiriy and coworkers[105] developed an elegant approach for preparing P3HT-based rod–coil block copolymers *via* NMP. They synthesized an external nickel initiator consisting of an NMP-initiating group. The subsequent Kumada CTP afforded macro-initiator **MI-4** for the NMP of styrene (Scheme 3.12d). This approach avoided multi-step macromolecular reactions in the preparation of the macro-initiator.[106] The NMP initiator was also prepared *via* click chemistry (Scheme 3.12e) and was used for the controlled synthesis of amphiphilic rod–coil P3HT-*b*-poly(4-vinylpyridine) (P4VP).

Various P3HT-based rod–coil block copolymers with the coil block functionalized with acceptor moieties have been synthesized and their OPV

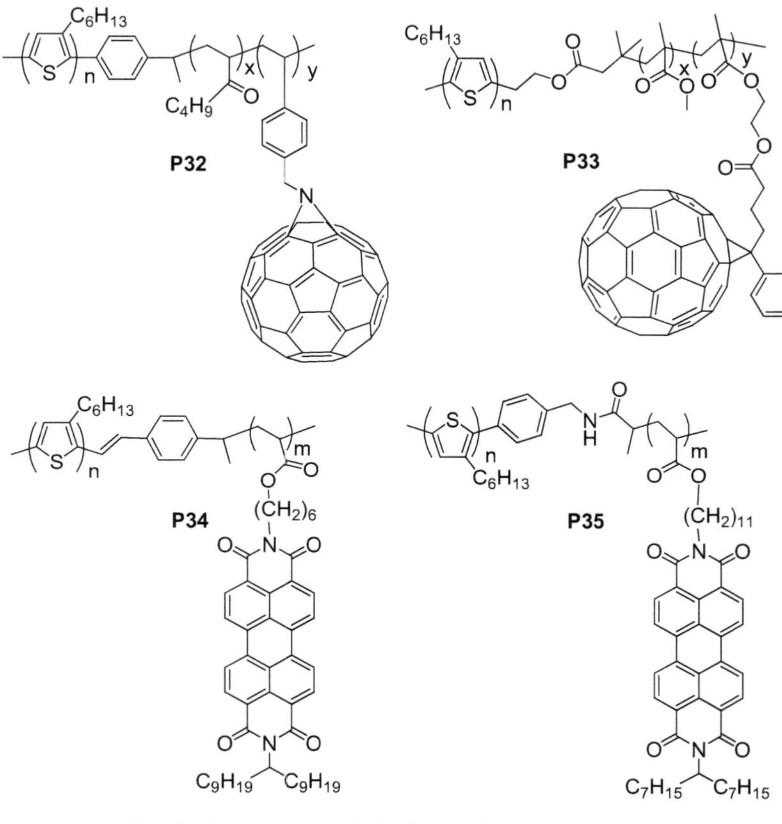

Scheme 3.13 Chemical structures of block copolymers **P32–P35**.

properties investigated, inspired by the construction of films consisting of well-defined D–A nanostructures.[107–112] Scheme 3.13 shows **P32–P35** as examples. **P33** was synthesized *via* ATRP with a macro-initiator prepared following Scheme 3.12a.[108] It can serve as a compatibilizer for the P3HT/PCBM blend to reduce the phase size and suppress macrophase separation during long-term thermal annealing because the copolymer preferentially locates at the interface between the P3HT and PCBM phases. With 2.5 wt% **P33** as an additive, the resulting OPV devices not only showed high efficiency, but also exhibited a smaller decrease in efficiency with long-term annealing at 150 °C.[109] The devices based on **P34** exhibited a PCE of 0.49% on annealing at 150 °C for 20 min.[110] This efficiency is much higher than that of the corresponding blend. Fréchet and coworkers[111] found that the efficiency of OPV devices based on the P3HT/perylene diimide blend could be significantly improved by adding **P35** as a compatibilizer. A PCE of 0.55% was demonstrated.

Hydroxyl-terminated P3HT can be used as a macro-initiator for the cationic ROP of lactide, 2-ethyl-2-oxazoline, tetrahydrofuran, and ethylene

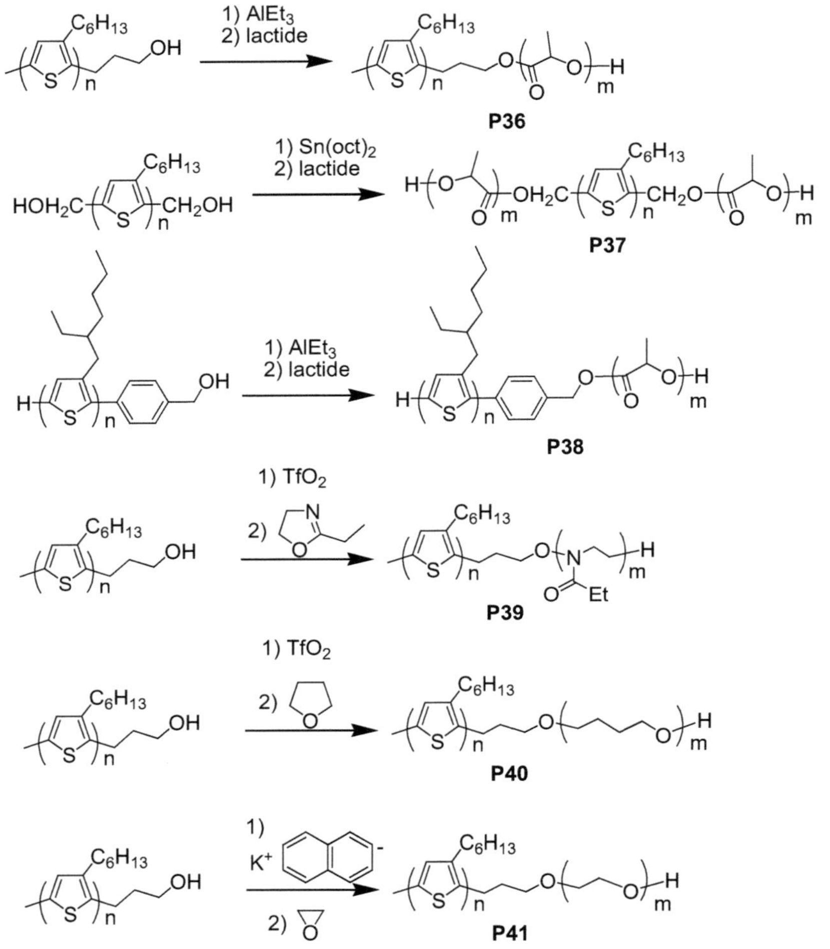

Scheme 3.14 Chemical structures and synthetic routes of block copolymers P36–P41.

oxide (Scheme 3.14).[113–119] Hillmyer and coworkers[113] synthesized the diblock copolymer **P36** *via* ROP of D,L-lactide (LA). Similar to the ATRP and NMP procedures described earlier, a small amount of P3HT homopolymer without an end-capping group always appeared in the resulting P3HT-*b*-PLA block copolymers. These workers also synthesized the PLA-*b*-P3HT-*b*-PLA triblock copolymer **P37** following a similar approach with hydroxyl-dicapped P3HT as the initiator.[114] The PLA segment could be etched with a strong base to afford nanoporous films. This phenomenon indicates the occurrence of phase separation and the resulting nanoporous films might be useful in the preparation of organic electronic devices.[113] The PLA block with an L-lactide repeating unit in the block copolymer P3HT-*b*-PLA is capable of forming lamellar nanostructures in the solid state *via* microphase

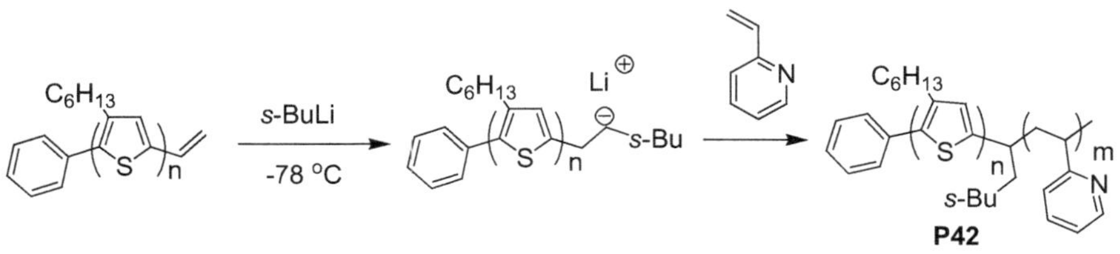

Scheme 3.15 Chemical structure and synthetic route of block copolymer **P42**.

separation. After the PLA block had been removed by etching with an aqueous solution of NaOH, fullerene-C60 could be filled into the empty spaces by dip-coating, resulting in bulk heterojunction nanostructured films.[115] Segalman and coworkers[116] synthesized a similar diblock co-polymer **P38** using 3-(2-ethylhexyl)thiophene instead of 3-hexylthiophene as the repeating unit in the polythiophene block. They found that the polymer could self-assemble into lamellar and hexagonally packed nanostructures as well as disordered micelles depending on the P3HT content. Diblock copolymers **P39–P41** were synthesized *via* the ROP of 2-ethyl-2-oxazoline, tet-rahydrofuran, and ethylene oxide, respectively.[117–119] They can self-assemble into various nanostructures. **P39** formed films with a nanofibrillar morph-ology, with the density of nanofibrills depending on the content of P3HT.[117] The field-effect mobility of both **P39** and **P40** decreased with the incorporation of an insulating coil block.[117,118] **P41** is capable of self-assembling into various well-defined core/shell-type nanostructures in solution when the solvent composition is altered. As a result, its photoluminescence colors can be re-versibly tuned from blue to red.[119]

The presence of a vinyl end-group also allows the preparation of a macro-initiator for anionic "living" polymerization. As shown in Scheme 3.15, vinyl-terminated P3HT reacts with s-BuLi. The resulting P3HT-based organo-lithium compound initiated the polymerization of 2-vinylpyridine (2VP) to afford P3HT-*b*-P2VP (**P42**).[120] About 20% P3HT homopolymer was contained in the product and could be removed by extraction. Depending on the content of P3HT, the copolymers could self-assemble into micelles or spheres, hexagonal close-packed cylinders, lamellar, filament-like, or nano-fiber structures.

3.4.3 Synthesis *via* One-pot Polymerization

The most convenient method by which to synthesize block copolymers consisting of conjugated and non-conjugated blocks is one-pot polymer-ization, similar to the preparation of D–D type fully conjugated block copolymers discussed in Section 3.2.1. However, the one-pot synthesis of rod–coil block copolymers or block copolymers with a non-conjugated block is very challenging because, in principle, two mechanically distinct poly-merizations have to be used. Geng and coworkers[121] found that an Ni catalyst could efficiently transfer across the bithienylmethylene derivative, a non-conjugated monomer, and then the P3HT-based block copolymer **P43** could be synthesized (Scheme 3.16a).

   According to the mechanism of CTP, a reactive catalyst species resides at the conjugated chain end after the monomer is consumed. This species may be able to catalyze the polymerization of some non-aromatic monomers. This idea was demonstrated for the first time by Wu and coworkers.[122–124] They found that the P3HT-Ni(dppp)Br produced by Kumada CTP could serve as a macro-initiator for the controlled polymerization of isocyanides. The block copolymer **P44** with a narrow molecular weight distribution and a well-defined block ratio and molecular weight was successfully synthesized *via* the subsequent addition of thiophene and isocyanine monomers (Scheme 3.16b). The resulting block copolymers were capable of microphase separation in the solid state and exhibited unique self-assembly properties in solution.[122–124] Depending on the substituents and solvent composition, they can self-assemble into nanofibrills, micelles, and vesicles.[123] P3HT-Ni(dppp)Br can also initiate the controlled polymerization of allenes to afford the block copolymer **P45** (Scheme 3.16c).[125]

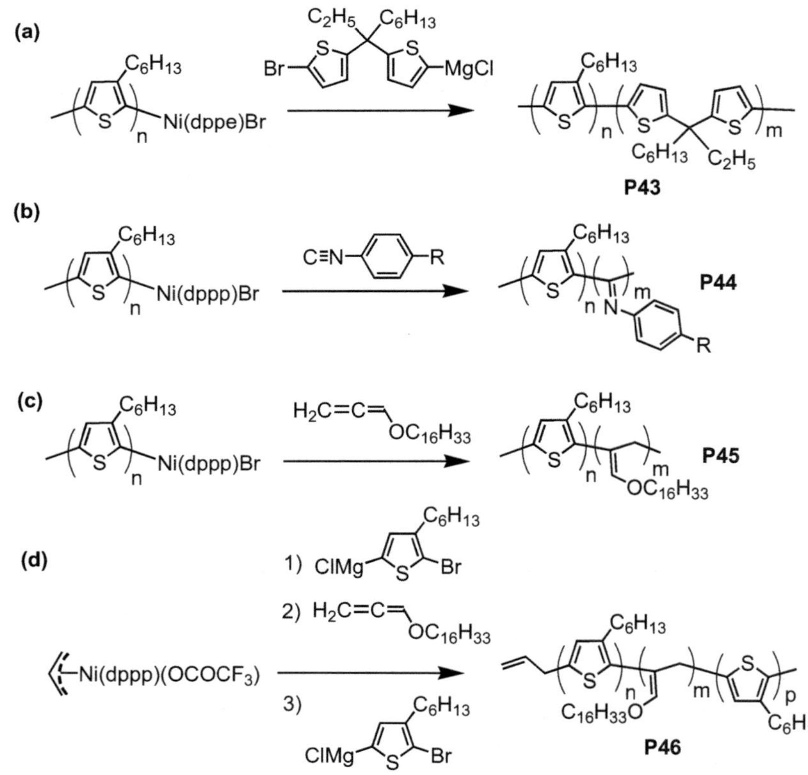

Scheme 3.16 One-pot synthesis of block copolymers **P43–P46** *via* sequential monomer addition.

Interestingly, a π-allylnickel complex can initiate the polymerization of thiophene, allene, styrene, and acrylate monomers. As a result, various rod–coil block copolymers, even the triblock copolymer **P46**, can be synthesized in a one-pot manner (Scheme 3.16d).[126]

3.5 Conjugated Polymer Brushes

Polymer brushes are attractive as a result of their various potential applications. Surface-initiated polymerization, a "graft-from" approach, is an efficient protocol with which to prepare polymer brushes. It needs the polymerization featured with "living" chain-growth characteristics. The development of CTP reactions has allowed the preparation of conjugated polymer brushes. As shown in Figure 3.4, the catalyst is first loaded onto the substrate *via* the reaction of an Ni^0 or Pd^0 complex with aryl-X (X = Cl, Br, or I) and the subsequent CTP affords the polymer brushes.

Efficiently loading the catalyst onto the substrate is the first step in the surface-initiated polymerization. It is known that $Ar–NiL_2–X$ complexes can be prepared by the oxidative addition of Ni^0 or Pd^0 to the aryl halide.[127–129] Therefore this type of initiator should be loaded onto the surface with the attached Ar–X moieties. Kiriy and coworkers[130,131] first demonstrated P3HT brushes on poly(4-bromostyrene) (PS(Br))-coated Si wafers. To avoid delamination on catalyst loading and subsequent polymerization, a 2 nm layer of poly(glycidyl methacrylate), which can react with the Si surface on thermal annealing (10 h at 150 °C in Ar) and with PS(Br) on the ultraviolet irradiation, was used as an adhesion layer. The PS(Br) layer was photo-cross-linked prior to polymerization. The substrate was then treated with $Ni(PPh_3)_4$ solution to afford films containing the $Ph-Ni(PPh_3)_2-Br$ initiator. About 16% C–Br reacted with the catalyst, as determined by X-ray photoelectron spectrometry. As the PS(Br) film can swell in the solvent used, the initiator species was also found inside the PS(Br) film as well as on the surface. As a result, the Kumada CTP followed an "interpenetrated network" model. This means that the polymerization proceeded, not only from the topmost layer, but also from the inner layer. Because of the significant chain termination in $Ph-Ni(PPh_3)_2-Br$ catalyzed polymerization, P3HT grafts of only about 10 nm

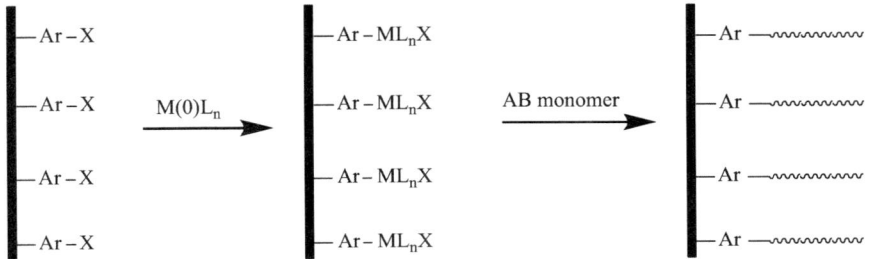

Figure 3.4 General "graft-from" approach for the synthesis of conjugated polymer brushes. M = Ni or Pd; L = ligand.

can be grown, corresponding to ∼30 repeating units. This approach is also applicable to other polymer-coated substrates. For example, P4VP can strongly absorb to a variety of polar substrates, such as Si wafers, glasses, and metal oxide surfaces. P4VP-*block*-poly(4-iodostyrene) (P4VP-*b*-PS(I)) was successfully used instead of PS(Br) as a polymer layer to grow P3HT brushes.[132] Micropatterned P4VP$_{25}$-*b*-PS(I)$_{350}$-*b*-P3HT brushes can be prepared by patterning P4VP$_{25}$-*b*-PS(I)$_{350}$ *via* colloidal lithography prior to catalyst loading.

To overcome the shortcomings of the Ph-Ni(PPh$_3$)$_2$-Br initiator, Kiriy and coworkers[133] developed an approach to fabricate organo-silica particles carrying a surface initiator based on the bidentate phosphorus ligand dppp, which allowed the preparation of hairy particles with densely grafted high molecular weight P3HT chains. Bulk heterojunction OPV devices with a PCE of 1.8–2.3% were prepared using this type of hairy particle as the donor material. Poly(9,9-dioctylfluorene) brushes can also be synthesized *via* surface-initiated Kumada and Suzuki–Miyaura CTP reactions.[134,135] As the surface grafting is very efficient, the conjugated polymer chains adopt planar and ordered conformations, even in solution.[133,134]

Locklin and coworkers[136,137] developed an alternative approach to loading initiator species onto the substrate. A gold surface was functionalized with thienyl bromide monolayers, which underwent oxidative addition with Ni(COD)$_2$ in the presence of PPh$_3$ to produce a substrate loaded with Th-Ni(COD)(PPh$_3$)$_2$-Br (Th = thienyl) initiator moieties.[136] Polythiophene and polyphenylene brushes with a thickness of 10–42 nm were then prepared *via* a surface-initiated Kumada CTP reaction. When 1,2-bis(diphenyl phosphino)ethane was used instead of PPh$_3$ as the phosphorous ligand,[137] the surface-initiated Kumada CTP on Si wafer and glass substrates was strongly influenced by the structure of the monomers. Polymer growth was not observed when 1-bromo-4-chloromagnesio-2,5-dimethoxybenzene was used as the monomer; however, polymerization proceeded well with 1-chloromagnesio-4-iodo-2,5-dialkyloxybenzene as the monomer. The film thickness that could be achieved decreased from 26 nm to 14, 8, and 1 nm on increasing the size of the side-chain from H to CH$_3$O, CH$_3$CH$_2$O, and C$_6$H$_{13}$O, respectively.

A surface-initiated Kumada CTP reaction was also used in the preparation of silica nanoparticles coated with a thiophene-based polyelectrolyte. The photophysical properties of the conjugated polyelectrolyte brushes were found to be dependent on the surrounding chemical environment and thus these particles are promising materials for sensor applications.[138] Conjugated microporous polymers have potential applications in various fields. However, this type of material is usually intractable and is therefore very difficult to process into thin films. Surface-initiated CTP provides an efficient way to overcome this problem. For instance, Kiriy and coworkers[139] made organo-silica microparticles covered by thin film (∼30 nm) layers of conjugated microporous polymers *via* a Kumada CTP of a tetrafunctional thiophene-based monomer. The obtained microporous polymers on the particles had a specific surface area of 463 m^2 g^{-1}. Poly(*p*-phenyleneethynylene) brushes on silica particles (~200 nm) was also successfully synthesized *via* surface-initiated Stille CTP.[8]

Some efforts have been devoted to the applications of conjugated polymer brushes in OPVs. Poly(3,4-ethylenedioxythiophene):poly(styrenesulfonate) (PEDOT:PSS), a water-soluble conducting polymer, is often used as a hole transport layer on top of an indium tin oxide anode. However, the acidic nature of PEDOT:PSS can corrode the indium tin oxide electrode, leading to chemical instability at the interface. PEDOT:PSS does not have sufficient electron-blocking capability, which could cause electron leakage at the anode to reduce the short-circuit current. Luscombe and coworkers[140] and Locklin and coworkers[141] independently synthesized poly(3-methylthiophene) (P3MT) brushes on indium tin oxide. Bulk heterojunction OPV devices, which exhibited a PCE >5%, have been demonstrated with 9 nm electrochemically doped P3MT as the hole transport layer. Most importantly, the P3MT interfacial layers offer superior stability in air, water, and organic solvents.[141]

Conjugated polymers with a wide absorption range are highly desirable for OPV applications. Covalently linking two p-type conjugated polymers with complementary absorption is an effective approach to realize this purpose. Scheme 3.17a shows an example. A carbazole monomer carrying a P3HT pendant is first synthesized *via* Kumada CTP and then a subsequent Suzuki–Miyaura polycondensation reaction affords a conjugated polymer (**P47**) consisting of a DPP-based main chain and P3HT brushes. As expected, this brush-like conjugated polymer exhibits a broad absorption.[142]

Another challenge in OPV devices is the construction of bulk heterojunction films consisting of alternating donor/acceptor domains on a nanoscale. Hawker and coworkers[143] synthesized P3HT with nitroxide-type initiators in the side-chain. The NMP of styrene afforded PS-grafted P3HT (**P48**) (Scheme 3.17b). Thin films of the polymer displayed an irregular microphase-separated structure with an average domain size of ~30 nm. Nanoporous P3HT thin films were obtained by the cleavage of the trityl ether linker, followed by the complete removal of PS. The films may have potential applications in diverse fields such as OPV devices.

3.6 Other Applications

Intramolecular catalyst transfer cross-couplings have also found applications in the synthesis of polymers with special topologies. Scheme 3.18 shows three examples.[144–146] Kiriy and coworkers[144] synthesized star-like P3HT (**P49**) *via* Kumada CTP with hexaphenylbenzene as a core and an Ar-Nibipy-Br species as the initiator. The oxidation addition of Et$_2$Nibipy to C–Br bonds was very efficient, allowing the preparation of star-like molecules with six initiator moieties at 95% purity. The resulting six-armed P3HT stars were studied by atomic force microscopy, revealing that each arm had ~27 repeating units. The synthesis of hyperbranched polymers with a high degree of branching is very challenging. By using Suzuki–Miyaura CTP with

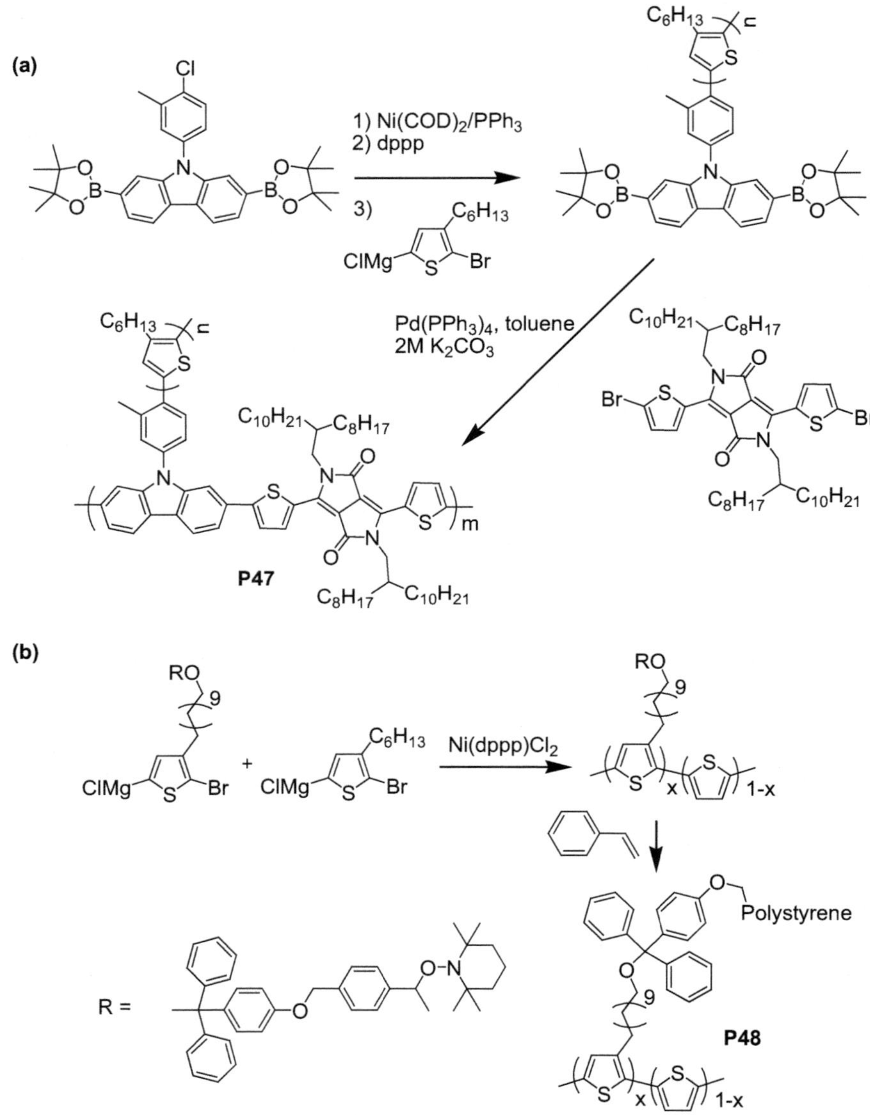

Scheme 3.17 Chemical structures and synthetic routes of conjugated polymer brushes (a) **P47** and (b) **P48**.

Pd$_2$(dba)$_3$/P(t-Bu)$_3$ as the catalyst, Bo and coworkers[145] successfully synthesized the carbazole-based hyperbranched polymer **P50** with a degree of branching of 100%. The fact that the catalyst species preferred intramolecular transfer instead of diffusion into the reaction mixture was believed to be the key to this success. They also found that macrocyclic conjugated molecules could be synthesized at high yields by using

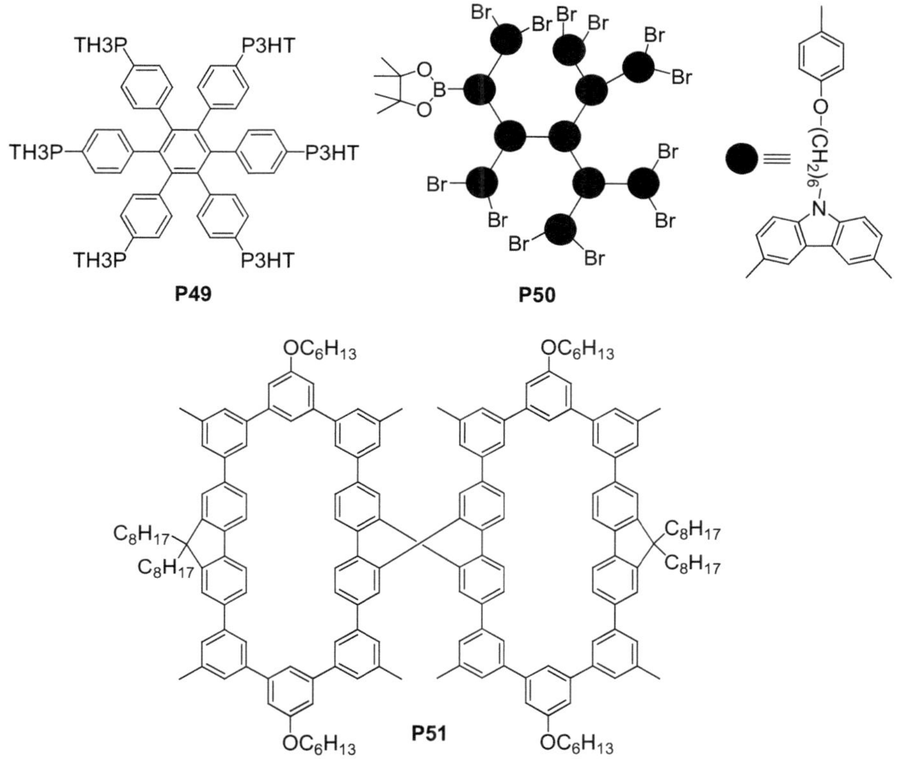

Scheme 3.18 Chemical structures of star-like P3HT (**P49**), the hyperbranched polymer **P50**, and the macrocyclic conjugated molecule **P51**.

intramolecular catalyst transfer coupling. Even for the macrocyclic conjugated molecule **P51**, which has a molecular weight of 2516 g mol^{-1}, a yield of 21% was achieved with Pd$_2$(dba)$_3$/P(t-Bu)$_3$ as the catalyst.[146]

3.7 Summary and Outlook

CTP reactions, particularly those following a "living" or quasi-"living" chain-growth polymerization mechanism, provide an efficient protocol to synthesize conjugated polymers with new chain or topological structures that are not accessible using conventional polycondensation reactions. These polymers exhibit unique properties and have potential applications in optoelectronic devices, in addition to providing opportunities to pursue new applications of conjugated polymers. However, several challenges still need to be overcome. First, block copolymers are promising for controlling film morphology and provide reliable methods to construct material systems with ambipolar transport properties and a good photovoltaic performance. This requires the block copolymers to have well-defined structures, such as a well-designed composition, a well-controlled block ratio, and low dispersity. Current protocols only allow the controlled synthesis of a few conjugated polymers consisting of electron-deficient aromatic moieties. Therefore new catalytic systems need to be developed to overcome this limitation. Second, the optoelectronic properties of multi-component material systems are greatly influenced by the interfacial nature of the components. Therefore (co)polymers with different chain and topological structures should be developed to establish a structure–property relationship. Third, more attention has been paid to the applications of the resulting materials. This needs solid collaboration between chemists, physicists, and material engineers.

References

1. R. Miyakoshi, A. Yokoyama and T. Yokozawa, *J. Am. Chem. Soc.*, 2005, **127**, 17542.
2. Z. J. Bryan and A. J. McNeil, *Macromolecules*, 2013, **46**, 8395.
3. E. E. Sheina, J. S. Liu, M. C. Iovu, D. W. Laird and R. D. McCullough, *Macromolecules*, 2004, **37**, 3526.
4. A. Yokoyama, R. Miyakoshi and T. Yokozawa, *Macromolecules*, 2004, **37**, 1169.
5. A. Yokoyama, H. Suzuki, Y. Kubota, K. Ohuchi, H. Higashimura and T. Yokozawa, *J. Am. Chem. Soc.*, 2007, **129**, 7236.
6. A. G. Sui, X. C. Shi, H. K. Tian, Y. H. Geng and F. S. Wang, *Polym. Chem.*, 2014, **5**, 7072.
7. Y. Qiu, J. Mohin, C.-H. Tsai, S. Tristram-Nagle, R. R. Gil, T. Kowalewski and K. J. T. Noonan, *Macromol. Rapid Commun.*, 2015, **36**, 840.
8. S. Kang, R. J. Ono and C. W. Bielawski, *J. Am. Chem. Soc.*, 2013, **135**, 4984.
9. T. Yokozawa and Y. Ohta, *Chem. Commun.*, 2013, **49**, 8281.
10. A. Kiriy, V. Senkovskyy and M. Sommer, *Macromol. Rapid Commun.*, 2011, **32**, 1503.
11. K. Okamoto and C. K. Luscombe, *Polym. Chem.*, 2011, **2**, 2424.
12. Y. H. Geng, L. Huang, S. P. Wu and F. S. Wang, *Sci. China: Chem.*, 2010, **53**, 1620.
13. S. P. Wu, L. J. Bu, L. Huang, X. H. Yu, Y. C. Han, Y. H. Geng and F. S. Wang, *Polymer*, 2009, **50**, 6245.
14. R. Tkachov, V. Senkovskyy, H. Komber, J.-U. Sommer and A. Kiriy, *J. Am. Chem. Soc.*, 2010, **132**, 7803.
15. X. H. Yu, H. Yang, S. P. Wu, Y. H. Geng and Y. C. Han, *Macromolecules*, 2012, **45**, 266.
16. W.-C. Chen, Y.-H. Lee, C.-Y. Chen, K.-C. Kau, L.-Y. Lin, C.-A. Dai, C.-G. Wu, K.-C. Ho, J.-K. Wang and L. Wang, *ACS Nano*, 2014, **8**, 1254.
17. A. Yokoyama, A. Kato, R. Miyakoshi and T. Yokozawa, *Macromolecules*, 2008, **41**, 7271.
18. R. Miyakoshi, A. Yokoyama and T. Yokozawa, *Chem. Lett.*, 2008, **37**, 1022.

19. K. Ohshimizu, A. Takahashi, T. Higashihara and M. Ueda, *J. Polym. Sci., Polym. Chem.*, 2011, **49**, 2709.
20. A. G. Sui, X. C. Shi, S. P. Wu, H. K. Tian, Y. H. Geng and F. S. Wang, *Macromolecules*, 2012, **45**, 5436.
21. A. G. Sui, X. C. Shi, Y. X. Wang, Y. H. Geng and F. S. Wang, *Polym. Chem.*, 2015, **6**, 4819.
22. Y. Zhang, K. Tajima, K. Hirota and K. Hashimoto, *J. Am. Chem. Soc.*, 2008, **130**, 7812.
23. Y. Zhang, K. Hirota and K. Hashimoto, *Macromolecules*, 2009, **42**, 7008.
24. J. Ge, M. He, F. Qiu and Y. L. Yang, *Macromolecules*, 2010, **43**, 6422.
25. M. He, L. Zhao, J. Wang, W. Han, Y. L. Yang, F. Qiu and Z. Q. Lin, *ACS Nano*, 2010, **4**, 3241.
26. M. He, W. Han, J. Ge, Y. L. Yang, F. Qiu and Z. Q. Lin, *Energy Environ. Sci.*, 2011, **4**, 2894.
27. P.-T. Wu, G. Q. Ren, C. X. Li, R. Mezzenga and S. A. Jenekhe, *Macromolecules*, 2009, **42**, 2317.
28. G. Q. Ren, P.-T. Wu and S. A. Jenekhe, *Chem. Mater.*, 2010, **22**, 2020.
29. M. Verswyvel, F. Monnaie and G. Koeckelberghs, *Macromolecules*, 2011, **44**, 9489.
30. K. Van den Bergh, I. Cosemans, T. Verbiest and G. Koeckelberghs, *Macromolecules*, 2010, **43**, 3794.
31. K. Van den Bergh, J. Huybrechts, T. Verbiest and G. Koeckelberghs, *Chem. – Eur. J.*, 2008, **14**, 9122.
32. K. Obshimizu and M. Ueda, *Macromolecules*, 2008, **41**, 5289.
33. J. Hollinger, A. A. Jahnke, N. Coombs and D. S. Seferos, *J. Am. Chem. Soc.*, 2010, **132**, 8546.
34. J. Hollinger, P. M. DiCarmine, D. Karl and D. S. Seferos, *Macromolecules*, 2012, **45**, 3772.
35. P.-T. Wu, G. Q. Ren, F. S. Kim, C. X. Li, R. Mezzenga and S. A. Jenekhe, *J. Polym. Sci., Polym. Chem.*, 2010, **48**, 614.
36. C. C. Chueh, T. Higashihara, J.-H. Tsai, M. Ueda and W.-C. Chen, *Org. Electron.*, 2009, **10**, 1541.
37. C.-C. Ho, Y.-C. Liu, S.-H. Lin and W.-F. Su, *Macromolecules*, 2012, **45**, 813.
38. Z.-Q. Wu, D.-F. Liu, Y. Wang, N. Liu, J. Yin, Y.-Y. Zhu, L.-Z. Qiu and Y.-S. Ding, *Polym. Chem.*, 2013, **4**, 4588.
39. T. Yokozawa, I. Adachi, R. Miyakoshi and A. Yokoyama, *High Perform. Polym.*, 2007, **19**, 684.
40. E. Lee, B. Hammer, J.-K. Kim, Z. Page, T. Emrick and R. C. Hayward, *J. Am. Chem. Soc.*, 2011, **133**, 10390.
41. J. Kim, I. Y. Song and T. Park, *Chem. Commun.*, 2011, **47**, 4697.
42. I. Y. Song, J. Kim, M. J. Im, B. J. Moon and T. Park, *Macromolecules*, 2012, **45**, 5058.
43. L. Ying, P. Zalar, S. D. Collins, Z. Chen, A. A. Mikhailovsky, T.-Q. Nguyen and G. C. Bazan, *Adv. Mater.*, 2012, **24**, 6496.

44. K. Umezawa, T. Oshima, M. Yoshizawa-Fujita, Y. Takeoka and M. Rikukawa, *ACS Macro Lett.*, 2012, **1**, 969.
45. A. Thomas, J. E. Houston, N. Van den Brande, J. De Winter, M. Chevrier, R. K. Heenan, A. E. Terry, S. Richeter, A. Mehdi, B. Van Mele, P. Dubois, R. Lazzaroni, P. Gerbaux, R. C. Evans and S. Clément, *Polym. Chem.*, 2014, **5**, 3352.
46. S. Miyanishi, Y. Zhang, K. Tajima and K. Hashimoto, *Chem. Commun.*, 2010, **46**, 6723.
47. S. Miyanishi, Y. Zhang, K. Hashimoto and K. Tajima, *Macromolecules*, 2012, **45**, 6424.
48. L. S. Li, J. Hollinger, N. Coombs, S. Petrov and D. S. Seferos, *Angew. Chem., Int. Ed.*, 2011, **50**, 8148.
49. F. Huang, H. B. Wu and Y. Cao, *Chem. Soc. Rev.*, 2010, **39**, 2500.
50. M. F. Wang and F. Wudl, *J. Mater. Chem.*, 2012, **22**, 24297.
51. A. Yassar, L. Miozzo, R. Gironda and G. Horowitz, *Prog. Polym. Sci.*, 2013, **38**, 791.
52. M. J. Robb, S.-Y. Ku and C. J. Hawker, *Adv. Mater.*, 2013, **25**, 5686.
53. Y. Nanashima, A. Yokoyama and T. Yokozawa, *J. Polym. Sci., Polym. Chem.*, 2012, **50**, 1054.
54. V. Senkovskyy, R. Tkachov, H. Komber, M. Sommer, M. Heuken, B. Voit, W. T. S. Huck, V. Kataev, A. Petr and A. Kiriy, *J. Am. Chem. Soc.*, 2011, **133**, 19966.
55. W. Liu, R. Tkachov, H. Komber, V. Senkovskyy, M. Schubert, Z. Wei, A. Facchetti, D. Neher and A. Kiriy, *Polym. Chem.*, 2014, **5**, 3404.
56. C. R. Bridges, T. M. McCormick, G. L. Gibson, J. Hollinger and D. S. Seferos, *J. Am. Chem. Soc.*, 2013, **135**, 13212.
57. C. R. Bridges, H. Yan, A. A. Pollit and D. S. Seferos, *ACS Macro Lett.*, 2014, **3**, 671.
58. J. Wang, M. Ueda and T. Higashihara, *ACS Macro Lett.*, 2013, **2**, 506.
59. J. Wang, M. Ueda and T. Higashihara, *J. Polym. Sci., Polym. Chem.*, 2014, **52**, 1139.
60. U. Scherf, S. Adamczyk, A. Gutacker and N. Koenen, *Macromol. Rapid Commun.*, 2009, **30**, 1059.
61. U. Scherf, A. Gutacker and N. Koenen, *Acc. Chem. Res.*, 2008, **41**, 1086.
62. R. Verduzco, I. Botiz, D. L. Pickel, S. M. Kilbey II, K. Hong, E. Dimasi and S. B. Darling, *Macromolecules*, 2011, **44**, 530.
63. M. Sommer, H. Komber, S. Huettner, R. Mulherin, P. Kohn, N. C. Greenham and W. T. S. Huck, *Macromolecules*, 2012, **45**, 4142.
64. R. C. Mulherin, S. Jung, S. Huettner, K. Johnson, P. Kohn, M. Sommer, S. Allard, U. Scherf and N. Greenham, *Nano Lett.*, 2011, **11**, 4846.
65. Y.-H. Lin, K. A. Smith, C. N. Kempf and R. Verduzco, *Polym. Chem.*, 2013, **4**, 229.
66. C. H. Gao, Y.-H. Lin, M. D. Witman, K. A. Smith, C. Wang, A. Hexemer, J. Strzalka, E. D. Gomez and R. Verduzco, *Nano Lett.*, 2013, **13**, 2957.
67. L. J. Bu, X. Y. Guo, B. Yu, Y. Qu, Z. Y. Xie, D. H. Yan, Y. H. Geng and F. S. Wang, *J. Am. Chem. Soc.*, 2009, **131**, 13242.

68. L. J. Bu, X. Y. Guo, B. Yu, Y. Y. Fu, Y. Qu, Z. Y. Xie, D. H. Yan, Y. H. Geng and F. S. Wang, *Polymer*, 2011, **52**, 4253.

69. J. F. Qu, B. R. Gao, H. K. Tian, X. J. Zhang, Y. Wang, Z. Y. Xie, H. Y. Wang, Y. H. Geng and F. S. Wang, *J. Mater. Chem. A*, 2014, **2**, 3632.

70. G. L. Tu, H. B. Li, M. Forster, R. Heiderhoff, L. J. Balk and U. Scherf, *Macromolecules*, 2006, **39**, 4327.

71. K. Nakabayashi and H. Mori, *Macromolecules*, 2012, **45**, 9618.

72. S.-Y. Ku, M. A. Brady, N. D. Treat, J. E. Cochran, M. J. Robb, E. J. Kramer, M. L. Chabinyc and C. J. Hawker, *J. Am. Chem. Soc.*, 2012, **134**, 16040.

73. K. B. Woody, B. J. Leever, M. F. Durstock and D. M. Collard, *Macromolecules*, 2011, **44**, 4690.

74. J. Kim, M. K. Gray, H. Y. Zhao, S. T. Nguyen and J. M. Torkelson, *Macromolecules*, 2005, **38**, 1037.

75. J. R. Locke and A. J. McNeil, *Macromolecules*, 2010, **43**, 8709.

76. E. F. Palermo and A. J. McNeil, *Macromolecules*, 2012, **45**, 5948.

77. E. F. Palermo, H. L. van der Laan and A. J. McNeil, *Polym. Chem.*, 2013, **4**, 4606.

78. E. F. Palermo, S. B. Darling and A. J. McNeil, *J. Mater. Chem. C*, 2014, **2**, 3401.

79. M. Jeffries-EL, G. Sauvé and R. D. McCullough, *Macromolecules*, 2005, **38**, 10346.

80. M. Jeffries-EL, G. Sauvé and R. D. McCullough, *Adv. Mater.*, 2004, **16**, 1017.

81. A. Smeets, K. Van den Bergh, J. D. Winter, P. Gerbaux, T. Verbiest and G. Koeckelberghs, *Macromolecules*, 2009, **42**, 7638.

82. K. A. Günay, P. Theato and H.-A. Klok, *J. Polym. Sci., Polym. Chem.*, 2013, **51**, 1.

83. Z.-Q. Wu, R. J. Ono, Z. Chen, Z. C. Li and C. W. Bielawski, *Polym. Chem.*, 2011, **2**, 300.

84. R. Peng, B. Pang, D. Q. Hu, M. J. Chen, G. B. Zhang, X. H. Wang, H. B. Lu, K. Cho and L. Z. Qiu, *J. Mater. Chem. C*, 2015, **3**, 3599.

85. C. N. Kempf, K. A. Smith, S. L. Pesek, X. Y. Li and R. Verduzco, *Polym. Chem.*, 2013, **4**, 2158.

86. C. A. Traina, R. C. Bakus II and G. C. Bazan, *J. Am. Chem. Soc.*, 2011, **133**, 12600.

87. M. Urien, H. Erothu, E. Cloutet, R. C. Hiorns, L. Vignau and H. Cramail, *Macromolecules*, 2008, **41**, 7033.

88. A. C. Kamps, M. Fryd and S.-J. Park, *ACS Nano*, 2012, **6**, 2844.

89. H. Yang, H. Xia, G. W. Wang, J. Peng and F. Qiu, *J. Polym. Sci., Polym. Chem.*, 2012, **50**, 5060.

90. A. Smeets, P. Willot, J. De Winter, P. Gerbaux, T. Verbiest and G. Koeckelberghs, *Macromolecules*, 2011, **44**, 6017.

91. K. A. Smith, Y.-H. Lin, D. B. Dement, J. Strzalka, S. B. Darling, D. L. Pickel and R. Verduzco, *Macromolecules*, 2013, **46**, 2636.

92. M. C. Iovu, M. Jeffries-EL, R. Zhang, T. Kowalewski and R. D. McCullough, *J. Macromol. Sci., Part A: Pure Appl. Chem.*, 2006, **43**, 1991.

93. T. Higashihara, K. Ohshimizu, A. Hirao and M. Ueda, *Macromolecules*, 2008, **41**, 9505.

94. H. C. Moon, A. Anthonysamy, Y. Lee and J. K. Kim, *Macromolecules*, 2010, **43**, 1747.

95. H. Q. Nguyen, M. P. Bhatt, E. A. Rainbolt and M. C. Stefan, *Polym. Chem.*, 2013, **4**, 462.

96. C. S. Fisher, M. C. Baier and S. Mecking, *J. Am. Chem. Soc.*, 2013, **135**, 1148.

97. M. C. Iovu, M. Jeffries-EL, E. E. Sheina, J. R. Cooper and R. D. McCullough, *Polymer*, 2005, **46**, 8582.

98. G. Sauvé and R. D. McCullough, *Adv. Mater.*, 2007, **19**, 1822.

99. C. R. Craley, R. Zhang, T. Kowalewski, R. D. McCullough and M. C. Stefan, *Macromol. Rapid Commun.*, 2009, **30**, 11.

100. Y. Lee, K.-I. Fukukawa, J. Bang, C. J. Hawker and J. K. Kim, *J. Polym. Sci., Polym. Chem.*, 2008, **46**, 8200.

101. H. T. Nguyen, O. Coulembier, J. De Winter, P. Gerbaux, X. Crispin and P. Dubois, *Polym. Bull.*, 2011, **66**, 51.

102. Y.-K. Fang, C.-L. Liu, C. Li, C.-J. Lin, R. Mezzenga and W.-C. Chen, *Adv. Funct. Mater.*, 2010, **20**, 3012.

103. M. C. Iovu, C. R. Craley, M. Jeffries-EL, A. B. Krankowski, R. Zhang, T. Kowalewski and R. D. McCullough, *Macromolecules*, 2007, **40**, 4733.

104. K. Palaniappan, N. Hundt, P. Sista, H. Nguyen, J. Hao, M. P. Bhatt, Y.-Y. Han, E. A. Schmiedel, W. E. Sheina, M. C. Biewer and M. C. Stefan, *J. Polym. Sci., Polym. Chem.*, 2011, **49**, 1802.

105. E. Kaul, V. Senkovskyy, R. Tkachov, V. Bocharova, H. Komber, M. Stamm and A. Kiriy, *Macromolecules*, 2010, **43**, 77.

106. R. H. Lohwasser and M. Thelakkat, *Macromolecules*, 2012, **45**, 3070.

107. F. Richard, C. Brochon, N. Leclerc, D. Eckhardt, T. Heiser and G. Hadziioannou, *Macromol. Rapid Commun.*, 2008, **29**, 885.

108. J. U. Lee, A. Cirpan, T. Emrick, T. P. Russell and W. H. Jo, *J. Mater. Chem.*, 2009, **19**, 1483.

109. J. U. Lee, J. W. Jung, T. Emrick, T. P. Russell and W. H. Jo, *Nanotechnology*, 2010, **21**, 105201.

110. Q. L. Zhang, A. Cirpan, T. P. Russell and T. Emrick, *Macromolecules*, 2009, **42**, 1079.

111. S. Rajaram, P. B. Armstrong, B. J. Kim and J. M. J. Fréchet, *Chem. Mater.*, 2009, **21**, 1775.

112. M. Sommer, A. S. Lang and M. Thelakkat, *Angew. Chem., Int. Ed.*, 2008, **47**, 7901.

113. B. W. Boudouris, C. D. Frisbie and M. A. Hillmyer, *Macromolecules*, 2008, **41**, 67.

114. B. W. Boudouris, C. D. Frisbie and M. A. Hillmyer, *Macromolecules*, 2010, **43**, 3566.

115. I. Botiz and S. B. Darling, *Macromolecules*, 2009, **42**, 8211.
116. V. Ho, B. W. Boudouris, B. L. McCulloch, C. G. Shuttle, M. Burkhardt, M. L. Chabinyc and R. A. Segalman, *J. Am. Chem. Soc.*, 2011, **133**, 9270.
117. M. G. Alemseghed, J. Servello, N. Hundat, P. Sista, M. C. Biewer and M. C. Stefan, *Macromol. Chem. Phys.*, 2010, **211**, 1291.
118. M. G. Alemseghed, S. Gowrisanker, J. Servello and M. C. Stefan, *Macromol. Chem. Phys.*, 2010, **211**, 2007.
119. S.-J. Park, S.-G. Kang, M. Fryd, J. G. Saven and S.-J. Park, *J. Am. Chem. Soc.*, 2010, **132**, 9931.
120. C. A. Dai, W. C. Yen, Y. H. Lee, C. C. Ho and W. F. Su, *J. Am. Chem. Soc.*, 2007, **129**, 11036.
121. S. P. Wu, Y. Q. Sun, L. Huang, J. W. Wang, Y. H. Zhou, Y. H. Geng and F. S. Wang, *Macromolecules*, 2010, **43**, 4438.
122. Z.-Q. Wu, R. J. Ono, Z. Chen and C. W. Bielawski, *J. Am. Chem. Soc.*, 2010, **132**, 14000.
123. N. Liu, C.-G. Qi, Y. Wang, D.-F. Liu, J. Yin, Y.-Y. Zhu and Z.-Q. Wu, *Macromolecules*, 2013, **46**, 7753.
124. Z.-Q. Wu, C.-G. Qi, N. Liu, Y. Wang, J. Yin, Y.-Y. Zhu, L. Z. Qiu and H.-B. Lu, *J. Polym. Sci., Polym. Chem.*, 2013, **51**, 2939.
125. Z.-Q. Wu, Y. Chen, Y. Wang, X.-Y. He, Y.-S. Ding and N. Liu, *Chem. Commun.*, 2013, **49**, 8069.
126. L.-M. Gao, Y.-Y. Hu, Z.-P. Yu, N. Liu, J. Yin, Y.-Y. Zhu, Y.-S. Ding and Z.-Q. Wu, *Macromolecules*, 2014, **47**, 5010.
127. N. Doubina, A. Ho, A. K. Y. Jen and C. K. Luscombe, *Macromolecules*, 2009, **42**, 7670.
128. N. Doubina, M. Stoddard, H. A. Bronstein, A. K. Y. Jen and C. K. Luscombe, *Macromol. Chem. Phys.*, 2009, **210**, 1966.
129. H. A. Bronstein and C. K. Luscombe, *J. Am. Chem. Soc.*, 2009, **131**, 12894.
130. V. Senkovskyy, N. Khanduyeva, H. Komber, U. Oertel, M. Stamm, D. Kuckling and A. Kiriy, *J. Am. Chem. Soc.*, 2007, **129**, 6626.
131. N. Khanduyeva, V. Senkovskyy, T. Beryozkina, V. Bocharova, F. Simon, M. Nitschke, M. Stamm, R. Grötzschel and A. Kiriy, *Macromolecules*, 2008, **41**, 7383.
132. N. Khanduyeva, V. Senkovskyy, T. Beryozkina, M. Horecha, M. Stamm, C. Uhrich, M. Riede, K. Leo and A. Kiriy, *J. Am. Chem. Soc.*, 2009, **131**, 153.
133. V. Senkovskyy, R. Tkachov, H. Komber, U. Oertel, M. Horecha, V. Bocharova, M. Stamm, S. A. Gevorgyan, F. C. Krebs and A. Kiriy, *J. Am. Chem. Soc.*, 2009, **131**, 16445.
134. R. Tkachov, V. Senkovskyy, M. Horecha, U. Oertel, M. Stamm and A. Kiriy, *Chem. Commun.*, 2010, **46**, 1425.
135. T. Beryozkina, K. Boyko, N. Khanduyeva, V. Senkovskyy, M. Horecha, U. Oertel, M. Stamm and A. Kiriy, *Angew. Chem., Int. Ed.*, 2009, **48**, 2695.
136. S. K. Sontag, N. Marshall and J. Locklin, *Chem. Commun.*, 2009, **45**, 3354.

137. N. Marshall, S. K. Sontag and J. Locklin, *Macromolecules*, 2010, **43**, 2137.
138. R. Tkachov, V. Senkovskyy, U. Oertel, A. Synytska, M. Horecha and A. Kiriy, *Macromol. Rapid Commun.*, 2010, **31**, 2146.
139. V. Senkovskyy, I. Senkovska and A. Kiriy, *ACS Macro Lett.*, 2012, **1**, 494.
140. N. Doubina, J. L. Jenkins, S. A. Paniagua, K. A. Mazzio, G. A. MacDonald, A. K. Y. Jen, N. R. Armstrong, S. R. Marder and C. K. Luscombe, *Langmuir*, 2012, **28**, 1900.
141. L. Q. Yang, S. K. Sontag, T. W. LaJoie, W. T. Li, N. E. Huddleston, J. Locklin and W. You, *ACS Appl. Mater. Interfaces*, 2012, **4**, 5069.
142. D. F. Zeigler, K. A. Mazzio and C. K. Luscombe, *Macromolecules*, 2014, **47**, 5019.
143. K. Sivanandan, T. Chatterjee, N. Treat, E. J. Kramer and C. J. Hawker, *Macromolecules*, 2010, **43**, 233.
144. V. Senkovskyy, T. Beryozkina, R. Tkachov, H. Komber, A. Lederer, M. Stamm, N. Severin, J. P. Rabe and A. Kiriy, *Macromol. Symp.*, 2010, **291–292**, 17.
145. W. G. Huang, L. J. Su and Z. S. Bo, *J. Am. Chem. Soc.*, 2009, **131**, 10348.
146. W. G. Huang, M. Wang, C. Du, Y. L. Chen, R. P. Qin, L. J. Su, C. Zhang, Z. P. Liu, C. H. Li and Z. S. Bo, *Chem. – Eur. J.*, 2011, **17**, 440.

CHAPTER 4

Controlled Synthesis of Chain End Functional, Block and Branched Polymers Containing Polythiophene Segments

TOMOYA HIGASHIHARA AND MITSURU UEDA*

Graduate School of Organic Materials Science, Yamagata University,
4-3-16 Jonan, Yonezawa, Yamagata 992-8510, Japan
*Email: ueda.m.ad@m.titech.ac.jp

4.1 Introduction

In principle, 'living'/controlled polymerization provides the most versatile methodology for the synthesis of well-defined macromolecules with a predictable molecular weight (MW) and a narrow molecular weight distribution (or dispersity, Đ). A wide range of other compositional and structural parameters can also be controlled, such as the chain end functionality, copolymer composition, regio-regularity, stereo-regularity and branching.[1–7] More complex macromolecular architectures, exploiting the benefits of 'living'/controlled polymerization and the efficient coupling/linking reactions of building blocks, have rapidly emerged in polymer chemistry and have been reviewed elsewhere.[8–21]

The low degree of compositional heterogeneity in macromolecules is important in clarifying the quantitative relationship between the primary chemical structure and properties of polymers. Developments in the

RSC Polymer Chemistry Series No. 21
Semiconducting Polymers: Controlled Synthesis and Microstructure
Edited by Christine Luscombe
© The Royal Society of Chemistry 2017
Published by the Royal Society of Chemistry, www.rsc.org

synthesis of model polymers by 'living'/controlled polymerization have been of interest for more than 50 years. Block copolymers (BCPs), in particular, have attracted attention as a result of their unique behaviour and effects in morphological studies, micelles, processing, materials science and functional device applications.[22-30] In a typical synthesis of well-defined BCPs, the second monomer is charged *in situ* after the polymerization of the first monomer, leading to an efficient crossover reaction, followed by polymerization of the second monomer in a 'living' manner. As the requirements for polymer properties and performances have become stricter and more versatile, both the monomer varieties and the combinations of polymer segments need to be broadened.

π-Conjugated polymers have emerged as intriguing and unique materials as a result of their stiff, rod-like structures in solution, as melts and/or in the solid state, their heterophase structures and their optoelectronic performance. Typical π-conjugated polymers include polyacetylenes, polyphenylenes, polyfluorenes, poly(phenylene vinylene)s and polythiophenes (PTs). Breakthroughs in condensative chain-growth polymerization techniques have allowed the synthesis of well-defined π-conjugated polymers and BCPs containing low-dispersity π-conjugated polymer segments via 'living'/controlled systems.[31-37] These advances may contribute to the accelerated evolution of potential areas of application, such as polymer photonics materials and optoelectronic devices.

We review here recent progress in the synthetic methodologies of well-defined chain end functional, block and branched polymers containing PT segments, especially those based on 'living'/controlled polymerization systems. As applications using BCPs containing PT segments have been reviewed elsewhere,[34,38] this chapter focuses on recent synthetic methodologies and examples.

4.2 Controlled Synthesis of Polythiophenes

Poly(3-alkylthiophene)s (P3ATs) are considered to be the best class of balanced high-performance p-type semiconductors among all the π-conjugated polymers in terms of solubility, chemical stability, charge mobility and commercial availability. Therefore much attention has been paid to the application of these materials in electronic devices such as organic field-effect transistors[39-44] and photovoltaic cells.[45-49]

PTs were initially synthesized by protic acid, Lewis acid or electrochemical oxidative polymerization. As a result of the strong intermolecular π–π stacking between the extended conjugation along the main chains, PTs are only poorly soluble and Jen and coworkers[50] first synthesized P3ATs to improve their solubility. Sugimoto *et al.* also succeeded in synthesizing P3ATs by FeCl₃-mediated oxidative coupling polymerization.[51] The substitution of an alkyl group at the 3-position of thiophene rings causes a significant regio-chemistry problem derived from the asymmetrical monomer structure. In addition, conventional condensative polymerization limits controllability in terms of the MW and *Đ*; theoretically, the *Đ* values reach 2 at the end of monomer conversion in condensative polymerization. Practical methodologies for addressing these issues are introduced in the following section.

4.3   Regio-regularity

P3ATs have regio-structures along the main chain as a result of the asymmetrical chemical structure of the monomer. Figure 4.1a shows the three different regio-regularities: head-to-tail (HT), head-to-head (HH) and tail-to-tail (TT). This results in four triad regio-isomers in the polymer chain: HT–HT, HT–HH, TT–HT and TT–HH (Figure 4.1b). McCullough and co-workers[52,53] reported the control of regio-regularity of P3ATs based on the Kumada coupling reaction, which gave a regio-regularity >98%. The synthetic routes are shown in Scheme 4.1a. At almost the same time, Rieke and coworkers[54,55] used the Negishi coupling reaction to provide HT–HT regio-regular P3ATs (Scheme 4.1b). The Stille coupling reaction (Scheme 4.1c)[56,57] and the Suzuki coupling reaction (Scheme 4.1d)[58,59] have also been used to synthesize HT–HT regio-regular P3ATs and related PTs.

McCullough and coworkers[60,61] further developed the inexpensive methodology of Grignard metathesis polymerization to give regio-regular P3ATs, which is also referred to as Kumada catalyst-transfer polymerization (KCTP). The intermediate Grignard monomer can be obtained at room temperature (or reflux temperature) *via* the Grignard exchange reaction from sp^3 to sp^2 orbital formation, which is much cheaper than the earlier technique using lithium diisopropylamide and low temperature conditions (Scheme 4.1e).

The regio-regularity can be calculated from ^1H NMR spectra by comparing the deconvoluted signal intensity of the 4-H thiophene protons (HT–HT 6.98 ppm; HT–HH 7.00 ppm; TT–HT 7.02 ppm; and TT–HH 7.05 ppm) as well as α-methylene protons (HT 2.80 ppm; HH 2.58 ppm) next to the thiophene ring.[54,55]

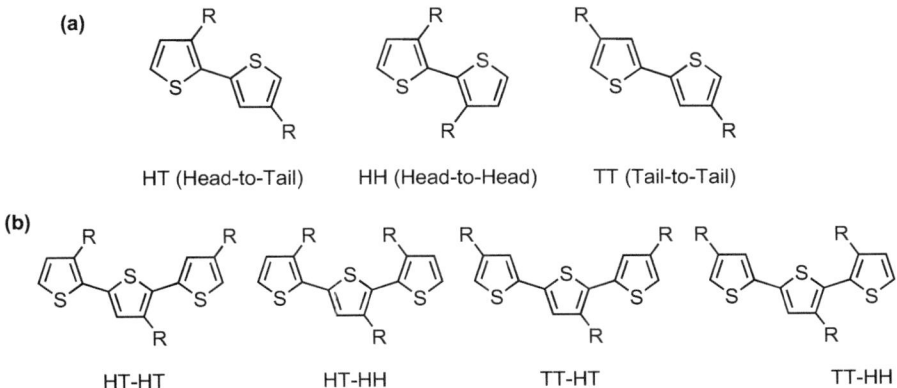

(a)

HT (Head-to-Tail)   HH (Head-to-Head)   TT (Tail-to-Tail)

(b)

HT-HT   HT-HH   TT-HT   TT-HH

Figure 4.1   Regio-regularity of poly(3-alkylthiophene) polymers.

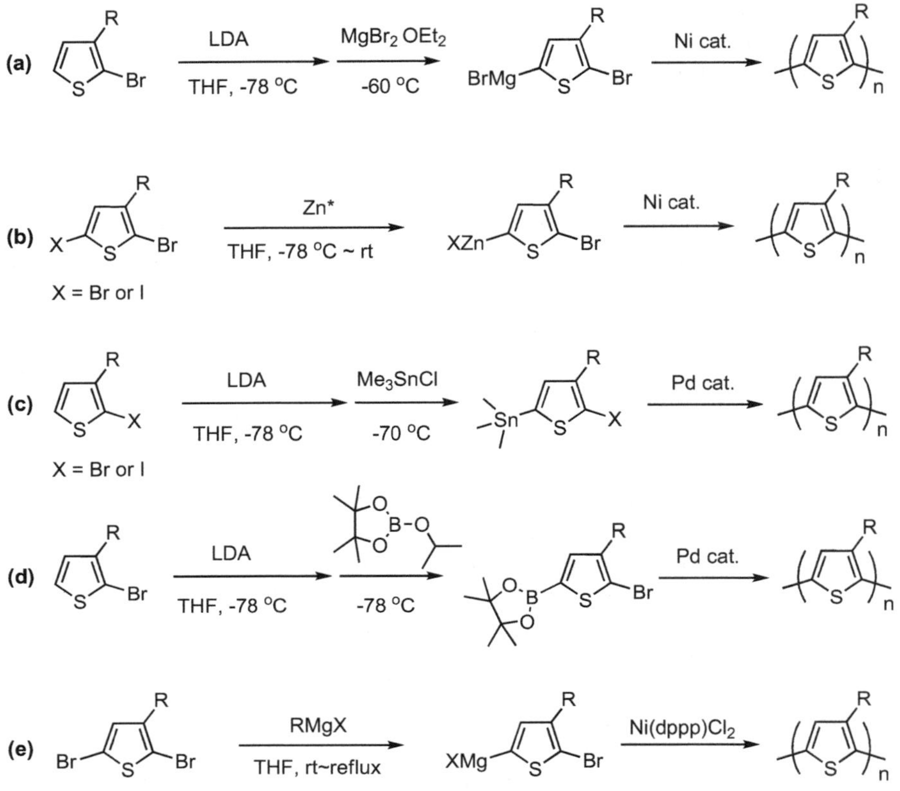

Scheme 4.1 Synthetic routes for regio-regular poly(3-alkylthiophene) structures.

4.4 Control of Molecular Weight and Dispersity

A quasi-living KCTP system was independently reported by McCullough and coworkers[60,61] shortly after Yokozawa and coworkers.[62,63] The mechanism proposed by Yokozawa and coworkers[63] is as follows (see Scheme 4.2). Two equivalents of the thiophene monomer, 2-bromo-5-chloromagnesio-3-hexylthiophene (**1**), are coupled by adding catalytic Ni(dppp)Cl$_2$ (dppp = 1,3-diphenylphosphinopropane) to give the TT dimer. The Ni species then transfers to the next C–Br bond by walking through the π-conjugation without diffusion of the Ni^0 species to afford the dimer (**2**). Dimer **2** then acts as a virtual initiator to start the polymerization of **1** by transmetallation. The polymerization proceeds in a chain-growth manner without diffusion of the Ni^0 species, in which the transmetallation reaction only takes place between the activated polymer chain end and the incoming monomer of **1**. When the initiation rate is faster or similar to the propagation rate, the resulting P3HT has a controlled MW with a feed ratio of **1** to Ni(dppp)Cl$_2$ and a low *Đ* < 1.2. McCullough and coworkers[61] proposed that the associate pairing between the generating polymer and the

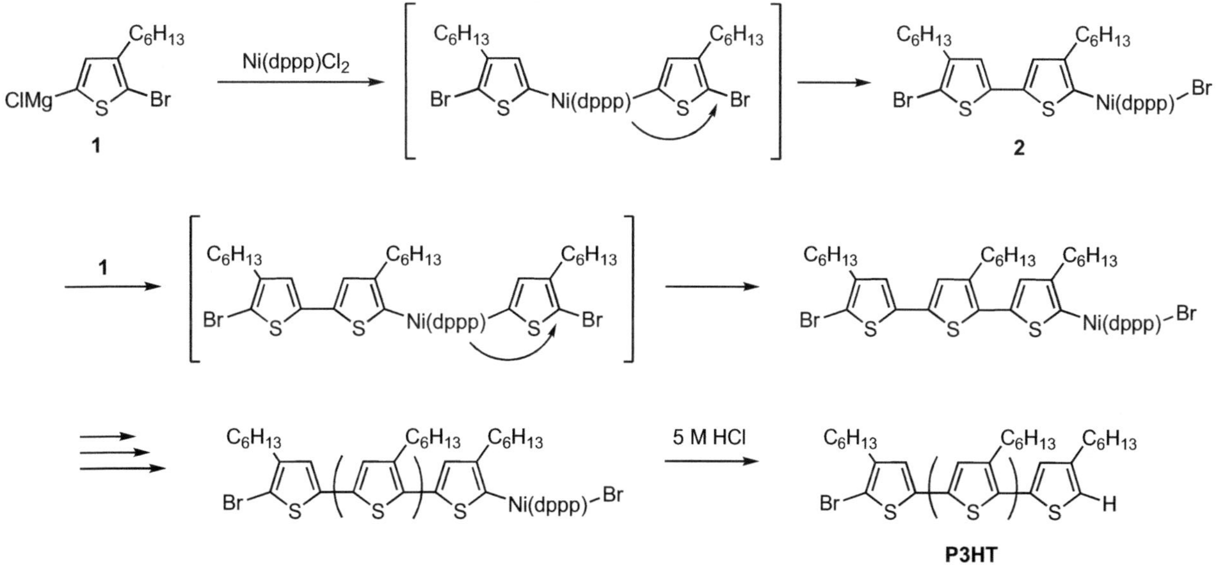

Scheme 4.2 Proposed mechanism for catalyst-transfer polycondensation in the synthesis of regio-regular P3HT. Reprinted with permission from T. Yokozawa and A. Yokoyama, *Chem. Rev.*, 2009, **109**, 5595–5619. Copyright (2009) American Chemical Society.[33]

Ni^0 species prevents the diffusion of the Ni^0 species into the polymerization media to achieve the chain-growth mechanism.

Thorough purification steps for the monomers and solvents are often necessary for the KCTP system, however, because highly reactive and moisture-sensitive Grignard reagents are used. In addition, many functional groups (*e.g.* hydroxy, aldehyde, ketone, ester and amide groups) cannot generally be tolerated in the presence of Grignard reagents. Uchiyama and coworkers[64] reported the anionic polymerization of *N*-isopropylacrylamide in aqueous media using the newly designed bulky zincate complex of $^tBu_4ZnLi_2$ with a low basicity as a highly selective anionic initiator. The results indicated that $^tBu_4ZnLi_2$ can be tolerated by hydroxyl functional groups. Regio-regular P3HT can be synthesized by a Negishi-type catalyst-transfer polymerization by replacing the Grignard-type monomer **1** by the zincate complex (Scheme 4.3a).[65]

The average number molecular weight (M_n) proportionally increased from 2500 to 30 700 by increasing the feed ratio of the monomer to Ni catalyst, while maintaining the relatively low $Đ < 1.2$. The quasi-livingness of Negishi-type catalyst-transfer polymerization has been shown by a monomer addition experiment in which M_n increased after the second addition of the same monomer as the first. The regio-regularity of P3HT was >97% when M_n exceeded 10 000. In addition, this new system could be used in as-received tetrahydrofuran as well as in tetrahydrofuran containing artificially added protic impurities such as isopropanol, methanol and even 1000 ppm water (equal molar ratio to the monomer). Based on these successful results, the direct synthesis of poly[3-(6-hydroxyhexyl)thiophene] (P3HHT) could be performed without protection (Scheme 4.3b). An Ni catalyst with an optimized phosphine ligand of 1,2-bis(cyclohexyl)ethane successfully provided P3HTs with extremely low $Đ$ values (1.03–1.17).[66] The controlled synthesis of P3HT based on Negishi-type catalyst-transfer polymerization has been reviewed elsewhere.[36] Another strategy for replacing Grignard-type thiophene monomers is the Suzuki–Miyaura catalyst-transfer polymerization system. Yokozawa *et al.*[67] reported the $^tBu_3PPd(Ph)Br$-catalysed Suzuki–Miyaura coupling polymerization of 2-(4-hexyl-5-iodo-2-thienyl)-4,4,5,5-tetramethyl-1,3,2-dioxaborolane (Scheme 4.4).

4.5   Dehydrogenative Synthesis of Polythiophene

An atom-economical method involving the dehydrogenative polycondensation of 2-bromo-3-hexylthiophene or 2-chloro-3-hexylthiophene has been reported for the synthesis of regio-regular P3HT (regio-regularity >98%). As shown in Scheme 4.5a, Ozawa and coworkers[68] reported the dehydrogenative polycondensation of 2-bromo-3-hexylthiophene with Herrmann's catalyst and tris(2-dimethylaminophenyl)phosphine as catalyst precursors, affording HT-P3HT with a high regio-regularity of 98%. Mori and coworkers[69,70] prepared HT-P3HT by dehydrogenative polycondensation using 2-bromo-3-hexylthiophene and the Knochel–Hauser base (TMPMgCl · LiCl) (Scheme 4.5b).

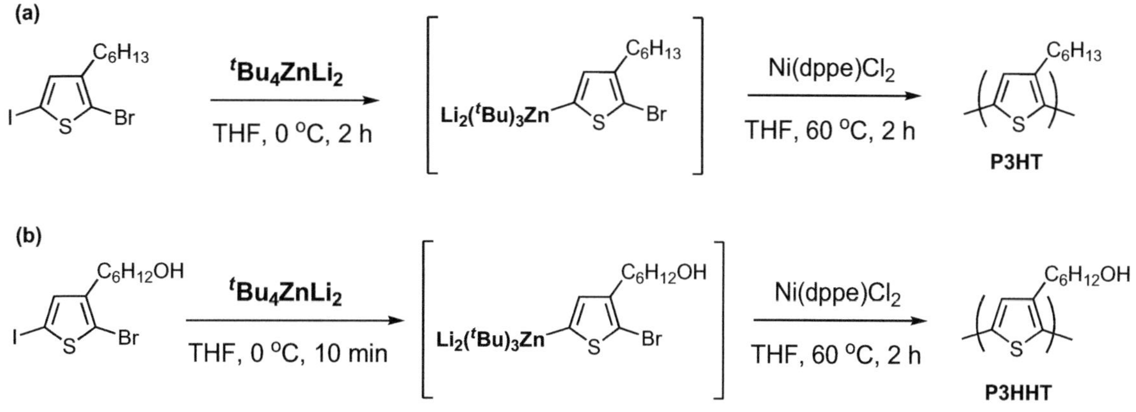

Scheme 4.3 Synthesis of regio-regular P3HT using the zincate complex ᵗBu₄ZnLi₂.

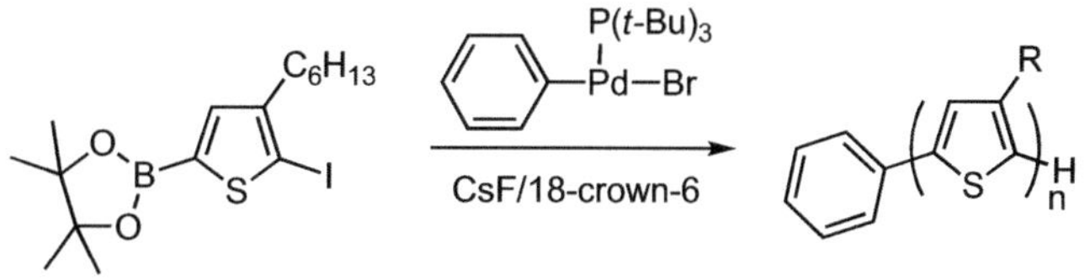

Scheme 4.4 Synthesis of regio-regular P3HT by catalyst-transfer Suzuki–Miyaura coupling polymerization.

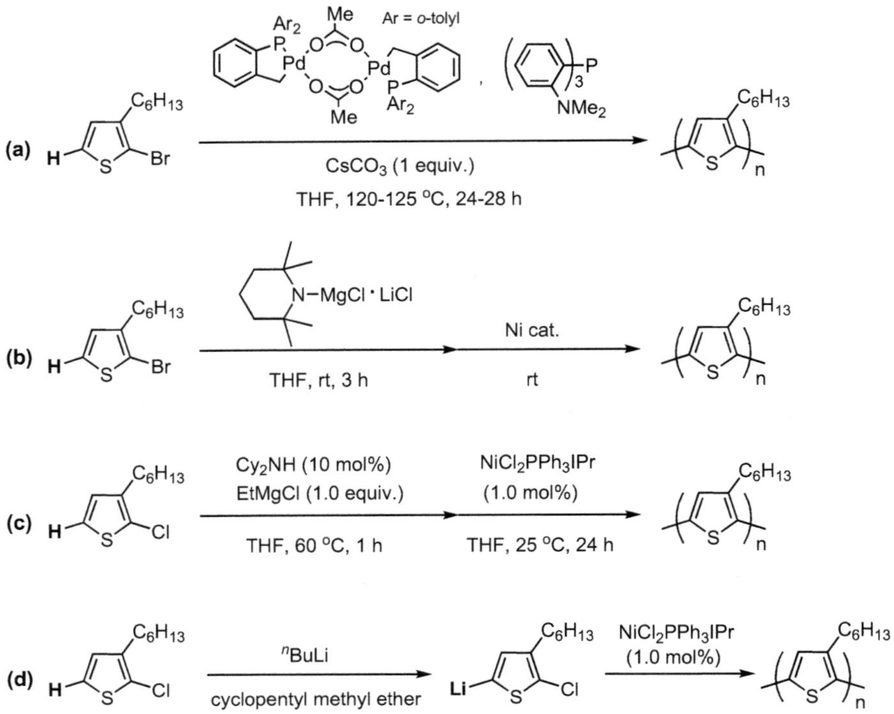

Scheme 4.5 Synthesis of regio-regular P3HT by dehalogenative polycondensation.

The use of a Grignard reagent and catalytic amine also enabled the dehydrogenative synthesis of HT-P3HT starting from 2-chloro-3-hexylthiophene, which is a more atom-economical monomer (Scheme 4.5c).[71] The same group has also utilized the Murahashi coupling reaction in the Ni–N-heterocyclic carbene complex catalysed polycondensation of organo-lithium species derived from 2-chloro-3-hexylthiophene (Scheme 4.5d).[72] Based on the quasi-living nature of the Ni-catalysed polycondensation, all-conjugated block copolythiophenes could also be synthesized by the dehydrogenative polycondensation reaction based on the sequential monomer addition technique.[69,73]

4.6 Chain End Functional Polythiophenes

The chain end functionalization of PTs is a very important step in preparing BCPs because the terminal functional groups are responsible for the introduction of the second block in the subsequent reactions or polymerizations. McCullough and coworkers[74–76] reported the chain end functionalization of P3HT by terminating quasi-living P3HT-Ni(dppp)Br with the Grignard reagent (RMgX) (Scheme 4.6). The functionality widely includes methyl,

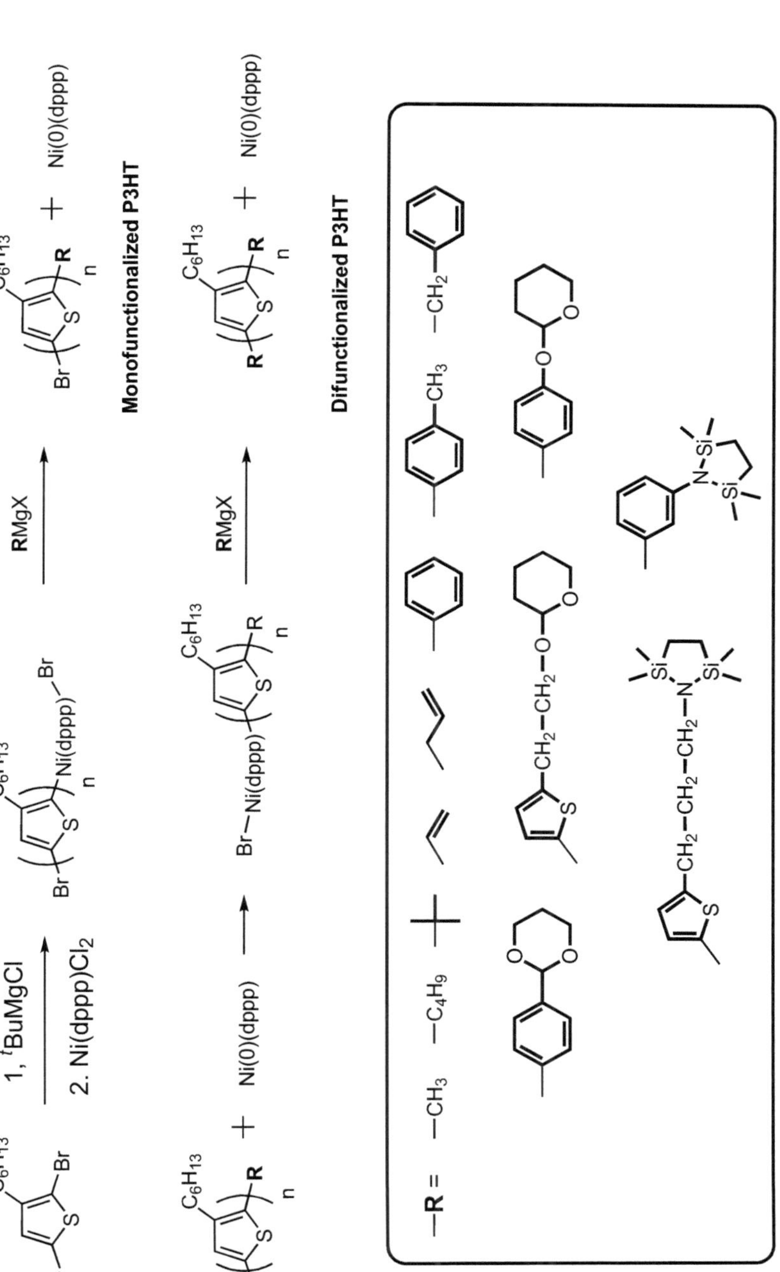

Scheme 4.6 Synthesis of chain end functional P3HTs by the termination method.

butyl, *t*butyl, vinyl, allyl, phenyl, tolyl, benzyl, formyl (protected by acetal), alcohol/phenol (protected by tetrahydropyran) and 3-aminopropyl/*p*-aminophenyl (protected by trimethylsilane) groups (Scheme 4.6).

However, the type of RMgX significantly affects the final product structures with respect to whether monofunctionalization at the ω-terminal or difunctionalization at the α,ω-termini takes place. As shown in Scheme 4.6, difunctional P3HT can be obtained by following mechanism: (1) reductive elimination of the Ni catalyst after monofunctionalization; (2) the Ni⁰ generated oxidatively inserts into the α-terminal; and (3) the metal-exchange reaction with excess RMgX eliminates Ni⁰ again. This mechanism illustrates the difficulty in selectively synthesizing either mono- or difunctional P3HT derivatives.

Indirect multistep end functionalization methodologies have also been reported. Thelakkat and Lohwasser[77] synthesized monocarboxylated P3HT by a post-functionalization method using the α-Br terminus (see Figure 4.2a). Tajima and coworkers[78] and Okamoto and Luscombe[79] independently used the post-functionalization method to afford α,ω-ditrimethylstannylated/ω-trimethylstannylated P3HTs and α,ω-dithiolated/ω-thiolated P3HTs, respectively (Figure 4.2b and 4.2c).

Bronstein and Luscombe[80] reported an efficient initiation method using an *o*-halotoluene precursor based on a ligand-exchange approach with the aim of synthesizing a defect-free regio-regular P3HT (Scheme 4.7). This system affords defect-free P3HTs with virtually 100% *o*-tolyl chain end functionality. They extended this approach to other types of initiators and obtained relatively good chain end functionalities for phenyl (89%) and thiophen-2-yl (72%) groups at the α-chain end of P3HTs.[81] Koeckelberghs and coworkers[82–84] reported the preparation of α-chain end functional P3HTs with functional initiators with OH (protected), SH (protected), COOH (protected), NH₂ (protected), alkyne (protected), phosphonic ester, pyridine and phenol (protected) groups based on the initiation method (Figure 4.3a). Kiriy and coworkers[85,86] synthesized a specially designed Ni initiator with an initiating fragment for nitroxy-mediated radial polymerization (NMRP) and another initiator with electronic functionality (Figure 4.3b).

Higashihara and coworkers[87] succeeded in the direct synthesis of the ω-chain end functional P3HTs with mono-bromobutyl groups without protection by the combination of initiation and termination methods in a one-pot process (Scheme 4.8). When Ni(dppp)Cl₂ was used as an initiator, difunctional P3HTs with 4-bromobutyl groups at both ends were formed. Matrix-assisted laser desorption/ionization time-of-flight mass spectrometry confirmed the selective preparation and quantitative chain end functionality of both the monobromo and dibromo functional P3HTs.

The bromoalkyl group can be transformed into a wide variety of other useful functionalities, such as azide, alkyne, amine, cyano, ether, alcohol and aldehyde groups. This system is also important for BCP synthesis using chain end functional P3HT as a precursor because the end functionality of P3HT is predominant for the sequence control of AB diblock or

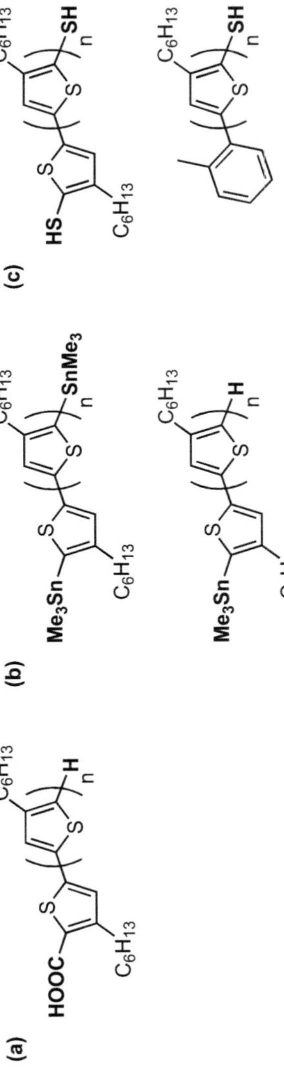

Figure 4.2   Chemical structures of chain end functional P3HTs.

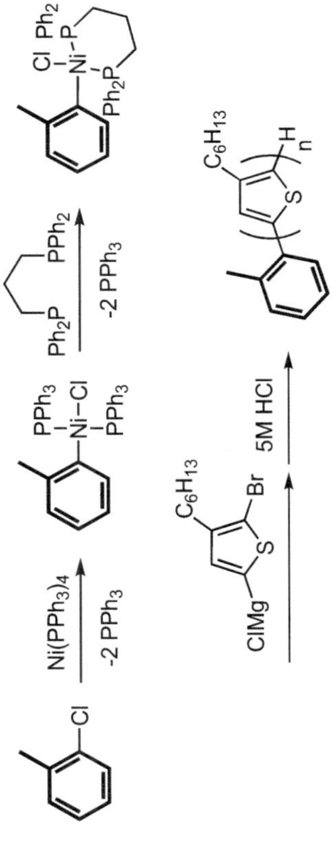

Scheme 4.7   Synthesis of α-chain end functional P3HTs by an initiation method.

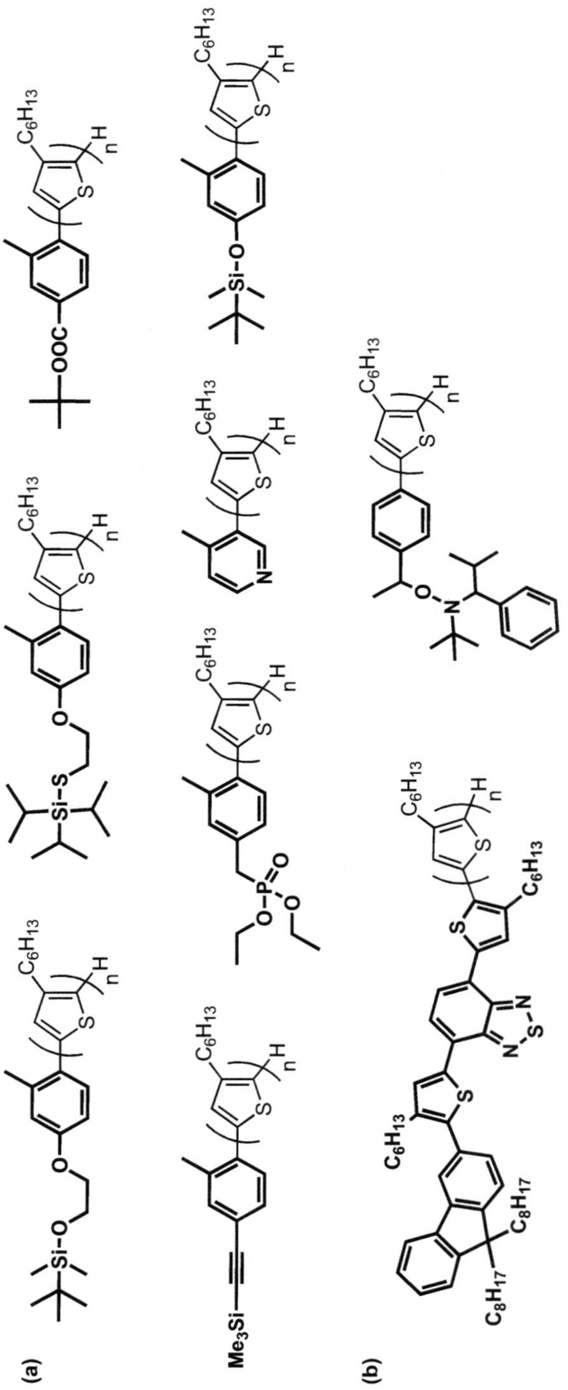

Figure 4.3 Chemical structures of chain end functional P3HTs prepared with functional initiators.

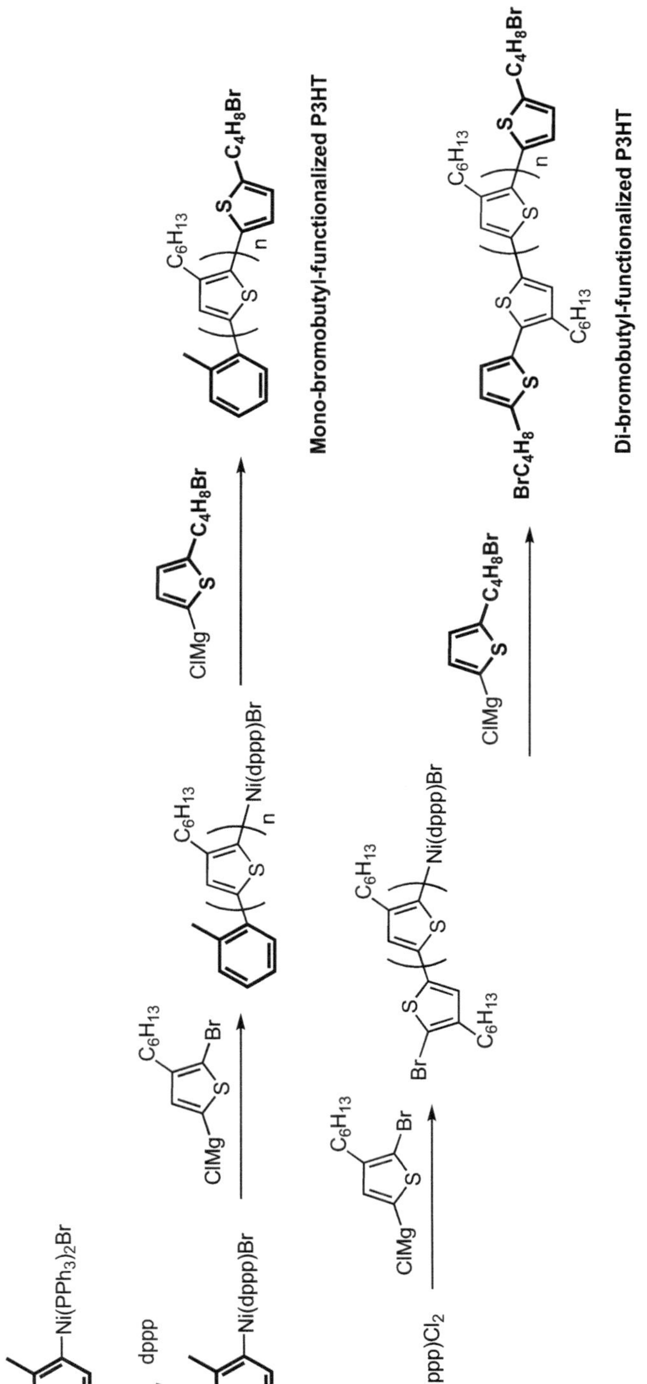

Scheme 4.8 Synthesis of mono- and dibromobutyl chain end functional P3HTs.

ABA triblock copolymers. Bao and coworkers[88] synthesized α,ω-chain end functional PTs with trimethylsilyl/thiol and azide moieties at the α- and ω-chain ends, respectively, based on a catalyst-transfer Suzuki–Miyaura coupling polymerization approach (Scheme 4.9). Recent progress in chain end functional P3HTs has been reviewed by Hawker and coworkers.[38]

4.7 Block Copolymers with Polythiophene Segments

4.7.1 All-conjugated Block Copolythiophenes

An all-conjugated block copolythiophene is generally defined as a BCP with a perfect π-conjugation along the main chain. Based on KCTP, an array of block copolythiophenes can be synthesized using a simple sequential monomer addition technique. McCullough and coworkers[61] first reported the sequential polymerization of 2-bromo-5-chloromagnesio-3-dodecylthiophene (**3**) just after the completion of the first block obtained by the polymerization of **1** to afford P3HT-*b*-poly(3-dodecylthiophene) (**4**) (Scheme 4.10). Yokozawa *et al.*[89] reported the synthesis of P3HT-*b*-poly(3-(2-(2-methoxyethoxy)ethoxy)thiophene) (**5**) based on a similar approach (Figure 4.4). Ueda and coworkers[90] and Tajima and coworkers[91] independently synthesized the crystalline–amorphous block copolythiophenes P3HT-*b*-poly(3-phenoxymethylthiophene) (**6**) and P3HT-*b*-poly(3-(2-ethylhexyl)thiophene) (**7**), respectively (Figure 4.4). Many other all-conjugated block copolythiophenes have also been reported (Figure 4.4), including P3HT-*b*-poly(3-(3,7-dimethyloctyloxy)thiophene) (**8**),[92] poly(3-butylthiophene)-*b*-poly(3-octylthiophene) (**9**),[93] P3HT-*b*-poly(3-(4,4,5,5,6,6,7,7,7-nonafluoroheptyl)thiophene) (**10**),[94] P3HT-*b*-poly(3-(2-(2-(2-(2-methoxyethoxy)ethoxy)ethoxy)ethoxy)thiophene) (**11**),[95] P3HT-*b*-poly(3-(3-aminopropyloxy)methylthiophene) (**12**),[96] P3HT-*b*-poly(hexyl (thien-3-yl)acetate) (**13**),[97] P3HT-*b*-poly(7-(2-(thien-3-yl)ethoxy)-heptanoic acid) (**14**),[98] P3HT-*b*-poly(3-(10-(4-(3,3,4,4,5,5,6,6,7,7,8,8,8-tridecafluorooctyloxy)phenoxy)decyl)thiophene) (**15**),[99] P3HT-*b*-poly(3-(10-(2,3,6,7,10-pentakis(hexyloxy)triphenylene)decyloxy)thiophene) (**16**)[100] and P3HT-*b*-poly(3-(6-(*N*-methylimidazolium)hexyl)thiophene) (**17**).[101,102] Higashihara and coworkers also reported the synthesis of P3HT-*b*-poly(3-(2-(2-(2-methoxyethoxy)ethoxy)ethoxy)thiophene) (**18**)[103] and P3HT-*b*-poly(3-(4-(3,7-dimethyloctyloxy)-3-pyridyl)thiophene) (**19**).[104]

During the course of our morphological studies on block copolythiophenes, it was found that a thin film of BCP **18**, which has an amphiphilic structure, displayed a hierarchical nanostructure with a periodic perpendicular lamellar morphology in the range 25–30 nm in which crystalline P3HT tends to show an edge-on rich orientation, as observed by atomic force microscopy, transmission electron microscopy and glazing incidence X-ray scattering.[103] Such a perpendicularly oriented morphology is important in organic photovoltaic devices as a result of the high speed of charge transportation before recombination of the separated charges occurs.

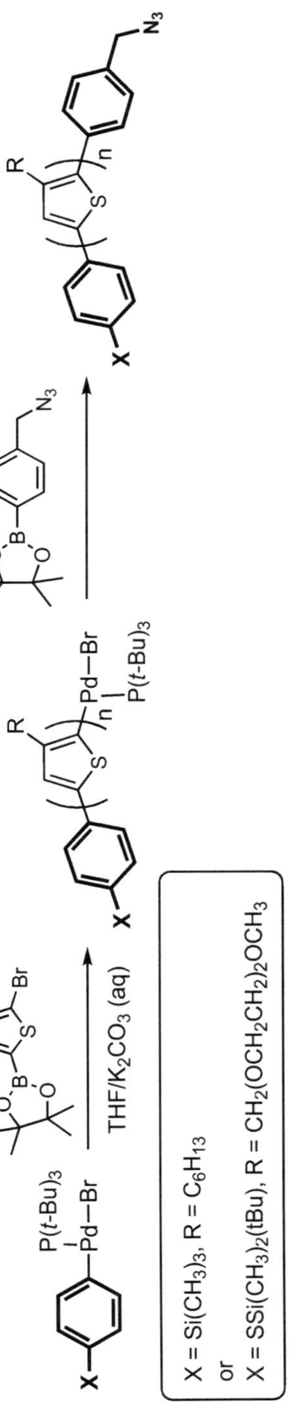

Scheme 4.9 Synthesis of α,ω chain end functional P3HTs.

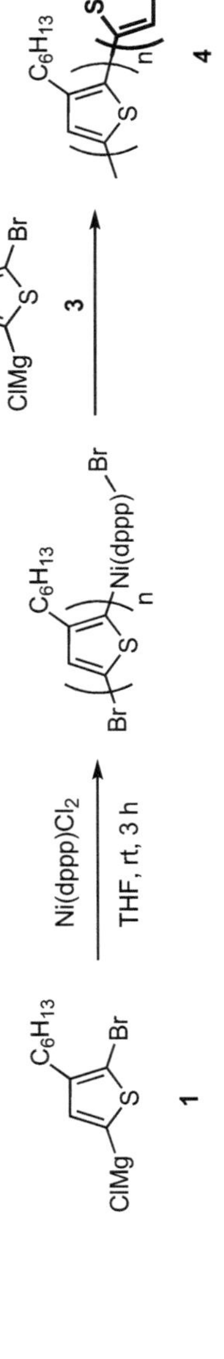

Scheme 4.10 Synthesis of all-conjugated block copolythiophene by the sequential monomer addition technique.

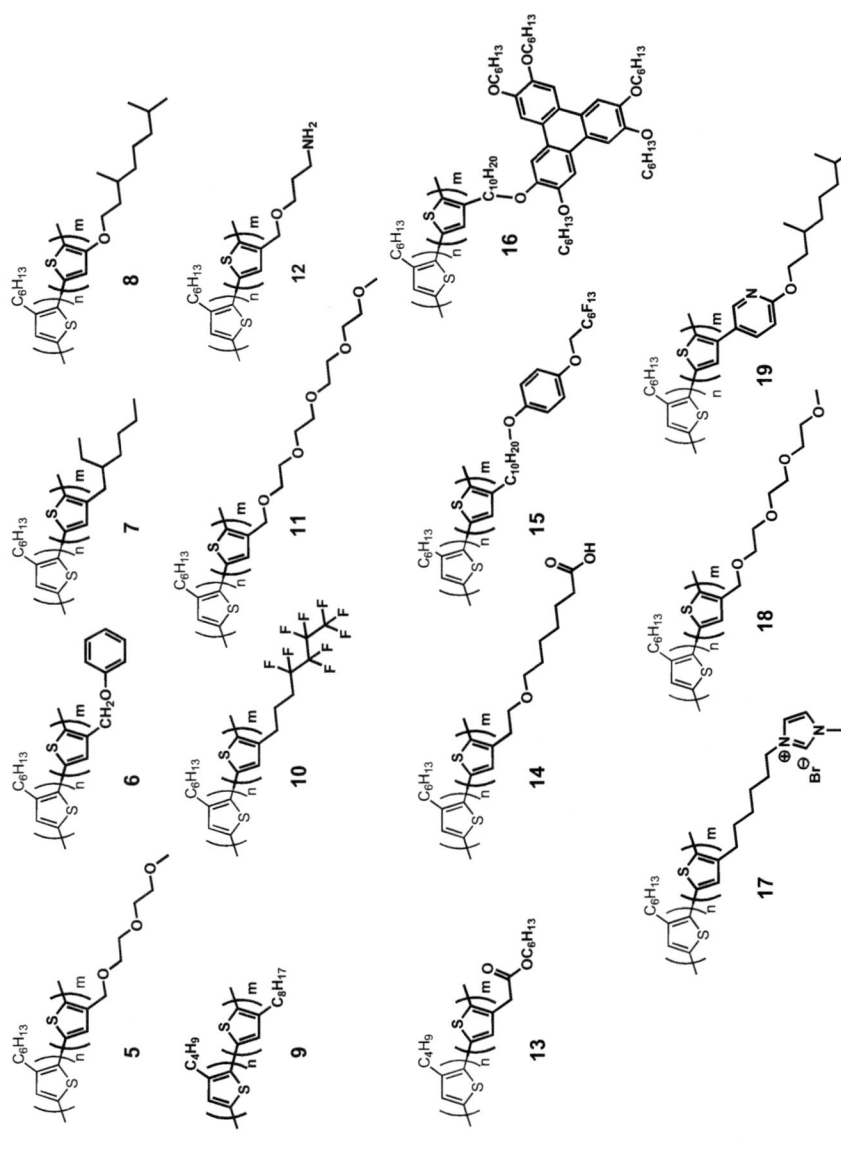

Figure 4.4 Chemical structures of all-conjugated block copolythiophenes.

4.7.2   All-conjugated Block Copolymers Containing Polythiophene and Other Polymer Segments

Scherf and coworkers[105] reported the synthesis of all-conjugated block copolymers containing PT and polyfluorene segments using H-/Br-terminated P3HT and AB type fluorine monomers (Scheme 4.11) and reviewed the synthesis and characterization of the related block copolymers.[106] Based on recent progress in the controlled synthesis of π-conjugated polymers by the chain-growth catalyst-transfer polymerization of various monomers other than thiophene-based monomers, a new type of all-conjugated BCP has emerged. For instance, Yokozawa and coworkers[107] reported that the sequential monomer addition technique is applicable to the synthesis of poly(2,5-dihexyloxy-1,4-phenylene)(PPP)-*b*-P3HT, although the order of addition should be carefully selected to guarantee an efficient crossover reaction from the first monomer to the second (Scheme 4.12). To synthesize well-defined PPP-*b*-P3HT, the phenylene-based monomer must first be polymerized. Other BCPs containing PT and other types of polymer segment include P3HT-*b*-poly(3-hexylselenophene) (**20**),[108] P3HT-*b*-poly(4,4-bis(2-ethylhexyl)dithieno[3,2-*b*:2′,3′-*d*]silole) (**21**)[109] and P3HT-*b*-poly(5,8-di-*p*-tolyl-quinoxaline-2,3-diyl) (**22**) (Figure 4.5).[110] Higashihara and coworkers[111] synthesized a novel soluble BCP with P3HT and graphene-like nanoribbon segment (**23**). Koeckelberghs and coworkers[112] reported an array of ABC-type BCPs containing P3HT, poly(3-octylselenophene) and poly(9,9-dioctyl-fluorene) segments (**24**) with all sequential combinations.

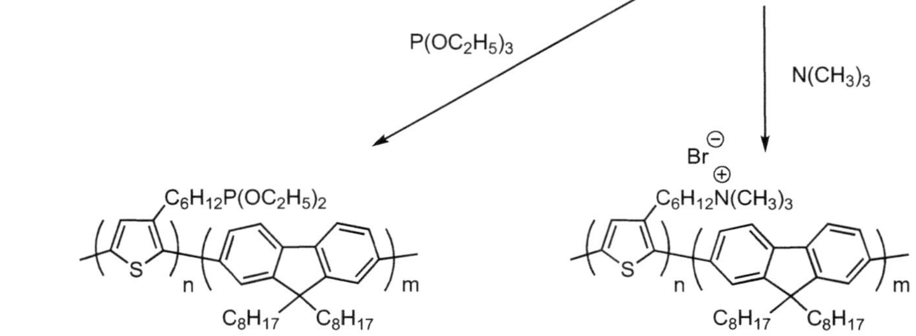

Scheme 4.11   Synthesis of all-conjugated BCPs containing polythiophene and poly-flurene segments.

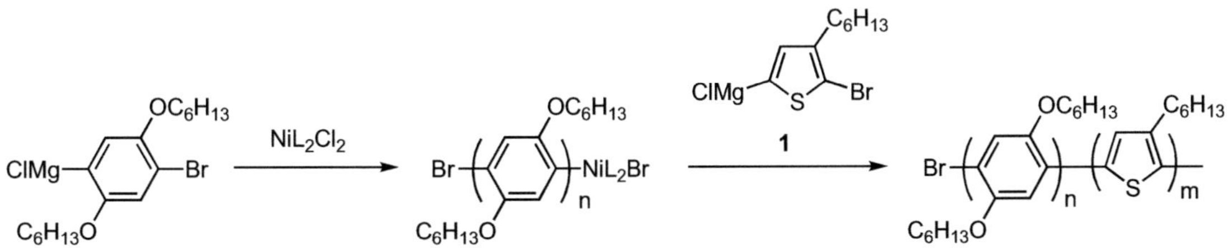

Scheme 4.12 Synthesis of poly(2,5-dihexyloxy-1,4-phenylene)-*b*-P3HT by the sequential monomer addition technique.

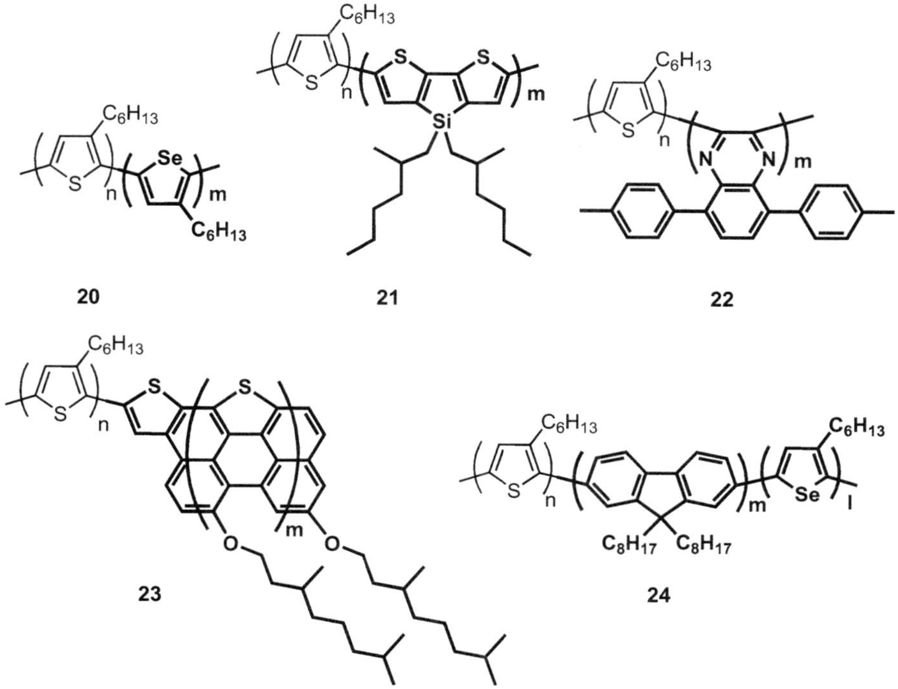

Figure 4.5 Chemical structures of all-conjugated BCPs containing polythiophene and other types of polymer segment.

4.7.3 All-conjugated Donor–Acceptor Block Copolymers

All-conjugated donor–acceptor BCPs are of interest because of their potential application to fullerene-free organic photovoltaic devices, especially those in which an active layer is composed of a single polymeric material so that consideration of the phase separation of the donor and acceptor is unnecessary. The prototype of all-conjugated block copolythiophenes with donor and acceptor units was reported by Dagron-Lartigau and coworkers;[113] Tajima and coworkers[114] modified the synthetic procedure using an alkyne-azide click reaction. They synthesized block copolythiophenes with P3HT and PT and pendent C_{60} units at the side-chains. Scherf and coworkers[115] reported the synthesis of fullerene-free all-conjugated donor–acceptor BCPs by the polycondensation of an AB type monomer with H-/Br-terminated P3HT based on the Stille coupling reaction (Scheme 4.13a). Hawker and coworkers[116] reported a donor–acceptor BCP consisting of P3HT and a diketopyrrolopyrrole-based narrow bandgap block segment (Scheme 4.13b).

Nakabayashi and Mori[117] reported a donor–acceptor BCP with P3HT and n-type naphthalene diimide (NDI)-based π-conjugated polymer segments (Scheme 4.14a) by the Yamamoto coupling between a dibrominated

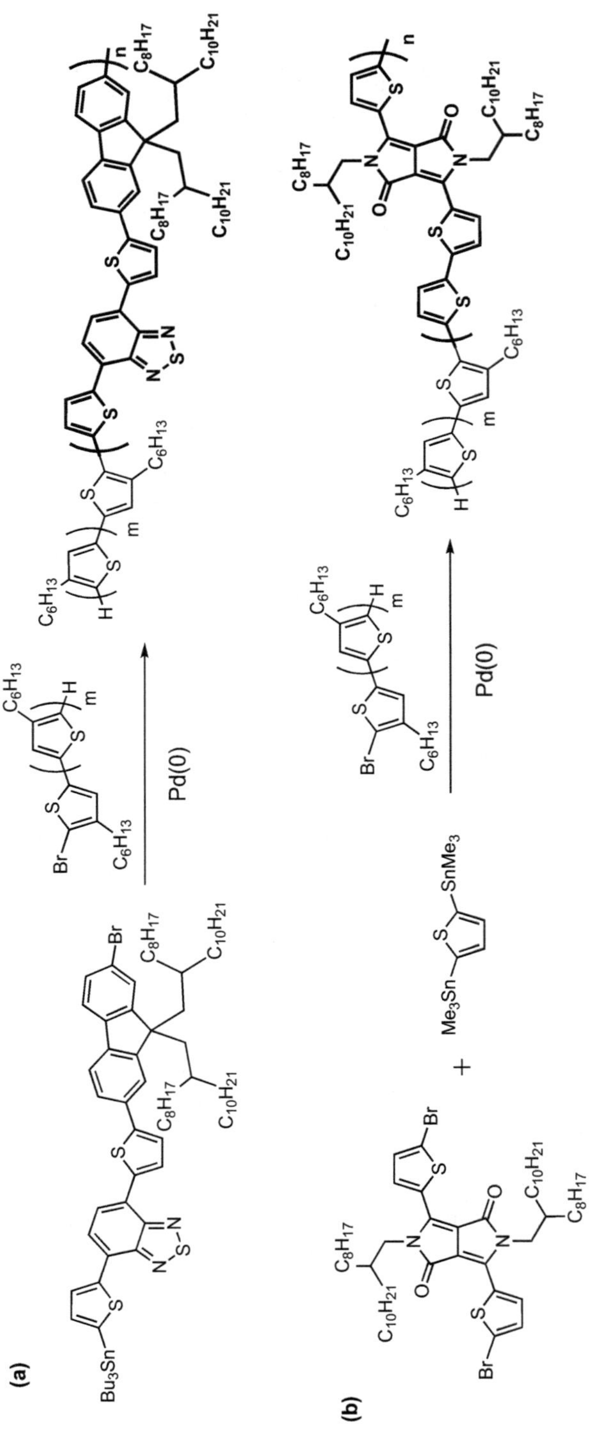

Scheme 4.13 Synthesis of all-conjugated donor–acceptor block copolymers.

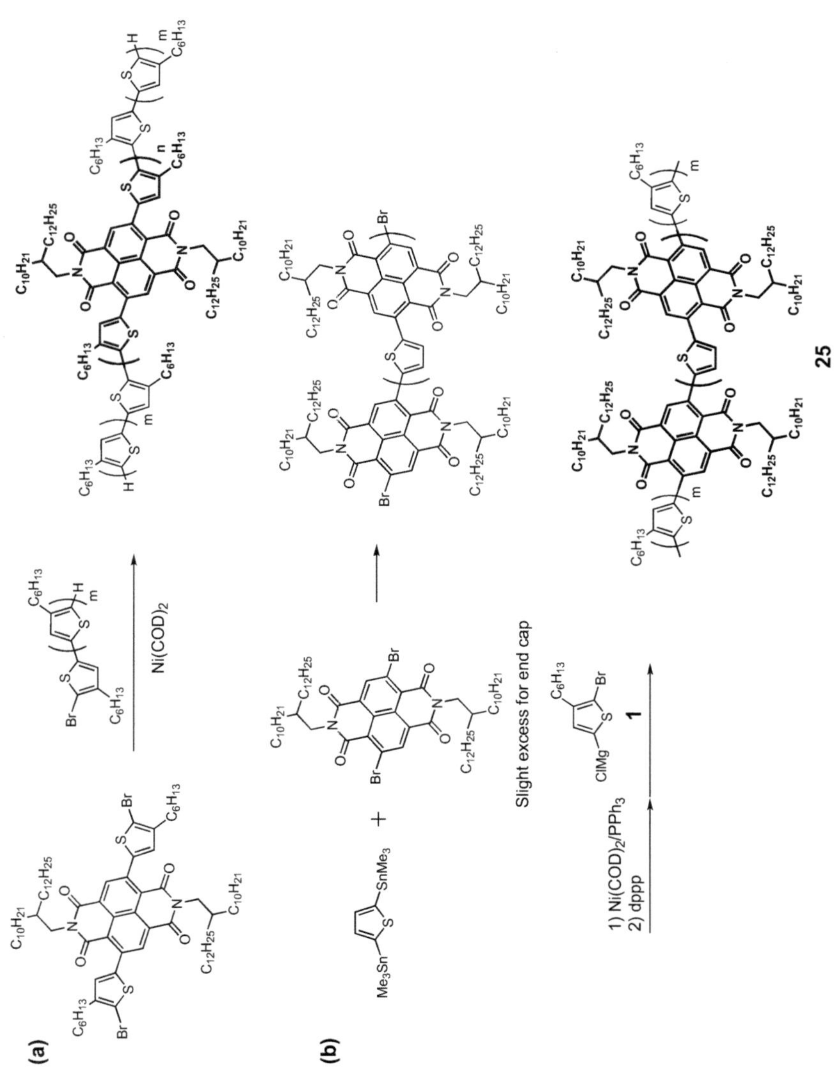

Scheme 4.14 Synthesis of all-conjugated donor–acceptor block copolymers containing naphthalene diimide based n-type polymer segments.

NDI monomer and Br-terminated P3HT. Wang *et al.*[118] synthesized a similar
BCP (**25**) (Scheme 4.14b). A Br-terminated NDI-based polymer was first
prepared by the Stille coupling polymerization of bis(trimethyl-
stannyl)thiophene and a slight excess of dibrominated NDI monomer. It was
then utilized as a macro-initiator for KCTP of the thiophene monomer
1 after insertion of an Ni catalyst with $Ni(COD)_2/PPh_3$, followed by ligand
exchange with dppp. The BCP **25** showed phase separation of two
distinct crystalline domains corresponding to P3HT (edge-on rich) and
NDI-based polymer segments (face-on rich) by glazing incidence wide-
angle X-ray scattering.[119] Koeckelberghs and coworkers[120] have reported
the one-pot synthesis of P3HT-*b*-poly(didodecylthieno[3,4-*b*]pyrazine)
(Scheme 4.15).

4.7.4 Semi-conjugated Block Copolymers Containing Polythiophene Segments

Discussion of the combination of block segments in the preceding sections
was limited to a π-conjugated system along the entire main chain. To extend
the application range, the second block segments connected to PT segments
should be more diverse. For example, to synthesize semi-conjugated BCPs
containing P3HT and poly(acrylate) (PMA) segments, McCullough and
coworkers[121] combined KCTP with an atom-transfer radical polymerization
(ATRP) system involving several site transformation steps (Scheme 4.16).
First, ω-chain end vinyl functional P3HT was synthesized by terminating
quasi-living P3HT with an ω-Ni-Br chain end with an excess of vinyl magnesium bromide. The introduced vinyl group was then transformed into a
hydroxyl group by a hydroboration/oxidation process. After transforming the
hydroxyl group into a 2-bromopropionate group by an esterification reaction,
ATRP of the acrylates was performed to afford P3HT-*b*-poly(methyl acrylate)
(PMA) and P3HT-*b*-poly(*t*butyl acrylate) (P*t*BMA). They also succeeded in
synthesizing P3HT-*b*-polyisoprene (**26**) and P3HT-*b*-polystyrene (PS) (**27**)
(Figure 4.6) using NMP and addition fragmentation chain transfer polymerization, respectively.[122]

New amphiphilic BCPs of P3HT-*b*-poly(acrylic acid) (PAA) (**28**) were synthesized by the hydrolysis of P3HT-*b*-P*t*BMA, which was prepared by a similar
method.[123] Zou *et al.*[124] synthesized a series of P3HT-containing BCPs with
PAA, poly(PEG acrylate) (**29**) and PMMA (**30**) and used these as efficient
dispersants for carbon nanotubes. Urien *et al.*[125] reported an alternative
synthetic approach via a copper(I)-catalysed Huisgen [3 + 2] dipolar cycloaddition reaction based on click chemistry, combined with KCTP and ATRP,
to afford AB and ABA BCPs containing P3HT and PS segments. Another
example based on click chemistry included P3HT-*b*-poly(styrene sulfonate)
(**31**).[126] A series of interesting donor–acceptor BCPs was synthesized based
on 'living'/controlled radical polymerization co-systems. Richard *et al.*[127]
synthesized P3HT-based rod–coil BCPs with fullerene-grafted PS segments

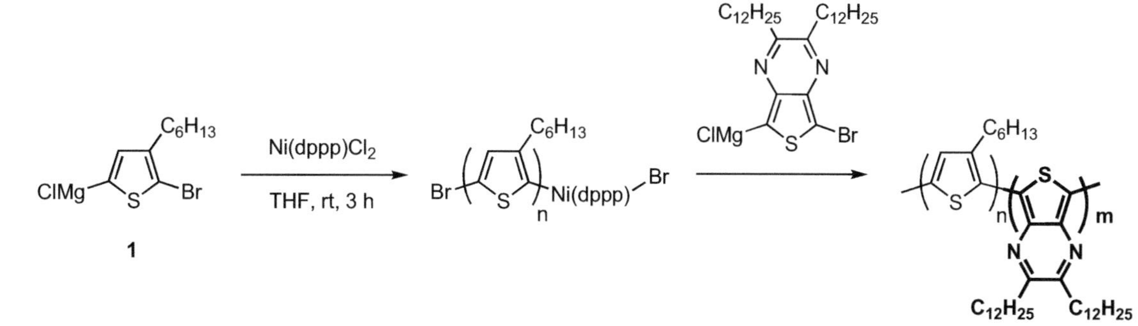

Scheme 4.15 One-pot synthesis of P3HT-*b*-poly(didodecylthieno[3,4-*b*]pyrazine) by the sequential monomer addition technique.

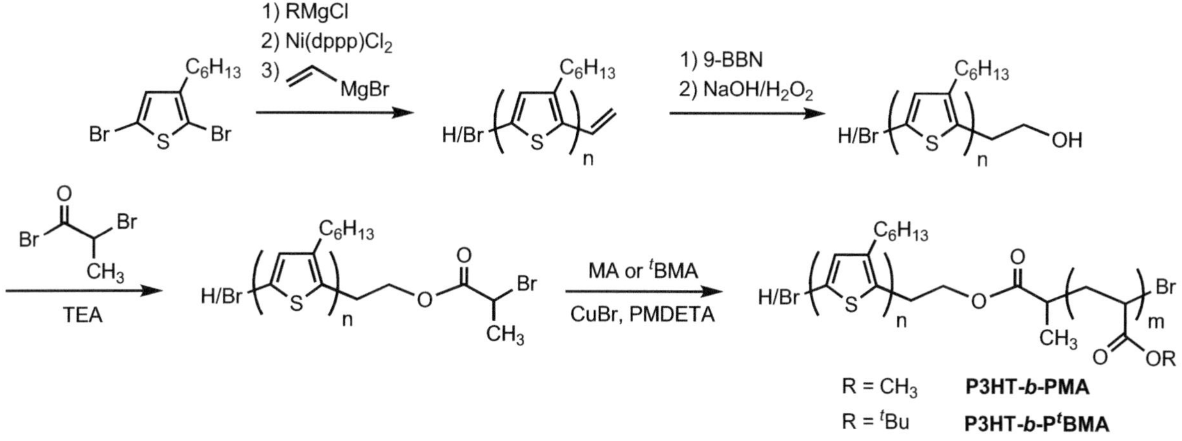

Scheme 4.16 Synthesis of P3HT-*b*-PMA and P3HT-*b*-P'BMA by the combination of KCTP and ATRP.

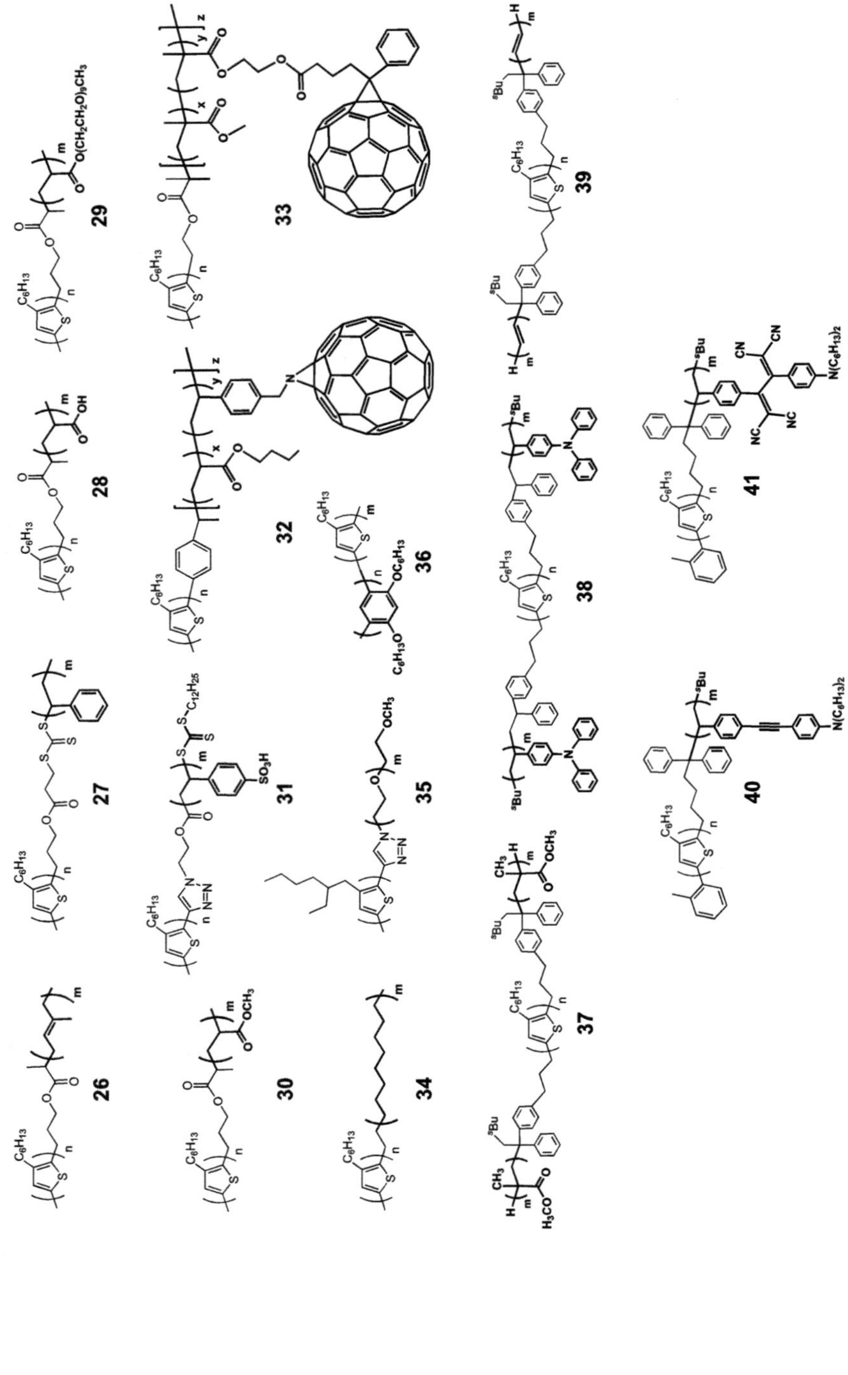

Figure 4.6 Chemical structures of semi-conjugated BCPs containing polythiophene segments.

(**32**) using the NMP system. Lee *et al.*[128] synthesized P3HT-based rod–coil BCPs with fullerene-grafted PMMA segments (**33**). Although many examples have been reported based on KCTP in conjunction with the 'living'/controlled radical polymerization system, tedious site transformation reactions as well as the limited control nature of radical polymerization often cause large Đ values. A thorough purification process is essential to remove the residual copper catalyst, if used, from products used in optoelectronic applications.

In addition to radical polymerization co-systems, Janssen and co-workers[129] reported the synthesis of crystalline–crystalline BCPs consisting of P3HT and polyethylene segments (**34**) by a combination of KCTP and ring-opening polymerization. Balsra and coworkers reported the synthesis and thermodynamic properties of P3HT-*b*-poly(ethylene oxide) (**35**)[130] based on click chemistry. Higashihara and coworkers[131] reported the synthesis of poly(4,6-dihexyloxy-1,3-phenylene)-*b*-P3HT with low dispersity (Đ = 1.15) by a sequential monomer addition technique (**36**). Dai *et al.*[132] first synthesized P3HT-*b*-poly(2-vinylpyridine) (P2VP) by a combination of KCTP and 'living' anionic polymerization. As shown in Scheme 4.17, ω-chain end vinyl functional P3HT was synthesized by the established method. Then the α-Br chain end was converted into a phenyl group so that unwanted side-reactions, such as Li–Br exchange, or single-electron transfer reactions could be eliminated. After lithiation of the terminal vinyl groups with ⁱBuLi, 'living' anionic polymerization of 2VP was carried out to synthesize the desired BCPs. Simply by changing the block compositional ratio, spherical, cylindrical, lamellar and nanofibril structures could clearly be observed by transmission electron microscopy and small-angle X-ray scattering. The Đ values of all the BCP samples were low due to the use of the 'living' anionic polymerization system (Đ = 1.18–1.28). However, multistep reactions were still required, especially for the modification of the α-Br chain end of P3HT. A possible

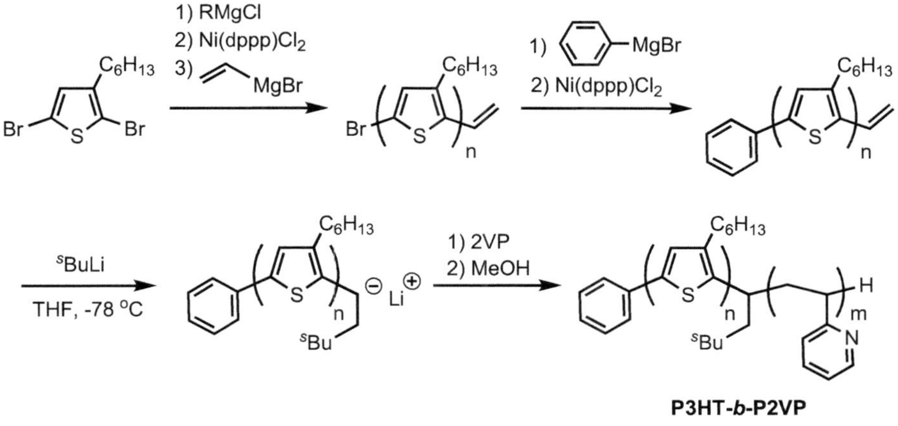

Scheme 4.17 Synthesis of P3HT-*b*-P2VP by the combination of KCTP and 'living' anionic polymerization.

homopolymerization of the terminal thiophenylvinyl units might generate coupling by-products in the lithiation reaction.

A facile synthesis of PS-*b*-P3HT-*b*-PS was been reported by Higashihara *et al.*[133] based on a novel approach involving only two steps: the *in situ* introduction of non-homopolymerizable 1,1-diphenylethylene (DPE) moieties at the α,ω-ends of P3HT and a linking reaction with a 'living' anionic polymer of styrene. The synthetic routes are shown in Scheme 4.18. The key step is an *in situ* introduction of DPE moieties simultaneously at the α,ω-ends of P3HT, which readily and quantitatively link with 'living' PS without homopolymerization of the DPE units as a result of steric hindrance and resonance effects. This spontaneous disubstitution allows the extra steps of the transformation reactions at the chain end to be skipped. The polymer–polymer linking method provides more reliable MWs and compositions of the BCPs than conventional macro-initiator methods because both segments can be pre-made and separately characterized. The purified PS-*b*-P3HT-*b*-PS has an MW of 18 900 and a *Đ* value of 1.15. The surface morphology of the BCP thin films was observed by atomic force microscopy and continuous nanofibril and nanowire structures were observed. The precursor α,ω-end DPE functional P3HT was also used to synthesize well-defined PMMA-*b*-P3HT-*b*-PMMA with a low *Đ* value (1.10) (37) (Figure 4.6).[134] Kim and coworkers[135] synthesized well-defined P3HT-*b*-PMMA based on the polymer–polymer linking approach using phenyl acrylate (PA) α-chain end functional P3HT.

Unfortunately, semi-conjugated BCPs show limited optoelectronic properties. Therefore the introduction of optoelectronically active segments into P3HT-based BCPs is highly desirable. Higashihara and Ueda[136] synthesized novel BCPs containing P3HT and optoelectronically active poly(4-vinyltriphenylamine) (PVTPA) by a similar approach. PVTPA-*b*-P3HT-*b*-PVTPA (38) may be a better hole-transporting material than PS-*b*-P3HT-*b*-PS or PMMA-*b*-P3HT-*b*-PMMA. The compatibilizing effect of 38 was discovered in the application to P3HT:[6,6]-phenyl-C$_{61}$-butyric acid methyl ester bulk heterojunction organic photovoltaic devices and showed a 13% improvement in the power conversion efficiency after only a 1.5 wt% addition of 38.[137] Our other examples of semi-conjugated BCPs include polyacetylene-*b*-P3HT-*b*-polyacetylene (non-conjugated linkage between π-conjugated segments) (39),[138] P3HT-*b*-poly(4-(4′-*N*,*N*-dihexylaminophenylethynyl)styrene) (40) and tetracyanoethylene adduct to 39 (41).[87]

4.8   Graft Copolymers with Polythiophene Segments

Graft copolymers with PT segments are categorized into two types. One type consists of PT side-chains and a backbone of another other polymer. Kiriy and coworkers[139,140] reported the grafting of P3HT from poly(4-bromostyrene) by externally initiated KCTP based on the grafting-from method (Scheme 4.19a). A different macromonomer method was reported by Kilbey and coworkers[141] in which a norbornenyl functional P3HT macromonomer

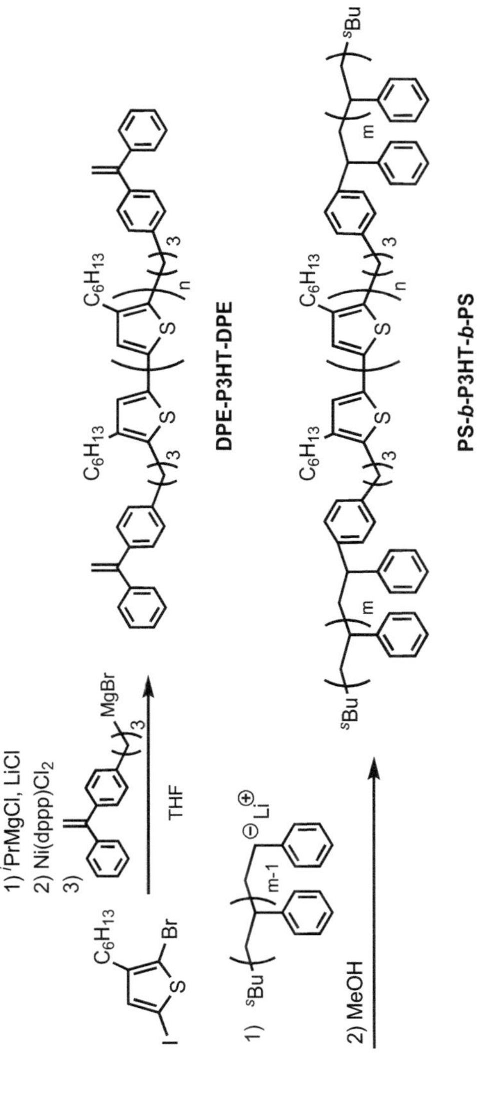

Scheme 4.18 Synthesis of PS-*b*-P3HT-*b*-PS by the combination of KCTP and 'living' anionic polymerization.

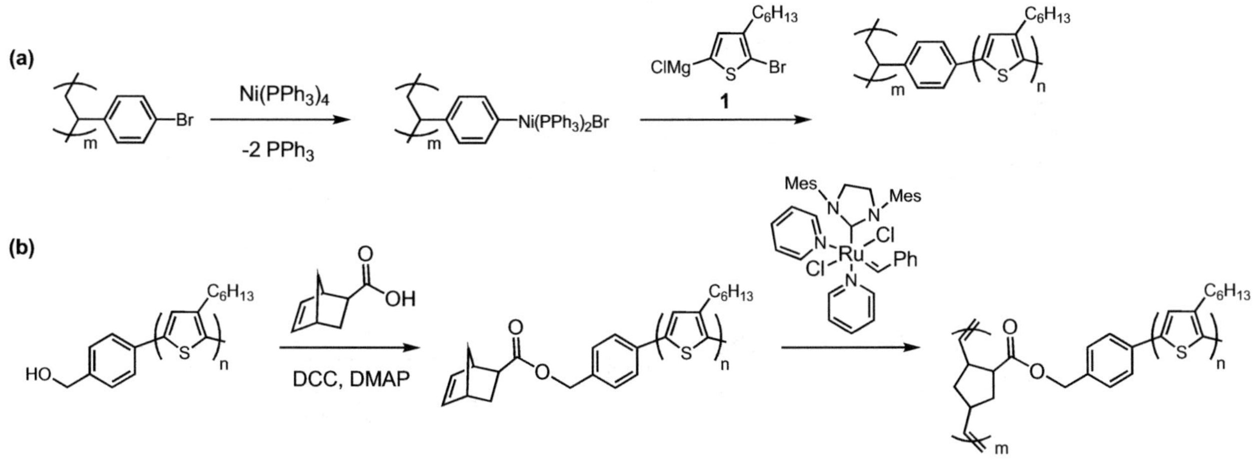

Scheme 4.19 Synthesis of semi-conjugated graft copolymers with P3HT segments.

was polymerized by ring-opening metathesis polymerization to synthesize a graft copolymer with a poly(norbornene) backbone and P3HT side-chains (Scheme 4.19b). This type of polymer consists of a PT backbone and side-chains of another polymer. For example, Hawker and coworkers[142] prepared P3HT grafted with PS segments *via* a trityl ether linkage by a macromonomer method. Nanoporous P3HT films were obtained after cleavage of the trityl ether units followed by selective solvent etching of the PS segments (Figure 4.7a). To obtain a better optoelectronic performance, an all-conjugated system is in high demand. Higashihara and coworkers[119,143] reported the synthesis of all-conjugated graft copolymers with a PT backbone and P3HT side-chains as well as an n-type NDI-based polymer backbone and P3HT side-chains using the grafting-from method (Scheme 4.20). Luscombe and coworkers[144] synthesized a p-type donor–acceptor polymer backbone and P3HT side-chains by the macromonomer method (Figure 4.7b). The resulting polymers showed a strong internal charge transfer peak in UV-visible spectra and ambipolar mobilities ($\mu_\text{h} = 5 \times 10^{-3}$ cm^2 V^{-1} s^{-1} and $\mu_\text{e} = 7 \times 10^{-4}$ cm^2 V^{-1} s^{-1}).

4.9   Star-branched Polymers with Polythiophene Segments

Star-branched polymers are classified as branched macromolecules with more than three arm segments generated from a core. The hydrodynamic volume of star polymers is smaller than that of their linear counterparts. In addition, less entanglement of the star-branched polymers leads to a lower viscosity compared with the linear analogues in the melt and solution states.[145] However, there are only a few reports of the synthesis of star P3HTs as a result of synthetic difficulties.[146,147] Kim *et al.*[148] synthesized multi-arm star P3HTs based on a divinyl compound technique and characterized their unique crystalline behaviour. However, this method causes ill-defined structures in terms of the arm number. Kiriy and coworkers[149] reported a core-first method based on externally initiated KCTP to synthesize a six-arm star P3HT using a core compound with six initiating sites. Luscombe and coworkers[150] reported the synthesis of a well-defined three-arm star P3HT with a low dispersity ($Đ = 1.15$) using a specially designed core compound (Scheme 4.21). Zhao and coworkers[151] reported the synthesis of six-arm star-block copolymers containing PS-*b*-P3HT arm segments based on ATRP and a 1,3-dipolar cycloaddition click reaction.

Asymmetrical star polymers (also called miktoarm star polymers), in which different types of arm segments are connected at the core, have the potential to create novel morphological nanostructures other than linear block copolymers or regular star polymers with the same arm segments. However, it has been more difficult to synthesize asymmetrical star polymers than regular star polymers for the following reasons: (1) there are strict requirements for the multistep reactions corresponding to chemically

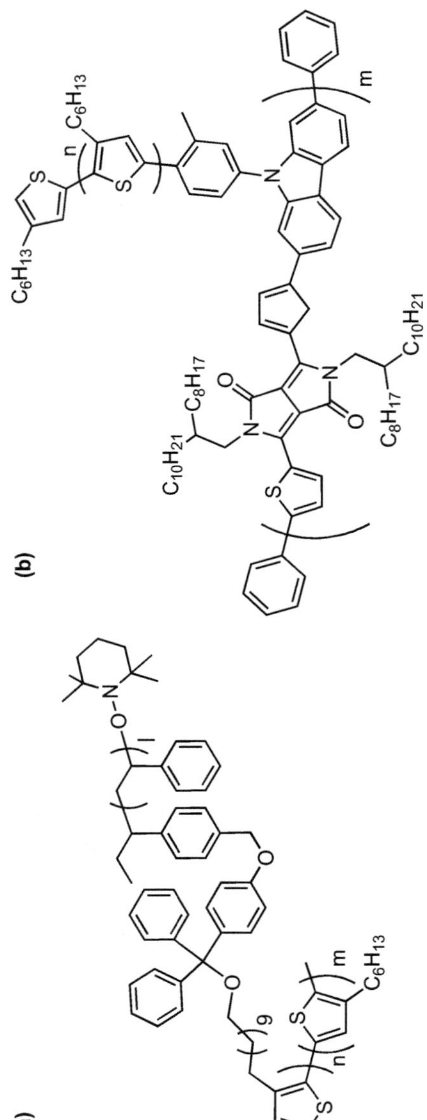

Figure 4.7 Chemical structures of graft copolymers containing polythiophene segments.

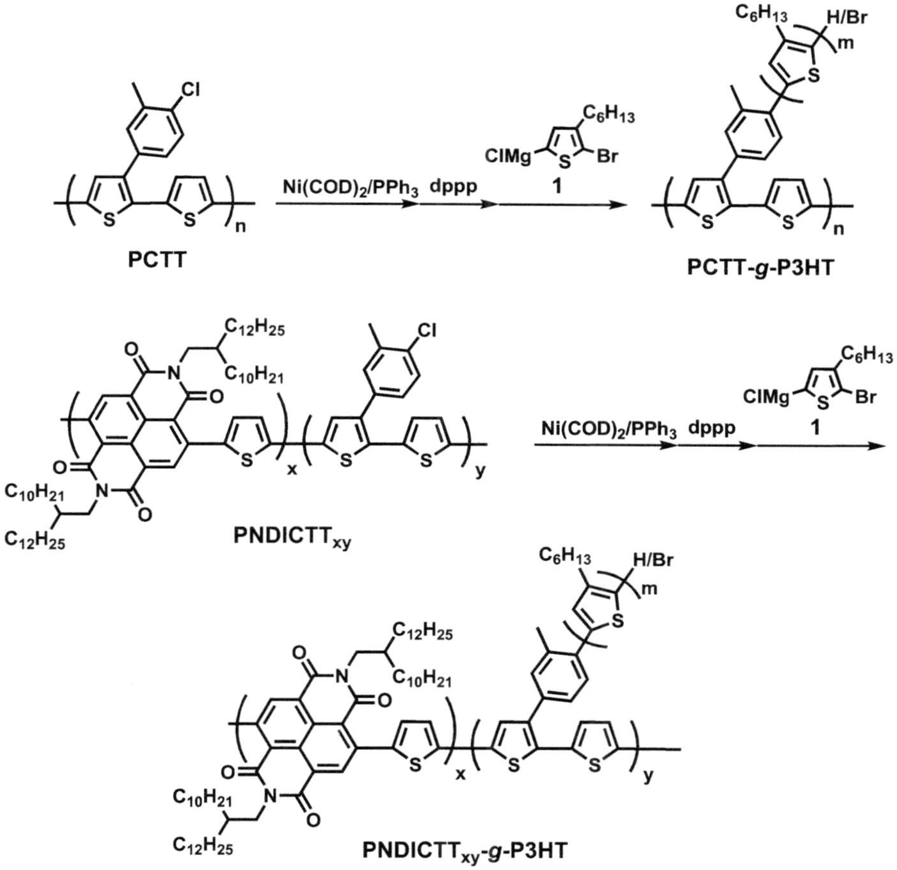

Scheme 4.20 Synthesis of all-conjugated graft copolymers with P3HT segments. Reprinted with permission from J. Wang, M. Ueda and T. Higashihara, *ACS Macro Lett.*, 2013, **2**, 506–510. Copyright (2013) American Chemical Society.[118]

different arms; (2) each polymer–polymer reaction should be nearly quantitative; and (3) the isolation and purification of intermediate polymers are often required. Kim and coworkers[152] reported a well-defined AB$_2$-type asymmetrical star polymer, where A and B were PMMA and P3HT segments, respectively, by the copper(ı)-catalysed Huisgen 1,3-dipolar cycloaddition click reaction of chain end functional PMMA with two azide moieties with monoethynyl chain end functional P3HT. Higashihara *et al.*[153] reported the synthesis of well-defined ABC-type asymmetrical star polymers, where A, B and C were P3HT, PS and P2VP, respectively, with controlled MWs, composition and extremely low *Đ* values (<1.05) based on catalyst-free anionic linking reactions in conjunction with KCTP (Scheme 4.22). The synthetic methodology involved the following three-stage reaction: (1) an anionic

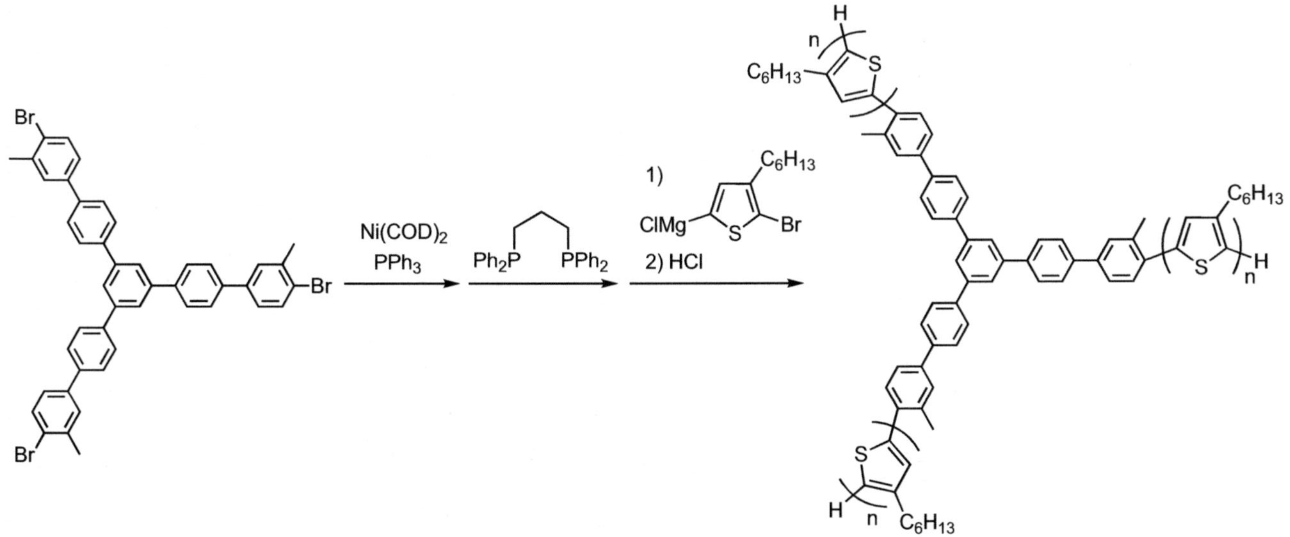

Scheme 4.21   Synthesis of well-defined three-armed star P3HT.
Reprinted with permission from M. Yuan, K. Okamoto, H. A. Bronstein and C. K. Luscombe, *ACS Macro Lett.*, 2012, **1**, 392–395. Copyright (2012) American Chemical Society.[150]

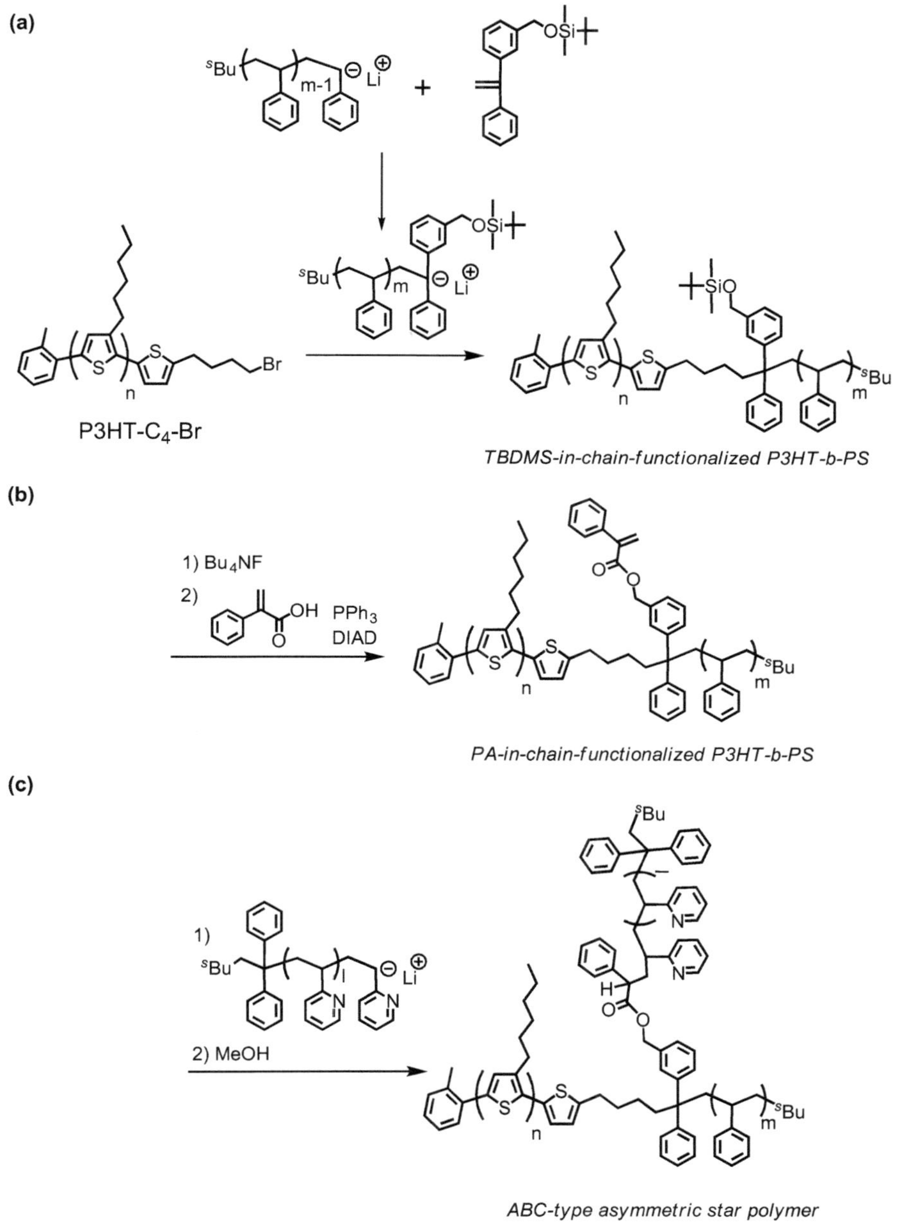
(a)
sBu
m-1
Li
+
OSi
sBu
m
Li
OSi
SiO
S
S
Br
n
P3HT-C4-Br
S
S
n
sBu
m
TBDMS-in-chain-functionalized P3HT-b-PS
(b)
1) Bu4NF
2)
OH
O
PPh3
DIAD
O
O
S
S
n
sBu
m
PA-in-chain-functionalized P3HT-b-PS
(c)
1)
sBu
l
Li
N
N
2) MeOH
sBu
N
N
H
O
O
S
S
n
m
sBu
ABC-type asymmetric star polymer

Scheme 4.22 Synthesis of well-defined ABC-type asymmetrical star polymer. Reprinted with permission from T. Higashihara, S. Ito, S. Fukuta, T. Koganezawa, M. Ueda, T. Ishizone and A. Hirao, *Macromolecules*, 2015, **48**, 245–255. Copyright (2015) American Chemical Society.[153]

linking reaction between ω-chain end functional P3HT with a bromobutyl moiety (P3HT-C$_4$-Br) and 'living' anionic PS end-capped with 1-(3-^{t}butyldimethylsilyloxymethylphenyl)-1-phenylethylene; (2) a transformation reaction of the 3-^{t}butyldimethylsilyloxymethylphenyl (TBDMS) moiety at the junction between the P3HT and PS segments into the PA moiety through the hydroxymethylphenyl moiety; and (3) an anionic linking reaction between the PA in the chain functional P3HT-*b*-PS and the 'living' anionic P2VP. The thermal, optical and morphological properties of the ABC-type star polymer were fully investigated. Glazing incidence X-ray scattering studies of the ABC-type star polymer thin films confirmed their hierarchical nanofibril structures in which the P3HT crystalline domains aligned with an edge-on orientation in the exclusive area of the microphase-separated domains from the PS + P2VP domains in the thin films. It was also found that the fibril period distance and the π–π stacking distance of the P3HT crystalline domains increased with increasing length of the P2VP arm segments.

4.10   Hyperbranched Polythiophene with a Controlled Degree of Branching

Hyperbranched polymers (HBPs) generally exhibit an irregular architecture with incompletely reacted branching points throughout the structure, leading to large *Đ* values and a low degree of branching (DB) in a theoretical sense, in contrast with dendrimers, which have a well-controlled size, shape and monodispersed structure. As the DB is the key parameter in the characterization and specific properties of HBPs, it is highly desirable to control the branched structures in HBPs.

Several groups have reported a synthetic approach to control the DB to 100% by the following strategies:[21,154] the first step is reversible and produces an unstable linear unit and the second step is irreversible and produces a stable dendritic unit;[155,156] the second nucleophilic substitution in an AB$_2$ monomer must be faster than the first substitution – in other words, the first reaction, *i.e.* the substitution reaction of an A group and a B group in an AB$_2$ monomer, facilitates the second reaction.[157–164]

Bo and coworkers[165] successfully applied this new method to prepare HBPs with a DB of 100%. Hyperbranched polyphenylene and polycarbazole with a DB of 100% were prepared by the catalyst-transfer Suzuki–Miyaura coupling reaction of AB$_2$ monomers with one phenyl boronic ester and dibromobenzene or dibromocarbazole. However, the synthesis of hyperbranched PTs with a nearly 100% DB has never been reported.

Ueda and coworkers[166] reported the synthesis of a practically defect-free hyperbranched PT *via* the catalyst-transfer Suzuki–Miyaura polycondensation of a specially designed AB$_2$ monomer containing phenyl boronic acid pinacol ester and dibromothiophene (Scheme 4.23) and characterized the product by ^{1}H and ^{13}C NMR spectrometry.

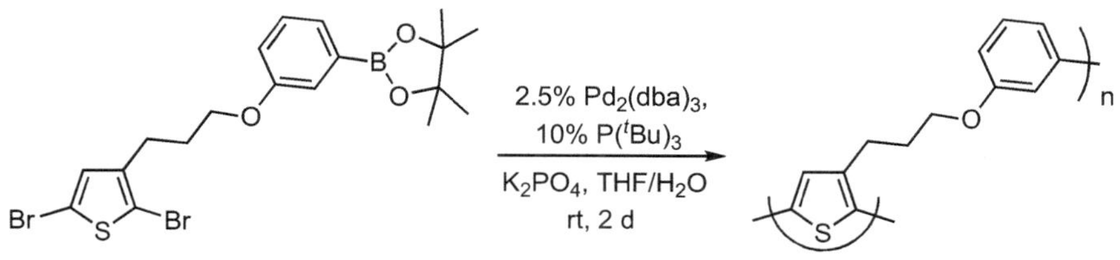

Scheme 4.23   Synthesis of hyperbranched polythiophene with a controlled degree of branching.

4.11   Conclusions

We have reviewed recent progresses in the controlled synthesis of chain end functional, block and branched polymers containing PT segments, focusing on advanced synthetic methodologies. The controlled synthesis of the P3HT homopolymer and combination with other 'living' polymerization systems are essential in designing and synthesizing these polymers with predictable MWs and low Đ values. Based on continuous efforts by many researchers, a wide variety of architectural polymers containing P3HT segments has been synthesized. Their highly ordered nanostructures related to the primary chemical structures have been well studied. We did not consider the optoelectronic properties of these P3HT-containing polymers in much detail because the application of BCPs with P3HT segments to organic electronic devices has been reviewed elsewhere.[34,38] The relationships among the primary polymer structures, morphologies, optoelectronic properties and device performances are still not fully understood and they should be systematically clarified using a library of well-defined architectural polymer samples to provide reliable and reproducible conclusions in the near future.

Acknowledgements

Financial support of this work from the Japan Science and Technology Agency (JST), PRESTO programme and Japan Society for the Promotion of Science (JSPS) and KAKENHI is gratefully acknowledged.

References

1. M. Morton, in *Anionic Polymerization: Principles and Practice*, Academic Press, New York, 1983.
2. S. Bywater, in *Encyclopedia of Polymer Science and Engineering*, ed. J. I. Kroschwitz, Wiley-Interscience, New York, 2nd edn, 1985, vol. 2, p. 1.
3. P. Rempp, E. Franta and J. E. Herz, *Adv. Polym. Sci.*, 1988, **86**, 145–173.
4. R. H. Grubbs and W. Tumas, *Science*, 1989, **243**, 907–915.
5. O. W. Webster, *Science*, 1991, **251**, 887–893.

6. J. P. Kenedy and B. Ivan, *Designed Polymers by Carbocationic Macro-molecular Engineering: Theory and Practice*, Hanser Publishers, Munich, 1992.
7. H. L. Hsieh and R. P. Quirk, in *Anionic Polymerization: Principles and Practical Applications*, ed. D. E. Hudgin, Marcel Dekker, Inc., New York, 1996, ch. 4, pp. 71–92.
8. M. Pitsikalis, S. Pispas, J. W. Mays and N. Hadjichristidis, *Adv. Polym. Sci.*, 1998, **135**, 1–137.
9. N. Hadjichristidis, *J. Polym. Sci., Part A: Polym. Chem.*, 1999, **37**, 857–871.
10. N. Hadjichristidis, S. Pispas, M. Pitsikalis, H. Iatrou and C. Vlahos, *Adv. Polym. Sci.*, 1999, **142**, 71–127.
11. N. Hadjichristidis, H. Iatrou, S. Pispas and M. Pitsikalis, *J. Polym. Sci., Part A: Polym. Chem.*, 2000, **38**, 3211–3234.
12. N. Hadjichristidis, M. Pitsikalis, S. Pispas and H. Iatrou, *Chem. Rev.*, 2001, **101**, 3747–3792.
13. N. Hadjichristidis, H. Iatrou, M. Pitsikalis, S. Pispas and A. Avgeropoulos, *Prog. Polym. Sci.*, 2005, **30**, 725–782.
14. N. Hadjichristidis, H. Iatrou, M. Pitsikalis and J. W. Mays, *Prog. Polym. Sci.*, 2006, **31**, 1068–1132.
15. A. Hirao, M. Hayashi, Y. Tokuda, T. Higashihara, N. Haraguchi and S. W. Ryu, *Polym. J.*, 2002, **34**, 633–658.
16. A. Hirao, M. Hayashi, S. Loykulnant, K. Sugiyama, S. W. Ryu, N. Haraguchi, A. Matsuo and T. Higashihara, *Prog. Polym. Sci.*, 2005, **30**, 111–182.
17. A. Hirao, K. Inoue, T. Higashihara and M. Hayashi, *Polym. J.*, 2008, **40**, 923–941.
18. T. Higashihara, K. Sugiyama, H. S. Yoo, M. Hayashi and A. Hirao, *Macromol. Rapid Commun.*, 2010, **31**, 1031–1059.
19. T. Higashihara, M. Hayashi and A. Hirao, *Prog. Polym. Sci.*, 2011, **36**, 323–375.
20. A. Hirao, M. Hayashi, T. Higashihara and N. Hadjichristidis, in *Complex Macromolecular Architecture*, ed. N. Hadjichristidis, A. Hirao, Y. Tezuka and F. D. Prez, John Wiley & Sons (Asia) Pte Ltd., Singapore, 2011, ch. 4, pp. 97–167.
21. T. Higashihara, Y. Segawa, W. Sinananwanich and M. Ueda, *Polym. J.*, 2012, **44**, 14–29.
22. G. Krausch and R. Magerle, *Adv. Mater.*, 2002, **14**, 1579–1583.
23. K. Ishizu, K. Tsubaki, A. Mori and S. Uchida, *Prog. Polym. Sci.*, 2002, **28**, 27–54.
24. H. Mori and A. H. E. Mueller, *Prog. Polym. Sci.*, 2003, **28**, 1403–1439.
25. C. Park, J. Yoonand and E. L. Thomas, *Polymer*, 2003, **44**, 6725–6760.
26. M. A. Hillmyer, *Adv. Polym. Sci.*, 2005, **190**, 137–181.
27. D. A. Olson, L. Chen and M. A. Hillmyer, *Chem. Mater.*, 2008, **20**, 869–890.
28. F. J. M. Hoeben, P. Jonkheijm, E. W. Meijer and A. P. H. J. Schenning, *Chem. Rev.*, 2005, **105**, 1491–1546.

29. W. H. Tseng, C. K. Chen, Y. W. Chiang, R. M. Ho, S. Akasaka and H. Hasegawa, *J. Am. Chem. Soc.*, 2009, **131**, 1356–1357.
30. R. M. Ho, C. K. Chen and Y. W. Chiang, *Macromol. Rapid Commun.*, 2009, **30**, 1439–1456.
31. R. Miyakoshi, A. Yokoyama and T. Yokozawa, *J. Polym. Sci., Part A: Polym. Chem.*, 2008, **46**, 753–765.
32. T. Yokozawa, N. Ajioka and A. Yokoyama, *Adv. Polym. Sci.*, 2008, **217**, 1–77.
33. T. Yokozawa and A. Yokoyama, *Chem. Rev.*, 2009, **109**, 5595–5619.
34. T. Higashihara and M. Ueda, *Macromol. Res.*, 2013, **21**, 257–271.
35. J. Wang and T. Higashihara, *Polym. Chem.*, 2013, **4**, 5518–5526.
36. T. Higashihara and E. Goto, *Polym. J.*, 2014, **46**, 381–390.
37. T. Higashihara and M. Ueda, in *Complex Macromolecular Architecture*, ed. N. Hadjichristidis, A. Hirao, Y. Tezuka and F. D. Prez, John Wiley & Sons (Asia) Pte Ltd., Singapore, 2011, ch. 13, pp. 395–429.
38. N. V. Handa, A. V. Serrano, M. J. Robb and C. J. Hawker, *J. Polym. Sci., Part A: Polym. Chem.*, 2015, **53**, 831–841.
39. J. Roncali, *Chem. Rev.*, 1992, **92**, 711–738.
40. H. Sirringhaus, N. Tessler and R. H. Friend, *Science*, 1998, **280**, 1741–1744.
41. H. Sirringhaus, P. J. Brown, R. H. Friend, M. M. Nielsen, K. Bechgaard, B. M. W. Langeveld-Voss, A. J. H. Spiering, R. A. J. Janssen, E. W. Meijer, P. Herwig and D. M. de Leeuw, *Nature*, 1999, **401**, 685–688.
42. J. Zaumseil and H. Sirringhaus, *Chem. Rev.*, 2007, **107**, 1296–1323.
43. M. Mas-Torrent and C. Rovira, *Chem. Rev.*, 2011, **111**, 4833–4856.
44. C. Wang, H. Dong, W. Hu, Y. Liu and D. Zhu, *Chem. Rev.*, 2012, **112**, 2208–2267.
45. G. Li, V. Shrotriya, J. Huang, Y. Yao, T. Moriarty, K. Emery and Y. Yang, *Nat. Mater.*, 2005, **4**, 864–868.
46. J. Y. Kim, K. Lee, N. E. Coates, D. Moses, T.-Q. Nguyen, M. Dante and A. J. Heeger, *Science*, 2007, **317**, 222–225.
47. S. Günes, H. Neugebauer and N. S. Sariciftci, *Chem. Rev.*, 2007, **107**, 1324–1338.
48. C. H. Woo, B. C. Thompson, B. J. Kim, M. F. Toney and J. M. J. Fréchet, *J. Am. Chem. Soc.*, 2008, **130**, 16324–16329.
49. M. T. Dang, L. Hirsch and G. Wantz, *Adv. Mater.*, 2011, **23**, 3597–3602.
50. K. Y. Jen, G. G. Miller and R. L. Elsembaumer, *Chem. Commun.*, 1986, **17**, 1346–1347.
51. R. Sugimoto, S. Takeda, H. B. Gu and K. Yoshino, *Chem. Express*, 1986, **1**, 635–638.
52. R. D. McCullough and R. D. Lowe, *J. Chem. Soc., Chem. Commun.*, 1992, 70–72.
53. R. D. McCullough, R. D. Lowe, M. Jayaraman and D. L. Anderson, *Org. Chem.*, 1993, **58**, 904–912.
54. T. A. Chen and R. D. Rieke, *J. Am. Chem. Soc.*, 1992, **114**, 10087–10088.
55. T. A. Chen, X. Wu and R. D. Rieke, *J. Am. Chem. Soc.*, 1995, **117**, 233–244.

56. R. D. McCullough, P. C. Ewbank and R. S. Loewe, *J. Am. Chem. Soc.*, 1997, **119**, 633–634.

57. A. Iraqi and G. W. Barker, *J. Mater. Chem.*, 1998, **8**, 25–29.

58. S. Guillerez and G. Bidan, *Synth. Met.*, 1998, **93**, 123–126.

59. I. A. Liversedge, S. J. Higgins, M. Giles, M. Heeney and I. McCulloch, *Tetrahedron Lett.*, 2006, **47**, 5143–5146.

60. E. E. Sheina, J. Liu, M. C. Iovu, D. W. Laird and R. D. McCullough, *Macromolecules*, 2004, **37**, 3526–3528.

61. M. C. Iovu, E. E. Sheina, R. R. Gil and R. D. McCullough, *Macromolecules*, 2005, **38**, 8649–8656.

62. A. Yokoyama, R. Miyakoshi and T. Yokozawa, *Macromolecules*, 2004, **37**, 1169–1171.

63. R. Miyakoshi, A. Yokoyama and T. Yokozawa, *J. Am. Chem. Soc.*, 2005, **127**, 17542–17547.

64. M. Kobayashi, Y. Matsumoto, M. Uchiyama and T. Ohwada, *Macromolecules*, 2004, **37**, 4339–4341.

65. T. Higashihara, E. Goto and M. Ueda, *ACS Macro Lett.*, 2012, **1**, 167–170.

66. E. Goto, S. Nakamura, S. Kawauchi, H. Mori, M. Ueda and T. Higashihara, *J. Polym. Sci., PartA: Polym. Chem.*, 2014, **52**, 2287–2296.

67. T. Yokozawa, R. Suzuki, M. Nojima, Y. Ohta and A. Yokoyama, *Macromol. Rapid Commun.*, 2011, **32**, 801–806.

68. Q. Wang, R. Takita, Y. Kikuzaki and F. Ozawa, *J. Am. Chem. Soc.*, 2010, **132**, 11420–11421.

69. S. Tamba, S. Tanaka, Y. Okubo, S. Okamoto, H. Meguro and A. Mori, *Chem. Lett.*, 2011, **40**, 398–399.

70. S. Tamba, K. Fuji, H. Meguro, S. Okamoto, T. Tendo, R. Komobuchi, A. Sugie, T. Nishino and A. Mori, *Chem. Lett.*, 2013, **42**, 281–283.

71. S. Tamba, K. Shono, A. Sugie and A. Mori, *J. Am. Chem. Soc.*, 2011, **133**, 9700–9703.

72. K. Fuji, S. Tamba, K. Shono, A. Sugie and A. Mori, *J. Am. Chem. Soc.*, 2013, **135**, 12208–12211.

73. K. Nakamura, S. Tamba, A. Sugie and A. Mori, *Chem. Lett.*, 2013, **42**, 1200–1202.

74. J. Liu and R. D. McCullough, *Macromolecules*, 2002, **35**, 9882–9889.

75. M. Jeffries-EL, G. Sauvé and R. D. McCullough, *Adv. Mater.*, 2004, **16**, 1017–1019.

76. M. Jeffries-EL, G. Sauvé and R. D. McCullough, *Macromolecules*, 2005, **38**, 10346–10352.

77. R. H. Lohwasser and M. Thelakkat, *Macromolecules*, 2010, **43**, 7611–7616.

78. L. Zhang, K. Hashimoto and K. Tajima, *Polym. J.*, 2012, **44**, 1145–1148.

79. K. Okamoto and C. K. Luscombe, *Chem. Commun.*, 2014, **50**, 5310–5312.

80. H. A. Bronstein and C. K. Luscombe, *J. Am. Chem. Soc.*, 2009, **131**, 12894–12895.

81. N. Doubina, A. Ho, A. K. Y. Jen and C. K. Luscombe, *Macromolecules*, 2009, **42**, 7670–7677.

82. A. Smeets, K. Van den Bergh, J. D. Winter, P. Gerbaux, T. Verbiest and G. Koeckelberghs, *Macromolecules*, 2009, **42**, 7638–7641.
83. A. Smeets, P. Willot, J. D. Winter, P. Gerbaux, T. Verbiest and G. Koeckelberghs, *Macromolecules*, 2011, **44**, 6017–6025.
84. F. Monnaie, W. Brullot, T. Verbiest, J. D. Winter, P. Gerbaux, A. Smeets and G. Koeckelberghs, *Macromolecules*, 2013, **46**, 8500–8508.
85. E. Kaul, V. Senkovskyy, R. Tkachov, V. Bocharova, H. Komber, M. Stamm and A. Kiriy, *Macromolecules*, 2010, **43**, 77–81.
86. V. Senkovskyy, M. Sommer, R. Tkachov, H. Komber, W. T. S. Huck and A. Kiriy, *Macromolecules*, 2010, **43**, 10157–10161.
87. H. Fujita, T. Michinobu, M. Tokita, M. Ueda and T. Higashihara, *Macromolecules*, 2012, **45**, 9643–9656.
88. J. K. Lee, S. Ko and Z. Bao, *Macromol. Rapid Commun.*, 2012, **33**, 938–942.
89. T. Yokozawa, I. Adachi, R. Miyakoshi and A. Yokoyama, *High Perform. Polym.*, 2007, **19**, 684–699.
90. K. Ohshimizu and M. Ueda, *Macromolecules*, 2008, **41**, 5289–5294.
91. Y. Zhang, K. Tajima, K. Hirota and K. Hashimoto, *J. Am. Chem. Soc.*, 2008, **130**, 7812–7813.
92. K. Van den Bergh, J. Huybrechts, T. Verbiest and G. Koeckelberghs, *Chem. – Eur. J.*, 2008, **14**, 9122–9125.
93. P. T. Wu, G. Ren, C. Li, R. Mezzenga and S. A. Jenekhe, *Macromolecules*, 2009, **42**, 2317–2320.
94. I. Yamada, K. Takagi, Y. Hayashi, T. Soga, N. Shibata and T. Toru, *Int. J. Mol. Sci.*, 2010, **11**, 5027–5039.
95. J. Kim, I. Y. Song and T. Park, *Chem. Commun.*, 2011, **47**, 4697–4699.
96. B. A. G. Hammer, F. A. Bokel, R. C. Hayward and T. Emrick, *Chem. Mater.*, 2011, **23**, 4250–4256.
97. C. C. Ho, Y. C. Liu, S. H. Lin and W. F. Su, *Macromolecules*, 2012, **45**, 813–820.
98. C. Suspène, L. Miozzo, J. Choi, R. Gironda, B. Geffroy, D. Tondelier, Y. Bonnassieux, G. Horowitz and A. Yassar, *J. Mater. Chem.*, 2012, **22**, 4511–4518.
99. K. Yao, L. Chen, X. Chen and Y. Chen, *Chem. Mater.*, 2013, **25**, 897–904.
100. X. Chen, L. Chen, K. Yao and Y. Chen, *Appl. Mater. Interfaces*, 2013, **5**, 8321–8328.
101. T. Ghoos, N. Van den Brande, M. Defour, J. Brassinne, C. A. Fustin, J. F. Gohy, S. Hoeppener, U. S. Schubert, W. Vanormelingen, L. Lutsen, D. J. Vanderzande, B. V. Mele and W. Maes, *Eur. Polym. J.*, 2014, **53**, 206–214.
102. A. Thomas, J. E. Houston, N. Van den Brande, J. D. Winter, M. Chevrier, R. K. Heenan, A. E. Terry, S. Richeter, A. Mehdi and B. V. Mele, *Polym. Chem.*, 2014, **5**, 3352–3362.
103. T. Higashihara, K. Ohshimizu, Y. Ryo, T. Sakurai, A. Takahashi, S. Nojima, M. Ree and M. Ueda, *Polymer*, 2009, **52**, 3687–3695.

104. Y. C. Lai, K. Ohshimizu, A. Takahashi, J. C. Hsu, T. Higashihara, M. Ueda and W. C. Chen, *J. Polym. Sci., Part A: Polym. Chem.*, 2011, **49**, 2577–2587.
105. G. Tu, H. Li, M. Forster, R. Heiderhoff, L. J. Balk, R. Sigel and U. Scherf, *Small*, 2007, **3**, 1001–1006.
106. U. Scherf, A. Gutacker and A. N. Koenen, *Acc. Chem. Res.*, 2008, **41**, 1086–1097.
107. R. Miyakoshi, A. Yokoyama and T. Yokozawa, *Chem. Lett.*, 2008, **137**, 1022–1023.
108. J. Hollinger, A. A. Jahnke, N. Coombs and D. S. Seferos, *J. Am. Chem. Soc.*, 2010, **132**, 8546–8547.
109. T. Erdmann, J. Back, R. Tkachov, A. Ruff, B. Voit, S. Ludwigs and A. Kiriy, *Polym. Chem.*, 2014, **5**, 5383–5390.
110. Y. Y. Zhu, T. T. Yin, J. Yin, N. Liu, Z. P. Yu, Y. W. Zhu, Y. S. Ding, J. Yin and Z. Q. Wu, *RSC Adv.*, 2014, **4**, 40241–40250.
111. A. Takahashi, C. J. Lin, K. Ohshimizu, T. Higashihara, W. C. Chen and M. Ueda, *Polym. Chem.*, 2012, **3**, 479–485.
112. M. Verswyvel, J. Steverlynck, S. H. Mohamed, M. Trabelsi, B. Champagne and G. Koeckelberghs, *Macromolecules*, 2014, **47**, 4668–4675.
113. F. Ouhib, A. Khoukh, J. B. Ledeuil, H. Martinez, J. Desbrieres and C. Dagron-Lartigau, *Macromolecules*, 2008, **41**, 9736–9743.
114. S. Miyanishi, Y. Zhang, K. Hashimoto and K. Tajima, *Macromolecules*, 2012, **45**, 6424–6437.
115. R. C. Mulherin, S. Jung, S. Huetter, K. Johnson, P. Kohn, M. Sommer, S. Allard, U. Scherf and N. C. Greenham, *Nano Lett.*, 2011, **11**, 4846–4851.
116. S. Y. Ku, M. A. Brady, N. D. Treat, J. E. Cochran, M. J. Robb, E. J. Kramer, M. L. Chabinyc and C. J. Hawker, *J. Am. Chem. Soc.*, 2012, **134**, 16040–16046.
117. K. Nakabayashi and H. Mori, *Macromolecules*, 2012, **45**, 9618–9625.
118. J. Wang, M. Ueda and T. Higashihara, *ACS Macro Lett.*, 2013, **2**, 506–510.
119. J. Wang, M. Ueda and T. Higashihara, *J. Polym. Sci., Part A: Polym. Chem.*, 2014, **52**, 1139–1148.
120. P. Willot, D. Moerman, P. Leclere, R. Lazzaroni, Y. Baeten, M. Van der Auweraer and G. Koeckelberghs, *Macromolecules*, 2014, **47**, 6671–6678.
121. M. C. Iovu, M. Jeffries-EL, E. E. Sheina, J. R. Cooper and R. D. McCullough, *Polymer*, 2005, **46**, 8582–8586.
122. M. C. Iovu, C. R. Craley, M. Jeffries-EL, A. B. Krankowski, R. Zhang, T. Kowalewski and R. D. McCullough, *Macromolecules*, 2007, **40**, 4733–4735.
123. C. R. Craley, R. Zhang, T. Kowalewski, R. D. McCullough and M. Stefan, *Macromol. Rapid Commun.*, 2009, **30**, 11–16.
124. J. Zou, S. I. Khondaker, Q. Huo and L. Zhai, *Adv. Funct. Mater.*, 2009, **19**, 479–483.

125. M. Urien, H. Erothu, E. Cloutet, R. C. Hiorns, L. Vignau and H. Cramail, *Macromolecules*, 2008, **41**, 7033–7040.

126. H. Erothu, J. Kolomanska, P. Johnston, S. Schumann, D. Deribew, D. T. W. Toolan, A. Gregori, C. Dagron-Lartigau, G. Portale, W. Bras, T. Arnold, A. Distler, R. C. Hiorns, P. Mokarian-Tabari, T. W. Collins, J. R. Howse and P. D. Topham, *Macromolecules*, 2015, **48**, 2107–2117.

127. F. Richard, C. Brochon, N. Leclerc, D. Eckhardt, T. Heiser and G. Hadziioannou, *Macromol. Rapid Commun.*, 2008, **29**, 885–891.

128. J. U. Lee, A. Cirpan, T. Emrick, T. P. Russell and W. H. Jo, *J. Mater. Chem.*, 2009, **19**, 1483–1489.

129. C. P. Radano, O. A. Scherman, N. Stingelin-Stutzmann, C. Mueller, D. W. Breiby, P. Smith, R. A. J. Janssen and E. W. Meijer, *J. Am. Chem. Soc.*, 2005, **127**, 12502–12503.

130. S. N. Patel, A. E. Javier, K. M. Beer, J. A. Pople, V. Ho, R. A. Segalman and N. P. Balsra, *Nano Lett.*, 2012, **12**, 4901–4906.

131. K. Ohshimizu, A. Takahashi, T. Higashihara and M. Ueda, *J. Polym. Sci., Part A: Polym. Chem.*, 2011, **49**, 2709–2714.

132. C. A. Dai, W. C. Yen, Y. H. Lee, C. C. Ho and W. F. Su, *J. Am. Chem. Soc.*, 2007, **129**, 11036–11038.

133. T. Higashihara, K. Ohshimizu, A. Hirao and M. Ueda, *Macromolecules*, 2008, **41**, 9505–9507.

134. T. Higashihara and M. Ueda, *React. Funct. Polym.*, 2009, **69**, 457–462.

135. H. C. Moon, A. Anthonysamy, J. K. Kim and A. Hirao, *Macromolecules*, 2011, **44**, 1894–1899.

136. T. Higashihara and M. Ueda, *Macromolecules*, 2009, **42**, 8794–8800.

137. J. H. Tsai, Y. C. Lai, T. Higashihara, C. J. Lin, M. Ueda and W. C. Chen, *Macromolecules*, 2010, **43**, 6085–6091.

138. T. Higashihara, C. L. Liu, W. C. Chen and M. Ueda, *Polymers*, 2011, **3**, 236–251.

139. N. Khanduyeva, V. Senlovskyy, T. Beryozkina, V. Bocharova, F. Simon, M. Nitschke, M. Stamm, R. Grötzschel and A. Kiriy, *Macromolecules*, 2008, **41**, 7383–7389.

140. N. Khanduyeva, V. Senlovskyy, T. Beryozkina, M. Horecha, M. Stamm, C. Uhrich, M. Riede, K. Leo and A. Kiriy, *J. Am. Chem. Soc.*, 2009, **131**, 153–161.

141. S. K. Ahn, D. L. Pickel, W. M. Kochemba, J. Chen, D. Uhrig, J. P. Hinestrosa, J. M. Carrillo, M. Shao, C. Do, J. M. Messman, W. M. Brown, B. G. Sumpter and S. M. Kilbey II, *ACS Macro Lett.*, 2013, **2**, 761–765.

142. K. Sivanandan, T. Chatterjee, N. Treat, E. J. Kramer and C. J. Hawker, *Macromolecules*, 2010, **43**, 233–241.

143. J. Wang, C. Lu, T. Mizobe, M. Ueda, W. C. Chen and T. Higashihara, *Macromolecules*, 2013, **46**, 1783–1793.

144. D. F. Zeigler, K. A. Mazzio and C. K. Luscombe, *Macromolecules*, 2014, **47**, 5019–5028.

145. H. L. Hsieh and R. P. Quirk, in *Anionic Polymerization: Principles and Practical Applications*, ed. D. E. Hudgin, Marcel Dekker, Inc., New York, 1996, ch. 13, pp. 333–368.

146. F. Wang, R. D. Rauh and T. L. Rose, *J. Am. Chem. Soc.*, 1997, **119**, 11106–11107.

147. J. Bras, S. Guillerez and B. Pépin-Donat, *Chem. Mater.*, 2000, **12**, 2372–2384.

148. H. Kim, Y. Lee, S. Hwang, D. Choi, H. Yang and K. Baek, *J. Polym. Sci., Part A: Polym. Chem.*, 2011, **49**, 4221–4226.

149. V. Senkovskyy, T. Beryozkina, V. Bocharova, R. Tkachov, H. Komber, A. Lederer, M. Stamm, N. Sevrin, J. P. Rabe and A. Kiriy, *Macromol. Symp.*, 2010, **291–292**, 17–25.

150. M. Yuan, K. Okamoto, H. A. Bronstein and C. K. Luscombe, *ACS Macro Lett.*, 2012, **1**, 392–395.

151. D. Han, X. Tong and Y. Zhao, *J. Polym. Sci., Part A: Polym. Chem.*, 2012, **50**, 4198–4205.

152. J. Park, H. C. Moon and J. K. Kim, *J. Polym. Sci., Part A: Polym. Chem.*, 2013, **51**, 2225–2232.

153. T. Higashihara, S. Ito, S. Fukuta, T. Koganezawa, M. Ueda, T. Ishizone and A. Hirao, *Macromolecules*, 2015, **48**, 245–255.

154. Y. Segawa, T. Higashihara and M. Ueda, *Polym. Chem.*, 2013, **4**, 1746–1759.

155. G. Maier, C. Zech, B. Voit and H. Komber, *Macromol. Chem. Phys.*, 1998, **199**, 2655–2664.

156. S. Chatterjee and S. Ramakrishnan, *ACS Macro Lett.*, 2012, **1**, 593–598.

157. M. Smet, E. Schacht and W. Dehaen, *Angew. Chem., Int. Ed.*, 2002, **41**, 4547–4550.

158. Y. Fu, A. Vandendriessche, W. Dehaen and M. Smet, *Macromolecules*, 2006, **39**, 5183–5186.

159. S. Kono, W. Sinananwanich, Y. Shibasaki, S. Ando and M. Ueda, *Polym. J.*, 2007, **39**, 1150–1156.

160. W. Sinananwanich and M. Ueda, *J. Polym. Sci., Part A: Polym. Chem.*, 2008, **46**, 2689–2700.

161. W. Sinananwanich, T. Higashihara and M. Ueda, *Macromolecules*, 2009, **42**, 994–1001.

162. W. Sinananwanich, Y. Segawa, T. Higashihara and M. Ueda, *Macromolecules*, 2009, **42**, 8718–8724.

163. Y. Segawa, T. Higashihara and M. Ueda, *J. Am. Chem. Soc.*, 2010, **132**, 11000–11001.

164. Y. Segawa, W. Sinananwanich and M. Ueda, *Macromolecules*, 2008, **41**, 8309–8311.

165. W. G. Huang, L. J. Su and Z. S. Bo, *J. Am. Chem. Soc.*, 2009, **131**, 10348–10349.

166. Y. Segawa, T. Higashihara and M. Ueda, *Polym. Chem.*, 2013, **4**, 1208–1215.

Section II: Microstructure of Semiconducting Polymers

CHAPTER 5

Characterization of Polymer Semiconductors by Neutron Scattering Techniques

GREGORY M. NEWBLOOM,[a] KIRAN KANEKAL,[a]
JEFFREY J. RICHARDS[b] AND LILO D. POZZO*[a]

[a] Chemical Engineering, 105 Benson Hall, Seattle, WA 98195, USA;
[b] NIST Center for Neutron Research, 100 Bureau Drive, Gaithersburg, MD 20899, USA
*Email: dpozzo@uw.edu

5.1 Introduction

Neutron scattering encompasses a set of non-destructive experimental techniques that enable the molecular analysis of complex systems over a broad range of length and time scales. Fundamentally, all scattering experiments utilize collimated beams of light, X-rays or neutrons to probe the spatial distribution of atoms within a sample (structure) and/or the relative movement of these atoms (dynamics). Fortunately, the fundamental equations describing these techniques are almost identical, with the exception of a few important differences. Neutrons interact with atomic nuclei, whereas X-rays and light interact with electron clouds. This makes the different scattering techniques highly complementary because they have different sensitivities towards different parts of the same molecular system. Light and X-ray scattering techniques are fairly accessible and sources are found in many laboratories. However, for conjugated polymers, a large level of light

RSC Polymer Chemistry Series No. 21
Semiconducting Polymers: Controlled Synthesis and Microstructure
Edited by Christine Luscombe
© The Royal Society of Chemistry 2017
Published by the Royal Society of Chemistry, www.rsc.org

absorption and/or low scattering contrast often limits their use. The main advantage of neutron scattering is the large difference in scattering contrast between isotopes such as hydrogen and deuterium. Because of this, selective deuterium substitution is used to enhance the scattering contrast of specific parts of molecular systems so that more detail can be extracted.[1,2] This process changes the isotopic composition and the neutron contrast, but does not affect the chemical makeup and behaviour of most samples. Inelastic and quasi-elastic neutron scattering (QENS) experiments are also highly sensitive to all major forms of molecular motion found in complex polymer systems. This chapter gives a brief introduction to the primary neutron scattering techniques that have been applied to the study of molecular structure and dynamics in conjugated polymer systems.

5.2   Small-angle Neutron Scattering

Small-angle neutron scattering (SANS) instruments are used to study structural features in the range 1–500 nm. Larger structures (from 500 nm to 10 μm) can be studied with ultra-small-angle neutron scattering (USANS) instruments. This means that five orders of magnitude in length scale can be probed with the same technique in two complementary instruments. User facilities with SANS and USANS instruments are available worldwide. Figure 5.1 is a basic diagram of a SANS instrument. Neutrons of wavelength (λ), generated from a reactor or spallation source, are collimated into a beam that illuminates the sample. The scattered neutrons are registered as counts (*i.e.* intensity) on a two-dimensional detector. The detector is often moved to different distances relative to the sample to capture different ranges of scattering angles with good resolution.

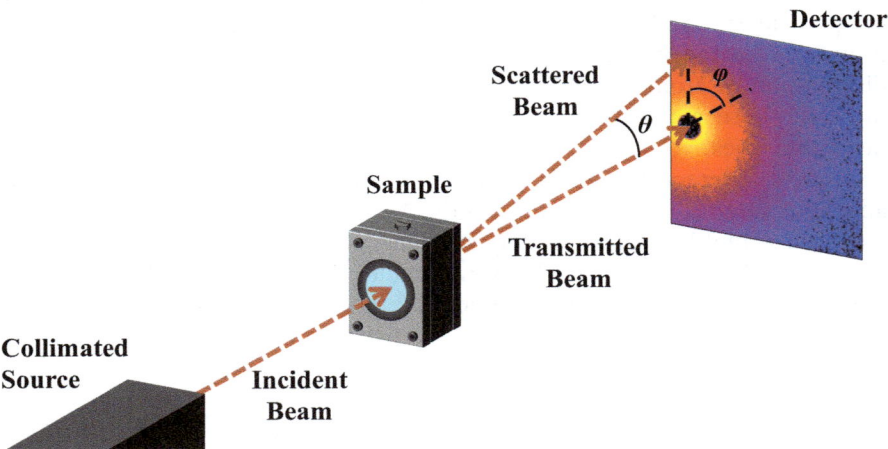

Figure 5.1   Standard elements of a SANS instrument. The scattering (θ) and azimuthal (φ) angles are both displayed.

SANS and USANS data are collected in reciprocal space with respect to the scattering vector Q, which is directly related to the scattering angle and the incident neutron wavelength, as shown in eqn (5.1).[1]

$$Q = 4\pi \frac{\sin(\theta/2)}{\lambda} \qquad (5.1)$$

Here, λ is the neutron wavelength and θ is defined as the angle between the incident neutron beam and the scattered neutron (Figure 5.1). The Q vector typically has units of Å^{-1} or nm^{-1}. The real space equivalent length scale (d) of the Q vector can be estimated using Bragg's law:

$$d = \frac{2\pi}{Q} \qquad (5.2)$$

The relationship between $I(Q)$ and Q is usually analysed with standard plots and models to quantify the morphological features. The absolute scale scattering intensity is dependent on the five terms in eqn (5.3):[1]

$$I(Q) = \phi \ (\Delta\rho)^2 \ V_P \ P(Q) \ S(Q) + Bkg \qquad (5.3)$$

For a simple two-component system, eqn (5.3) describes the scattering such that ϕ is the volume fraction of the dispersed particles, $\Delta\rho$ is the scattering length density (SLD) contrast between the two different phases (*e.g.* the polymer and the solvent), V_P is the volume of the particle, $P(Q)$ is the form factor of the object accounting for shape and $S(Q)$ is the structure factor accounting for inter-particle correlations in concentrated or interacting systems. The scattering background (Bkg) is the result of incoherent scattering. The SLD of conjugated polymers is calculated from the isotopic composition and the mass density.[2] Contrast variation experiments can also measure the SLDs of polymers of interest.[3,4]

It is important to note that SANS measurements also have limitations. In particular, experiments typically require large sample volumes (>400 μL) and deuterated materials. Neutron sources also have a much lower flux than X-ray sources and this leads to longer scattering times for reasonable statistics. Access to SANS facilities is competitive due to their high operating costs. It is highly recommended that first-time users collaborate with instrument scientists at user facilities.

5.2.1   SANS of Dissolved Conjugated Polymers

SANS is routinely used to characterize the conformation of polymers in solution. In fact, many of the fundamental scaling laws proposed by de Gennes, Debye, Kuhn, Flory and other researchers were experimentally confirmed using SANS techniques.[5,6] It is therefore natural to extend this work to conjugated polymers given the importance of the solution processing of electronic devices.[7] Some conjugated polymers also undergo conformational transitions (*e.g.* coil to rod) in solution before forming microcrystals.[8,9] There has been strong interest in probing the solution phase conformation to better understand and predict crystallization.[8,10,11]

In contrast with other polymers, delocalized electrons and steric hindrance of side-chains in conjugated polymers usually result in more rigid conformations in solution.[9,12–16] This is characterized by larger Kuhn lengths (*b*) and radii of gyration (R_g) compared with other polymers of similar contour length. As a result of this rigidity, conjugated polymers are often used above the critical overlap concentration (*c**).[16]

SANS analyses of dissolved conjugated polymers can be divided into two categories: dilute systems (*c* ≪ *c**) and concentrated systems (*c* ≥ *c**). The critical overlap concentration is determined from the scattering of solutions with increasing concentration. When these scattering profiles are normalized by the volume fraction (*φ*), overlapping curves are obtained in the dilute limit.[13,16] Above *c**, the normalized profiles do not overlap due to the emergence of inter chain correlations.

Scattering profiles of dilute conjugated polymers typically exhibit two regions: a low-*Q* Guinier region related to the size and conformation of the chain and a high-*Q* power law region related to the rigidity of the chain. To extract useful information, a form factor model is fitted with the input of known parameters (*e.g.* SLD and *φ*). A model for polymer chains with excluded volume interactions was found to be appropriate for dissolved poly(3-alkylthiophene)s (P3ATs).[13] The model accounts for a variable polymer–solvent affinity (*e.g.* Flory parameter) while fitting the size and stiffness of the chain.[17] McCulloch *et al.*[12] also fitted the scattering profiles of dissolved P3ATs using the Debye model for random coils (Figure 5.2) and a worm-like chain model. Aime *et al.*[9] used asymptotic laws to calculate the Kuhn length for a chain with a known contour length. In all of these cases, the Kuhn lengths of most regio-regular P3ATs were in the range 3–7 nm.[9,12,13]

Useful information is also extracted from power law fits near and above *c**. In SANS, power law exponents of −1, −1.7, −2 and −3 correspond to a rigid rod, a random walk polymer with excluded volume, a Gaussian chain and a collapsed chain structure, respectively. Researchers have used power law analysis in SANS to confirm the presence of rigid rod segments at high-*Q* values.[14,16,18,19] The low-*Q* structure of dissolved polymers has also been investigated in an attempt to understand the type of chain aggregation, *e.g.* the mesh-like structure observed in polyfluorenes.[20,21]

5.2.2 SANS of Colloidal Polymer Nanostructures

Conjugated polymers are known to self-assemble into colloidal structures (*e.g.* nanowires) depending on the solution conditions.[3,14,15,22–25] SANS and USANS are particularly useful in obtaining information on these systems. Figure 5.3 shows the combined SANS and USANS results for poly(3-hexythiophene) (P3HT) nanofibres that contain structural information from ∼1 nm to 10 μm.[23] Over these length scales, SANS/USANS is sensitive to

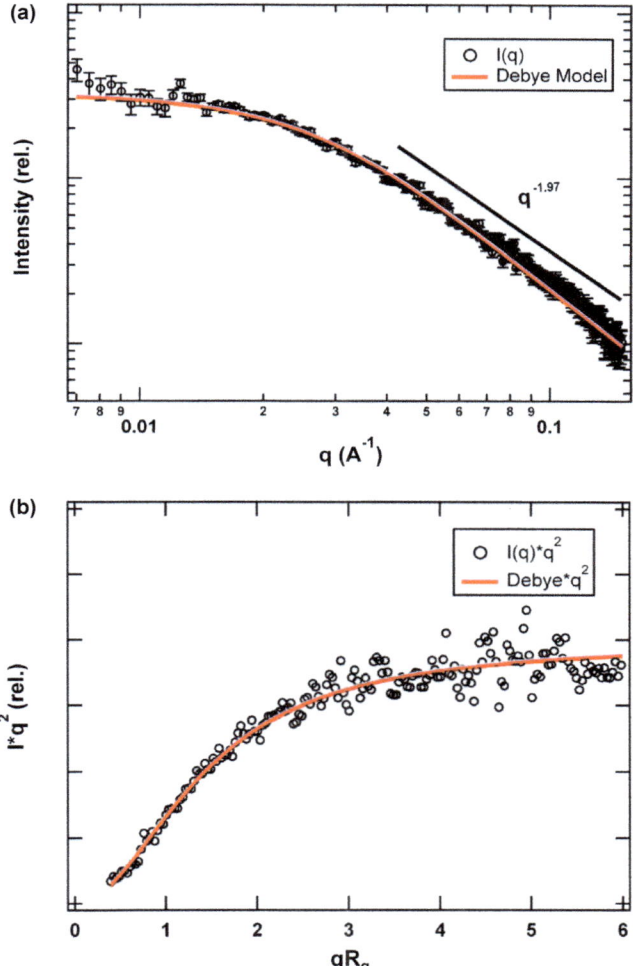

Figure 5.2   (a) SANS of poly(3-hexylthiophene) (P3HT) in dichlorobenzene shows that P3ATs adopt a random coil shape with a scaling of $I \approx q^{-2}$ over a large region corresponding to a Gaussian coil. (b) A Kratky plot also shows a plateau that indicates a random coil. The lower-case q is the same scattering vector Q described throughout the text.

Reprinted with permission from B. McCulloch *et al.*, *Macromolecules*, 2013, **46**, 1899–1907. Copyright 2013 American Chemical Society.

the inter-chain polymer packing, individual fibre structure, inter-fibre interactions and the overall size of the colloidal structure. Data analysis often requires the use of multiple models that are carefully delineated to extract relevant information over each length scale.[23]

Fits using eqn (5.3) provide information about both the size and shape of the nanostructure. Polythiophenes are known to form fibre-like nanostructures that have been successfully fitted with randomly oriented parallelepiped[3,13,23,26] and cylinder[25] models. Knaapila *et al.*[15] have fitted SANS data for polyfluorenes to both sheet and cylinder models. SANS data for polyphenylene vinylene have been fitted with a randomly oriented disc model.[19] These and other form factors (*e.g.* spheres and lamella) are available in the SASView package.[27] Constrained fits require a knowledge of the concentration and SLD to determine the nanoparticle shape using SANS.

Figure 5.3 shows the fit of a parallelepiped form factor at high-*Q*, but at low-*Q* the model no longer fits the SANS profile. This is because the analytical form factor assumes isolated fibres in solution and cannot describe the network structure. Shape-independent models are commonly used in cases where analytical models are not available to accurately describe a complex structure. Power law relationships (Figure 5.3) are common in SANS and the value of the power law exponent can be related to the fractal dimension of a network or to the shape of isolated scattering objects. Power law functions have also been used to characterize the structure of colloidal polyphenylene vinylene.[14] Porod scattering from sharp interfaces, a

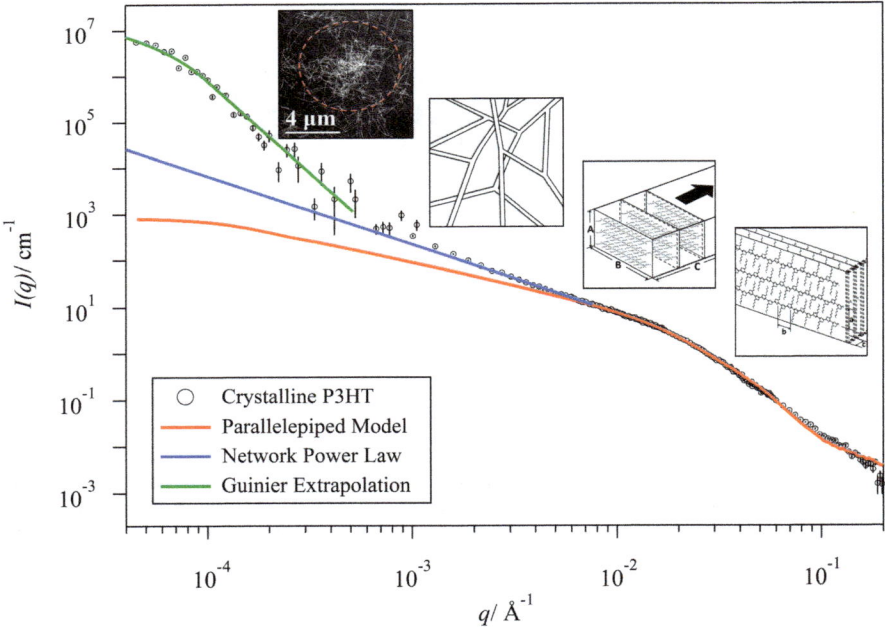

Figure 5.3   Combined SANS and USANS data from 0.2 wt% P3HT colloidal nanofibre networks in *p*-xylene. Fits (solid lines) describe the structures at different length scales. The insets are schematic representations of the structural features that are characterized at each *q*-range and a TEM image at low-*q*. The lower-case *q* is the same scattering vector *Q* described throughout the text.
Reprinted with permission from G. M. Newbloom *et al.*, *Macromolecules*, 2011, **44**, 3801–3809. Copyright 2011 American Chemical Society.

dependence of $I(Q) \approx Q^{-4}$, can be used to measure the specific surface area (S_v) of a nanostructure even without a knowledge of its shape. Specific surface area is an especially useful metric for organic photovoltaic devices and SANS has been utilized to obtain this parameter for nanostructured gels of polythiophenes and polyfluorenes.[3,22,24] Determining S_v requires a knowledge of the concentration, SLD and mass density.[3,24] The Guinier model is also useful for obtaining the radius of gyration (R_g), an average measure of size for all kinds of shapes, which can then be used to calculate specific feature sizes (*e.g.* radius) if the shape is known or assumed.[22,24] Figure 5.3 shows a Guinier fit at low-Q to extract the R_g of a colloidal P3HT network. In general, the use of a Guinier model requires a clear plateau to be visible in the scattering data and that $Q_{min}R_g < 1$, where Q_{min} is the minimum Q value used for fitting.

Standard plots have also been used to extract information from SANS data for other colloidal conjugated polymer nanostructures. Chen *et al.*[14] utilized a Kratky plot to determine the power law dependencies of Poly[2-methoxy-5-(2-ethylhexyloxy)-1,4-phenylenevinylene] (MEH-PPV) aggregates. The Kratky plot utilizes $I(Q) \times Q^{-x}$ *versus* Q to determine the Q-range (*i.e.* size range) over which a power law dependency exists (Figure 5.2). The Debye–Bueche plot has also been used to determine the correlation length (*e.g.* mesh size) of a polyfluorene gel.[20] The Stuhrmann plot was used to analyse the spatial distribution of P3HT and [6,6]-phenyl C61-butyric acid methyl ester (PCBM) within composite nanoparticles.[28]

SANS has also been applied to study the time-resolved self-assembly kinetics of conjugated polymers under different temperatures and solvent conditions.[3,4,26] The transparency of neutrons to metals also enabled the use of combined SANS, rheology and dielectric spectroscopy experiments for simultaneous structural, mechanical and electrical property analyses.[3,4,26] Although SANS is especially useful in characterizing nanostructured conjugated polymers, there are some constraints to consider. Solvent molecules can become trapped within self-assembled polythiophenes, leading to differences in the SLD, which affects form factor fits and specific surface area analyses.[3] It was also found that nanostructured polythiophenes may contain large fractions of dissolved polymers requiring the use of multiple form factors for accurate fitting.[3]

5.3 Neutron Scattering in Thin Films

The application of conjugated polymers to optoelectronics often requires solution-casting into thin films. Although low-temperature processing promises to enable low-cost manufacturing, it also imposes a number of significant challenges as a result of the semi-crystalline nature of conjugated polymers. For these materials, the distribution and orientation of the crystalline phases play important parts in the transport and lifetime of charge carriers and therefore the performance of functional devices.[29] These properties are determined by many details of the process history (*e.g.* the solvent, casting method, drying time and temperature) such that gaining insights into the relationship between processing and performance can be very difficult.[30]

This fact has motivated many active research efforts to focus on linking the mesoscale structure within conjugated polymer thin films to performance so that polymers can be designed to meet application requirements. Although electron and atomic force microscopy techniques can provide a direct window into this information, they have difficulties probing 'buried' materials and are often affected by experimental anomalies.[30,31] Particularly difficult to ascertain is the phase purity and degree of crystallinity.[32] Scattering measurements have thus been applied as complementary analyses for thin films. Many of the benefits of neutron scattering methods discussed in the preceding sections are also applicable to thin films. However, extending the measurement of microstructure to thin films (<1 μm thick) is not without challenges. In comparison with solution phase SANS, the interpretation of thin film scattering measurements is also more difficult. The goal of this section is to provide a starting point to understanding the methods necessary to both obtain high-quality neutron scattering measurements from thin films and to introduce the basics of the interpretation and analysis of data. The study of thin films of conjugated polymers using neutrons is generally achieved using transmission SANS, neutron reflectometry (NR) or grazing incidence small-angle neutron scattering (GISANS).

5.3.1   Transmission SANS

There are two inherent difficulties in the acquisition of high-quality SANS measurements from thin polymer films (typically <200 nm thick). The first, in traditional transmission geometry, is that the volume of the sample probed by the beam is small. Therefore the sample must scatter strongly if its structural features are to be detectable above the substrate's scattering and its inevitable absorption. In thin film composites, such as the polymer/fullerene thin films used in photovoltaics, contrast is provided by the chemistry due to the relative difference in hydrogen content between the conjugated polymer and fullerene (Figure 5.4). In contrast with bulk solution scattering, the total sample volume in thin films is vanishingly small. Therefore the scattering signal from the film is weak. To overcome this intrinsic limitation, Russell and co-workers[33,34] used stacks of up to 10 identical thin films to achieve high-quality SANS patterns within a reasonable amount of time. Using appropriate normalization methods, scattering patterns were collected and analysed for a number of thin film polymer/fullerene composites.

Examples of this approach include the work in which Russell and co-workers[33,35] measured mixtures of poly[2,6-(4,4-bis(2-ethylhexyl)-4H-cyclopenta[2,1-b;3,4-b′]-dithiophene)-alt-4,7-(2,1,3-benzothiadiazole)] (PCPDTBT) and PCBM with and without annealing and with or without a solvent additive. From fits of the SANS patterns, a correlation length was obtained from the Debye–Bueche relationship to quantify phase segregation due to

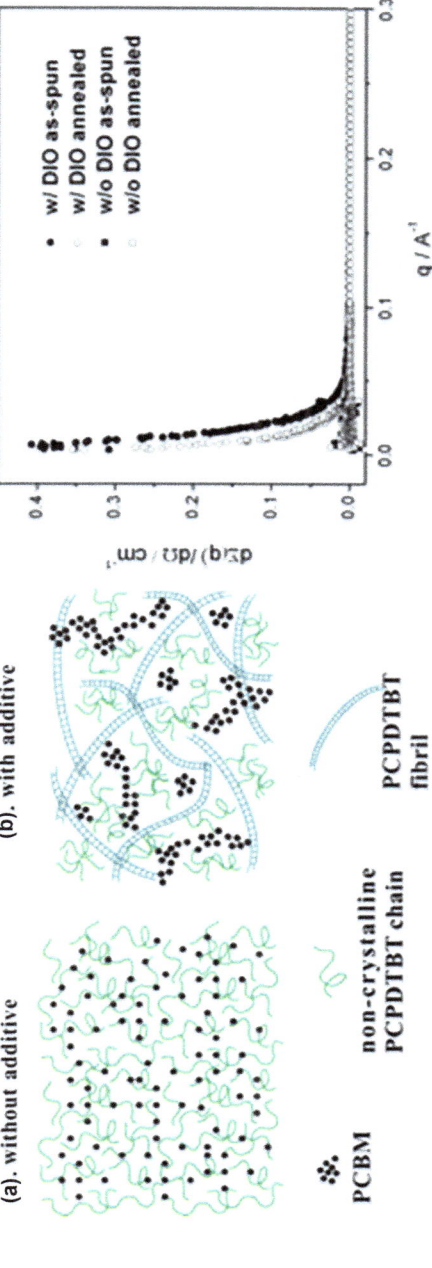

Figure 5.4 (Left) Morphology for PCPDTBT/PCBM thin films processed (a) without additives or (b) with additives. (Right) SANS profiles of PCPDTBT:PCBM blend films.
Reprinted from Y. Gu *et al.*, *Adv. Energy Mater.*, 2012, **2**, 683–690 with permission from John Wiley and Sons. Copyright © 2012 WILEY-VCH Verlag GmbH & Co. KGaA, Weinheim.

annealing. Using this correlation length, it was also possible to understand differences in performance within the context of the thin film micro-structure. Dadmun and co-workers[36] also reported SANS measurements from thin polymer/fullerene films. In their measurements of P3HT:PCBM thin films, they were able extract additional information by collecting data over a wide Q-range and fitting the SANS profiles to an elliptical cylinder that modelled the lamellar structure of P3HT crystals. A further example used transmission SANS to study the interface between polymer and fullerene domains. Using power law exponents at high-Q, Yin and Dadmun[37] dis-covered a fuzzy interface between the polymer and fullerene phases. They also found that the interfacial morphology changes as a function of fullerene loading and annealing time. A better understanding of the interfacial morphology is crucial in optimized organic electronics such as solar cells.

In principle, any of the analyses discussed in Section 5.2 could be utilized to extract information from thin film systems. However, the ability to per-form transmission SANS measurements on thin films is severely limited by long counting times. Only the in-plane structures of the film are probed in transmission SANS measurements. Because the beam propagation is normal to the stacking direction, structural variations through the thickness of the film cannot be investigated. Information about the distribution of material along the film thickness is more effectively studied with NR.

5.3.2 Neutron Reflectometry

Neutron reflectometry is a powerful probe of the vertical distribution of material in thin films. It has been used extensively in soft matter appli-cations to study interfacial phenomena such as polymer confinement, protein adsorption and the conformation of grafted polymers. In an NR measurement, the angle at which neutrons impinge on an interface is systematically varied and the intensity of the reflected neutrons is measured. As the angle increases above the critical angle of the film, the proportion of the beam that penetrates into the film increases, resulting in a decrease in intensity. Embedded interfaces within the film result in constructive and destructive interference of the incident wave that results in a complex dependence of the fraction of neutrons that reach the detector as a function of this angle. The intensity of reflected neutrons is normalized and the reflectivity is often plotted as R (reflectivity) *versus* Q_z.

From the angular dependence of the reflectivity profile and the known physical parameters of the sample (*e.g.* the film thickness, roughness), it is possible to use software packages to fit the reflectometry data to a known SLD profile as a function of depth (z).[38] For composite films, the known scattering length densities of the pure phases can also be used to convert the SLD profile into a volume fraction profile. For two-component systems, this solution is unique. This fact has been exploited to fit NR data from polymer/fullerene thin film composites.[32,39–41] These measurements have confirmed the often observed enrichment of fullerenes at air–film and substrate–film interfaces both before and after annealing. Mackay and co-workers[41] used enhanced phase-sensitive NR data with contrast variation of the air–film interface to refine the fit of the reflectivity data, demonstrating the importance of constraints on model quality. NR is an important complementary tool for the improvement of devices using conjugated polymers because subtle variations in the distribution of materials throughout the film, which would be easily missed in transmission SANS measurements, can have a significant effect on device performance.

5.3.3   Grazing Incidence Small-angle Neutron Scattering

Whereas NR and transmission SANS measurements probe specific scattering planes in a thin film, GISANS measurements can provide information about both in-plane and out-of-plane structures simultaneously. In grazing incidence measurements, a coated substrate is illuminated by a collimated beam oriented at a certain grazing angle (typically between 0.1 and 1°). In this configuration, the volume illuminated increases substantially due to the projected area of the beam and its complex interference with the thin film. In this way, stacking multiple samples is not necessary to obtain high-quality scattering profiles. Because of the similarity of the grazing incidence geometry to NR, the scattering profile will also be sensitive to the distribution of material through the sample thickness and in the plane of the film. In this way, it is possible to examine the orientation of structures with respect to the substrate surface and to also be sensitive to fluctuations in polymer density at the air–film or substrate–film interface. Although grazing incidence small-angle X-ray scattering (GISAXS) measurements are very commonly used to characterize conjugated polymer thin films, GISANS measurements have only recently become an attractive alternative. The benefits of GISANS over GISAXS are the same as in transmission SANS measurements: minimized sample damage, analysis through metals (*e.g.* electrodes), enhanced contrast and isotopic labelling.

In a GISANS experiment, collimated neutrons impinge on a sample at an incident angle, α_i, defined with respect to the plane of the substrate. Scattered neutrons are measured at a two-dimensional detector located a distance L from the sample. Two angles, α_f and $2\theta_f$, defined by the pixel position where the neutron is detected (Figure 5.5) are used to calculate the scattering vector q as shown in eqn (5.4).

$$q = \begin{pmatrix} Q_x \\ Q_y \\ Q_z \end{pmatrix} = \frac{2\pi}{\lambda} \begin{pmatrix} \cos(\alpha_f)\cos(2\theta) - \cos(\alpha_i) \\ \cos(\alpha_f)\sin(2\theta) \\ \sin(\alpha_f) + \sin(\alpha_i) \end{pmatrix} \qquad (5.4)$$

Once the GISANS data have been collected, the scattering intensity is corrected for dark current and normalized and the resulting pattern can be plotted as $I(Q_x, Q_y, Q_z)$. In this representation, horizontal line cuts, $I(Q_x)$ at a constant Q_z value, probe the in-plane structure and vertical line cuts, $I(Q_z)$ at a constant Q_x value, probe the out-of-plane structures.

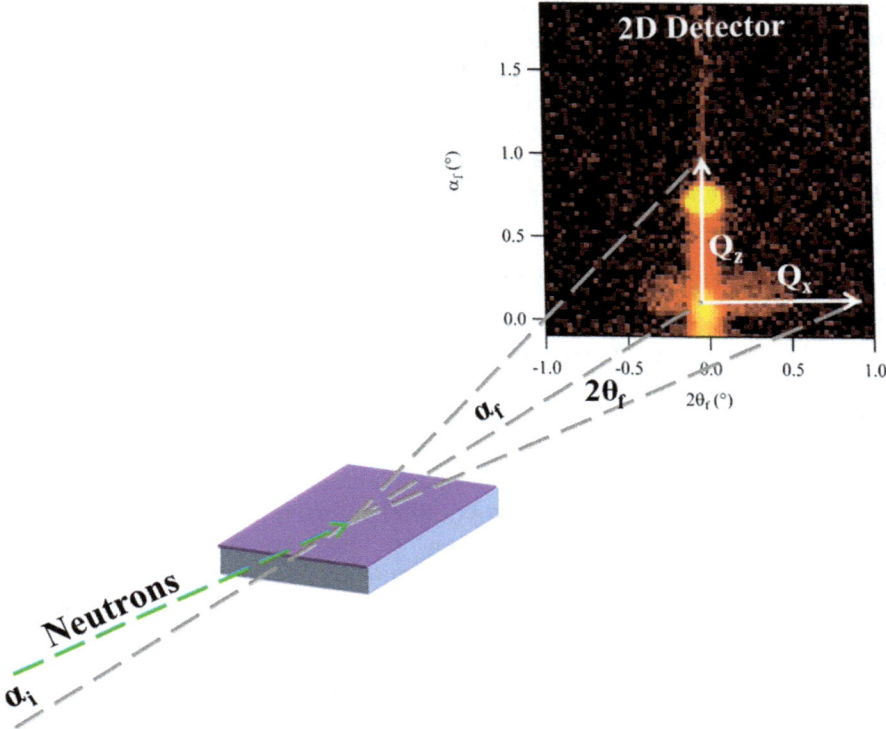

Figure 5.5 Schematic diagram of experimental GISANS measurement. The film is
oriented with an angle, α_i, relative to the substrate plane and scattering is
collected on a two-dimensional detector placed a distance L from the
sample, where angles α_f and $2\theta_f$ define the exiting angles of the scattered
neutrons from the sample. The resulting scattering pattern can be
converted to intensity as a function of Q_x and Q_z.

In the analysis of diffuse scattering contained within a GISANS dataset, the
approach varies depending on the nature of the structural features within
the film, the value of the incident angle with respect to the critical angle, and
the Q-range that is being fitted. Far from the critical edge, or at incident
angles much larger than the critical angle, the Born approximation can
adequately describe the scattering process. Unfortunately, at larger angles,
the diffuse scattering is often very weak, making it difficult to analyse.

To describe the entire scattering pattern and to reconstruct the in-plane
and out-of-plane density distribution within the film, it is necessary to model
the local potential imposed by the incident beam within each layer of the
film. Classically, the starting point for such an analysis was the full dy-
namical theory, which can describe scattering from arbitrarily complex
structures.[42] Although this theory is powerful, it can only be analytically
solved in certain special cases and requires significant computational power.
The distorted wave Born approximation provides an analytical framework
within which the scattering intensity from thin films can be treated

analytically.[43] Fortunately, there are now several software packages available for the analysis of GISAXS/GISANS data. These include IsGISAXS, FITGISAXS and BORNAgain.[44–46]

Although GISAXS experiments on thin films have been used for over a decade, GISANS measurements on conjugate polymer thin films have only recently been pioneered by Mueller-Buschbaum and co-workers.[47–50] GISANS probes large length scales and, for many conjugated polymer thin films, it is crystallinity and phase purity that is of primary interest and this is easily characterized with faster and more accessible GISAXS instruments. However, in polymer/fullerene thin films, it is often the mesoscopic length scales that principally determine the performance of the device. In this context, GISANS can be an ideal characterization tool.

Ruderer *et al.*[48] investigated annealed P3HT:PCBM blends using GISANS as a function of film composition. Quantitative fits using IsGISAXS and the distorted wave Born approximation to model structures embedded within the composite provided a quantitative measure of not only the size of fullerene domains, but also their spacing. To obtain adequate fits of the Yoneda peak in their data (arrows in Figure 5.6), molecularly dispersed PCBM had to be included in the P3HT-rich domains. The amount of fullerene dispersed within the P3HT phase of these thin films was found to be consistent with NR studies that showed intermixing of PCBM and P3HT under annealing conditions. These measurements demonstrate the power of quantitative fits to GISANS data. GISANS measurements have since been expanded to study phase separation in blends of PTB7:$PC_{71}BM$ thin films as a function of solvent treatment.[50]

Progress in the measurement of GISANS data on conjugated polymer thin films has also been catalysed by the development of time-of-flight (TOF) SANS instruments.[47] TOF-GISANS allows for patterns to be acquired in a fraction of the time required for traditional GISANS, expanding the utility of the technique and paving a path forward to new experiments that were previously limited by a low neutron flux. Improved software tools are also enabling a new generation of researchers interested in characterizing conjugated polymer thin films with GISANS.

5.4 Quasi-elastic Neutron Scattering

Although elastic neutron scattering provides a method of discerning the structural properties of polymers, QENS allows their dynamics to be probed on the molecular scale. This is because the energy range of neutrons used in this technique is similar to the activation energies of polymers (meV).[1,51–54] The dynamic disorder of polymer semiconductors is much higher than that of traditional semiconducting materials.[55,56] Molecular fluctuations also affect the degree to which π-orbital overlap is achieved, affecting both intra-chain and inter-chain charge transport.[57] In this section, the theory and techniques of QENS are briefly discussed before describing how QENS is being used to study polymer semiconductors.

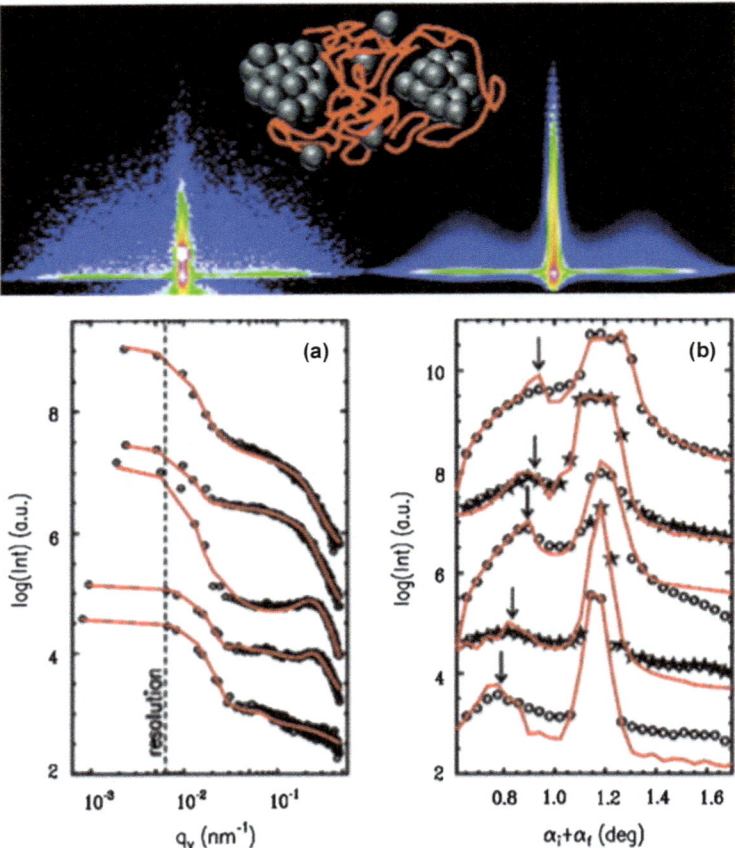

Figure 5.6 (Upper panel) Experimental and modelled GISANS data of P3HT:PCBM heterojunctions fitted with IsGISAXS. (Lower panel) (a) Horizontal and (b) vertical line cuts showing quantitative agreement.
Reprinted with permission from M. Ruderer *et al.*, *J. Phys. Chem. Lett.*, 2012, **3**, 683–688. Copyright 2012 American Chemical Society.

5.4.1 Theory

In QENS experiments, the quantity measured is the double differential scattering cross-section, which describes the probability of neutrons being scattered into a solid angle $\partial\Omega$ with a change in energy $\partial\hbar\omega$, as shown in eqn (5.5):[1]

$$\frac{\partial^2\sigma}{\partial\hbar\omega\partial\Omega} = \frac{1}{4\pi\hbar}\frac{K_f}{K_i}(\sigma_{\text{inc}}S_{\text{inc}}(Q,\omega) + \sigma_{\text{coh}}S_{\text{coh}}(Q,\omega)) \qquad (5.5)$$

If there is no change in energy, then the scattering is elastic, but if there is an energy transfer, then the scattering is inelastic. When the energy transfers are close to zero, the scattering is said to be quasi-elastic. K_f and K_i

correspond to the scattering vectors of the scattered and incident neutrons, the ratio of which approaches 1 for QENS. $S_{inc}(Q,\omega)$ and $S_{coh}(Q,\omega)$ are the incoherent and coherent dynamic scattering functions that link microscopic molecular motions with the observed scattering. $S_{inc}(Q,\omega)$ is related by double Fourier transform to the self-correlation function $G_{self}(R,t)$, which is the probability density of finding an atom at distance R away from itself after an elapsed time t. $S_{coh}(Q,\omega)$ is similarly related to $G_{self}(R,t)$, which is the probability density of finding an atom R away from another atom at an elapsed time t. Thus $S_{inc}(Q,\omega)$ describes the individual motion of atoms, whereas $S_{coh}(Q,\omega)$ describes the collective motion of atoms. QENS data are commonly represented using these scattering functions or their Fourier transforms in the time domain, known as the intermediate scattering functions $I_{inc}(Q,t)$ and $I_{coh}(Q,t)$. σ_{inc} and σ_{coh} are scattering cross-sections that are dependent on the scattering lengths characteristic of the nuclei of each atom. The incoherent scattering cross-section of hydrogen is almost two orders of magnitude higher than any of the cross-sections of other atoms found in polymers.[1] This means that the incoherent contribution to the double differential scattering cross-section will dominate for hydrogenated polymers and the data will largely probe the self-motions of hydrogen. For example, QENS measurements of fully hydrogenated P3HT highlight the molecular motions of the hexyl side-chains because they contain the majority of hydrogen atoms in each monomer. Deuterium, on the other hand, has a significantly lower σ_{inc}, almost equal to that of carbon. By substituting specific hydrogen atoms in a polymer with deuterium, the molecular motions of different parts of a polymer are 'highlighted'. With QENS, the dependence of the observed dynamics on Q can also be investigated. This makes it possible to ascertain the type of dynamic motion based on the value of Q (*i.e.* the length scale) where it is observed.

5.4.2 Instrumentation and Methods for Analysis

QENS experiments are usually conducted using three types of instruments: TOF spectrometers, back-scattering spectrometers and neutron spin echo (NSE) spectrometers. TOF and back-scattering spectrometers detect the changes in energy and momentum of scattered neutrons by measuring the neutron velocity, which is related to energy.[1] However, each instrument has a different method for measuring changes in energy, which results in different energy resolutions and time scales that can be probed.

Data from these instruments are measured as a function of energy transfer ($\hbar\omega$), where broadening of the elastic peak occurs due to molecular dynamics. QENS data in the energy domain can be fitted to linear combinations of Lorentzian functions, with the full width at half maximum of the Lorentzian peaks corresponding to the time scale of the observed relaxation.[1,54] For polymer systems, TOF spectrometers are sensitive to fast motions (*e.g.* methyl rotations), whereas back-scattering spectrometers are sensitive to slower segmental motions (*e.g.* segmental vibration). QENS experiments are usually slow (6–8 h are typical), but scans of the intensity of the elastic peak as a function of temperature are fast and can be used to identify the activation temperatures for different motions.[54] The intensity of the elastic peak decreases when molecular motions emerge because more intensity is scattered inelastically. This can be modelled by the Debye–Waller equation to obtain an estimate of the temperature-dependent mean square displacement $\langle r^2 \rangle(T)$.

$$S(Q, T, \omega \approx 0) = \exp\left(-\frac{1}{3}\langle r^2 \rangle(T)Q^2\right) \tag{5.6}$$

Unlike back-scattering and TOF spectrometers, NSE spectrometers are sensitive to energy transfers in the neV range, corresponding to time scales that can approach 1000 ns. The data are recorded directly in the time domain (*i.e.* intermediate scattering function) and are not convoluted by the instrument resolution. This technique is mainly used to monitor the segmental and chain dynamics of polymers over longer times (*i.e.* Rouse dynamics and reptation for polymers with very low friction coefficients). In the time domain, information is often extracted from QENS data by fitting to a Kolrausch–William–Watts stretched exponential function.[53,54,58]

$$I(Q, t) = \exp\left[-\left(\frac{t}{\tau}\right)^{\beta}\right] \tag{5.7}$$

Here, τ is the characteristic relaxation time of the dynamic process and β is a 'stretching' parameter $(0 < \beta < 1)$. Although τ varies greatly with both temperature and Q, a single β value is optimized for all values of Q (in many cases β is also fixed for multiple temperatures).[58] β values have been recorded within the range 0.2–0.8, with polymers usually having a β value of ~ 0.5. Fourier transforms are often used to convert data from TOF and back-scattering spectrometers into the time domain to combine with NSE data and to extract characteristic relaxation times using Kohlrausch–Williams–Watts (KWW) fits. The dynamic range of a single instrument is usually insufficient to completely capture all the relevant dynamics. A list of QENS spectrometers around the world has been compiled by Colmenero and Arbe.[52] To date, only TOF and back-scattering spectrometers have been used to characterize the dynamics of polymer semiconductors.[59–64]

5.4.3 QENS of Polymer Semiconductors

QENS studies conducted on polymer semiconductors have largely focused on P3HT and polyaniline (PANI). Obrzut and Page[59] measured the elastic intensity of P3HT as a function of temperature and compared these results with dielectric spectroscopy data. They observed that the motions of the side-chains are 'liberated' at ~ 175 K and there is also a corresponding decrease in the conductivity of bulk P3HT at this temperature. Using a modified version of eqn (5.6), they also calculated the activation energy for the side-chain motions to be ~ 9 kJ mol^{-1}. They concluded that local relaxations of the alkyl side-chains contributed to increased disorder in the polymer structure, causing an increase in the energy barriers to charge transport. Paterno *et al.*[60] characterized the dynamics of P3HT-PCBM blends cast from three different solvents (chloroform, chlorobenzene and *ortho*-dichlorobenzene). Interestingly, the solvent used had a negligible effect on the observed dynamics. They also noted that the addition of PCBM caused a decrease in the quasi-elastic broadening observed at temperatures >285 K. They hypothesized that this was due to an increase in domains that were not 'dynamically active', causing greater confinement of the P3HT side-chains.

Etampawala *et al.*[61] also investigated the dynamics of P3HT:PCBM blends. They assigned a fast relaxation process to P3HT and a slower process to PCBM and fitted their data as a sum of two KWW functions corresponding to each relaxation (see Figure 5.7). They confirmed that PCBM loading causes a frustration of P3HT side-chain dynamics and attributed this to the steric hindrance caused by PCBM on the P3HT side-chains in the amorphous phase. The Q dependence of the P3HT side-chain dynamics was also examined. For $Q < 0.9$ Å^{-1}, a near-linear relationship between the characteristic relaxation time and Q^2 was observed, indicating that simple diffusion dominated the observed dynamics at these length scales. For $Q > 1.1$ Å^{-1}, the relationship between the characteristic relaxation time was highly non-linear with respect to Q. These relaxation times were attributed to local motions within a side-chain, such as methyl rotations and crank-shaft motions.

Over the last decade, molecular dynamics (MD) simulations have also proved to be a highly useful complementary method for analysing QENS data.[52,53] This is because the coherent and incoherent intermediate scattering functions can be directly calculated with eqn (5.8) (with $i = j$ for incoherent scattering):[53]

$$I(Q,t) = \frac{1}{N} \sum_{i,j=1}^{N} \langle \exp(-iQ\mathbf{r}_{ij}(t)) \rangle \tag{5.8}$$

QENS data have been used to validate MD simulations by calculating intermediate scattering functions and comparing these with experimental $I(Q,t)$. The length scales and time scales accessible with MD simulations are the same as those accessible by QENS experiments. Validated MD simulations provide access to more detailed information than can be obtained from analytical models such as the KWW. For example, QENS can be used to validate force fields used in MD simulations. X-ray and neutron diffraction data can also be compared with MD simulations and used as further validation. Sniechowski and co-workers[62–64] used this approach to study the dynamics of PANI doped with the plasticizing dopant di-(2-butoxyethoxyethyl) ester of 4-sulfophthalic acid (DB3EPSA). The protons in the aliphatic tails of the dopant were the primary contributors to the observed scattering. They modelled the experimental data by assuming that the space explored by the aliphatic tail was a limited sphere in which protons could diffuse. They also performed MD simulations of PANI–DB3EPSA and validated their simulations with QENS. The simulations confirmed the proposed analytical model, but also revealed a dynamic heterogeneity

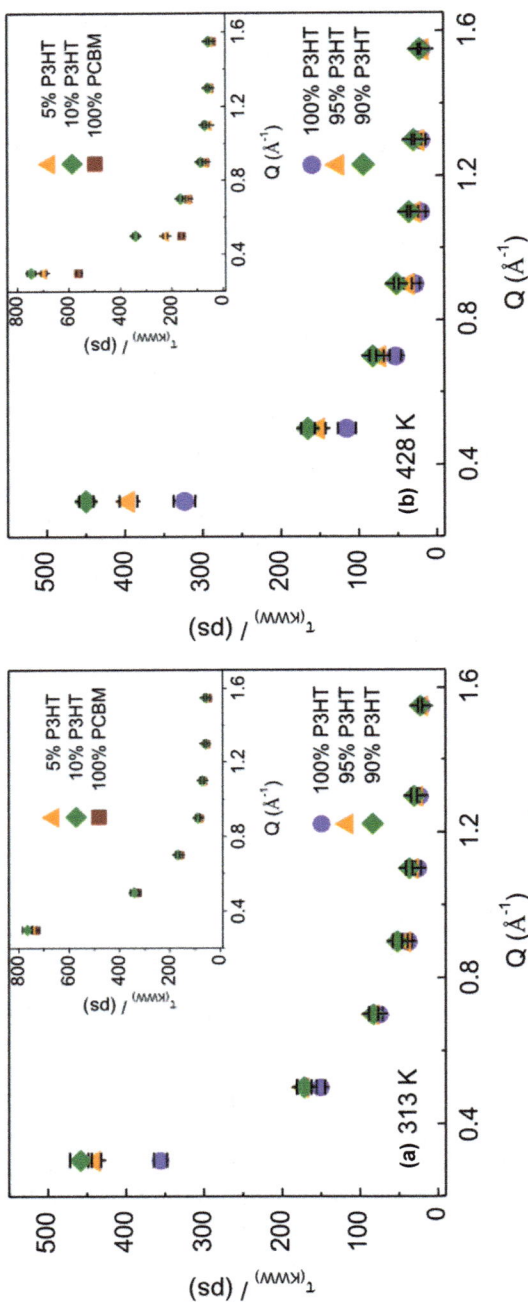

Figure 5.7 Relaxation times extracted from KWW fits to QENS spectra of P3HT:PCBM blends by Etampawala *et al.*[61] at 313 K (left) and 428 K (right). A single stretched exponential was fitted to QENS spectra for pure P3HT and PCBM. Blends were with two stretched exponentials (fast and slow). Reprinted with permission from T. Etampawala *et al.*, Monitoring the dynamics of miscible P3HT:PCBM blends: A quasi-elastic neutron scattering study of organic photovoltaic active layers, *Polymer*, 15, 155–162, Copyright 2015, with permission from Elsevier.

between the counter ions that was outside the scope of the model. Specifically, they observed a dramatic difference in the time evolution of the mean square displacement of protons in the dopant, which confirmed the relative immobility of the conjugated polymer compared with the dopant.

New efforts in our group are focused on using this approach to refine the MD force fields used to model the dynamics of P3ATs. Figure 5.8 shows that

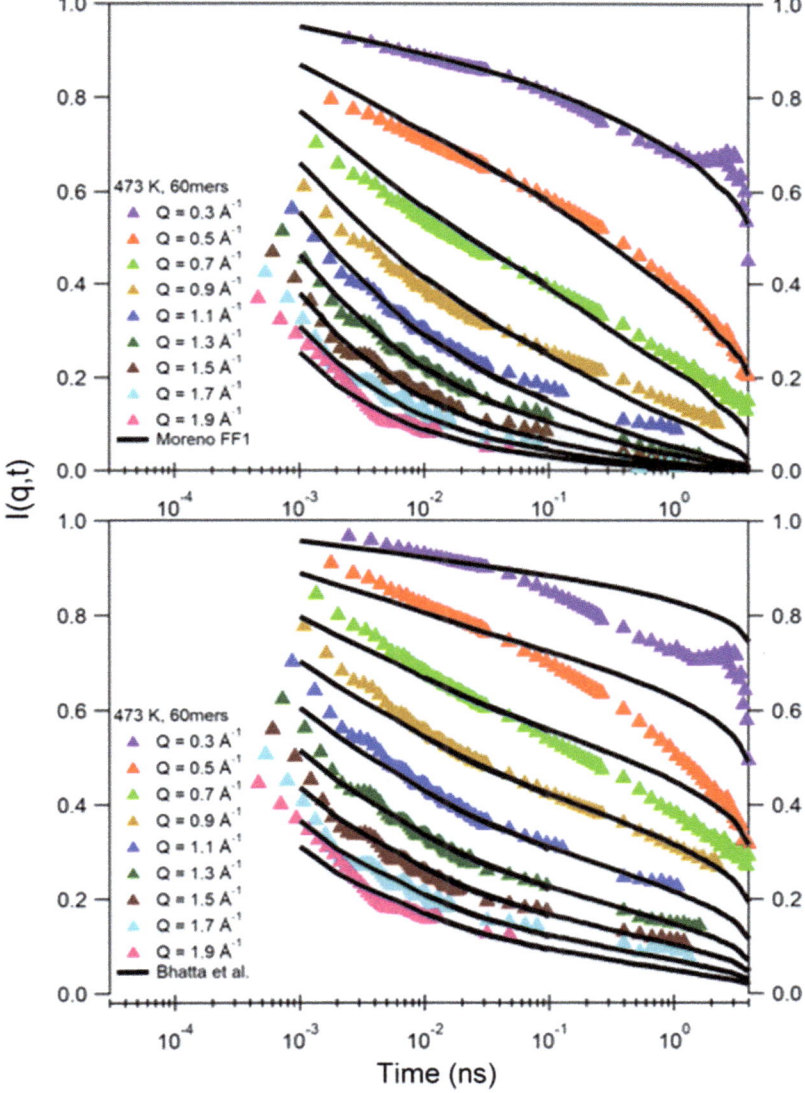

Figure 5.8 Comparison of MD simulations of melt phase regio-random P3HT with data obtained from DCS and HFBS at NIST as well as BASIS at ORNL. The upper plot shows a comparison with the OPLS-AA force field modified with a backbone torsion potential developed by Moreno *et al.*[65] The lower plot shows a comparison with the force field developed by Bhatta *et al.*[66]

two published torsion potentials for P3HT, both derived from *ab initio* density functional theory calculations, result in large discrepancies between simulations and experimental QENS data. The two torsion potentials overestimate or underestimate the time scales of the decay. This demonstrates that QENS of P3HT is very sensitive to torsional motions of both the backbone and side-chains. These parameters are also linked to π-orbital overlap and inter-chain charge transport. QENS coupled with MD simulations will probably continue to provide insights into the design of superior polymer semiconductors.

References

1. P. Lindner and T. Zemb, *Neutrons, X-rays and Light: Scattering Methods Applied to Soft Condensed Matter*, Elsevier Science, Amsterdam, 1st edn, 2002.
2. J. W. Kiel, B. J. Kirby, C. F. Majkrzak, B. B. Maranville and M. E. Mackay, *Soft Matter*, 2010, **6**, 641–646.
3. G. M. Newbloom, K. M. Weigandt and D. C. Pozzo, *Macromolecules*, 2012, **45**, 3452–3462.
4. G. M. Newbloom, P. de la Iglesia and L. D. Pozzo, *Soft Matter*, 2014, **10**, 8945–8954.
5. P.-G. de Gennes, *Scaling Concepts in Polymer Physics*, Ithaca, New York, United States, 1979.
6. P. J. Flory, *Principles of Polymer Chemistry*, Ithaca, New York, United States, 1953.
7. C. J. Brabec, *Sol. Energy Mater. Sol. Cells*, 2004, **83**, 273–292.
8. S. D. D. V. Rughooputh, S. Hotta, A. J. Heeger and F. Wudl, *J. Polym. Sci., Part B: Polym. Phys.*, 1987, **25**, 1071–1078.
9. J. P. Aime, F. Bargain, M. Schott, H. Eckhardt, G. G. Miller and R. L. Elsenbaumer, *Phys. Rev. Lett.*, 1989, **62**, 55–58.
10. G. Dufresne, J. Bouchard, M. Belletête, G. Durocher and M. Leclerc, *Macromolecules*, 2000, **33**, 8252–8257.
11. R. Potai, A. Kamphan and R. Traiphol, *J. Polym. Sci., Part B: Polym. Phys.*, 2013, **51**, 1288–1297.
12. B. McCulloch, V. Ho, M. Hoarfrost, C. Stanley, C. Do, W. T. Heller and R. A. Segalman, *Macromolecules*, 2013, **46**, 1899–1907.
13. G. M. Newbloom, S. M. Hoffmann, A. F. West, M. C. Gile, P. Sista, H.-K. C. Cheung, C. K. Luscombe, J. Pfaendtner and L. D. Pozzo, *Langmuir*, 2015, **31**, 458–468.
14. S. H. Chen, A. C. Su, C. S. Chang, H. L. Chen, D. L. Ho, C. S. Tsao, K. Y. Peng and S. A. Chen, *Langmuir*, 2004, **20**, 8909–8915.
15. M. Knaapila, L. Almásy, V. M. Garamus, M. L. Ramos, L. L. G. Justino, F. Galbrecht, E. Preis, U. Scherf, H. D. Burrows and A. P. Monkman, *Polymer*, 2008, **49**, 2033–2038.
16. Y.-C. Li, K.-B. Chen, H.-L. Chen, C.-S. Hsu, C.-S. Tsao, J.-H. Chen and S.-A. Chen, *Langmuir*, 2006, **22**, 11009–11015.

17. B. Hammouda, *Adv. Polym. Sci.*, 1993, **106**, 87–133.
18. J. H. Chen, C. S. Chang, Y. X. Chang, C. Y. Chen, H. L. Chen and S. A. Chen, *Macromolecules*, 2009, **42**, 1306–1314.
19. Y.-C. Li, C.-Y. Chen, Y.-X. Chang, P.-Y. Chuang, J.-H. Chen, H.-L. Chen, C.-S. Hsu, V. A. Ivanov, P. G. Khalatur and S.-A. Chen, *Langmuir*, 2009, **25**, 4668–4677.
20. L. L. G. Justino, M. L. Ramos, M. Knaapila, A. T. Marques, C. J. Kudla, U. Scherf, L. Almásy, R. Schweins, H. D. Burrows and A. P. Monkman, *Macromolecules*, 2011, **44**, 334–343.
21. M. Knaapila, D. W. Bright, B. S. Nehls, V. M. Garamus, L. Almasy, R. Schweins, U. Scherf and A. P. Monkman, *Macromolecules*, 2011, **44**, 6453–6460.
22. P. de la Iglesia and D. C. Pozzo, *Soft Matter*, 2013, **9**, 11214–11224.
23. G. M. Newbloom, F. S. Kim, S. A. Jenekhe and D. C. Pozzo, *Macromolecules*, 2011, **44**, 3801–3809.
24. J. J. Richards, K. M. Weigandt and D. C. Pozzo, *J. Colloid Interface Sci.*, 2011, **364**, 341–350.
25. C. Y. Chen, S. H. Chan, J. Y. Li, K. H. Wu, H. L. Chen, J. H. Chen, W. Y. Huang and S. A. Chen, *Macromolecules*, 2010, **43**, 7305–7311.
26. G. M. Newbloom, K. M. Weigandt and D. C. Pozzo, *Soft Matter*, 2012, **8**, 8854–8864.
27. P. Butler, G. Alina, R. C. Hernandez, M. Doucet, A. Jackson, P. Kienzle, S. Kline and J. Zhou, SASView for Small Angle Scattering Analysis, http://www.sasview.org/.
28. J. J. Richards, C. L. Whittle, G. Shao and L. D. Pozzo, *ACS Nano*, 2014, **8**, 4313–4324.
29. B. Zhao, C.-Z. Li, S.-Q. Liu, J. J. Richards, C.-C. Chueh, F. Ding, L. D. Pozzo, X. Li and A. K. Y. Jen, *J. Mater. Chem. A*, 2015, **3**, 6929–6934.
30. J. Zhao, A. Swinnen, G. Van Assche, J. Manca, D. Vanderzande and B. Van Mele, *J. Phys. Chem. B*, 2009, **113**, 1587–1591.
31. D. C. Coffey and D. S. Ginger, *Nat. Mater.*, 2006, **5**, 735–740.
32. A. Dokoutchaev, J. T. James, S. C. Koene, S. Pathak, G. K. S. Prakash and M. E. Thompson, *Chem. Mater.*, 1999, **11**, 2389–2399.
33. D. Chen, A. Nakahara, D. Wei, D. Nordlund and T. P. Russell, *Nano Lett.*, 2011, **11**, 561–567.
34. H. Lu, B. Akgun and T. P. Russell, *Adv. Energy Mater.*, 2011, **1**, 870–878.
35. Y. Gu, C. Wang and T. P. Russell, *Adv. Energy Mater.*, 2012, **2**, 683–690.
36. H. Chen, S. Hu, H. Zang, B. Hu and M. Dadmun, *Adv. Funct. Mater.*, 2013, **23**, 1701–1710.
37. W. Yin and M. Dadmun, *ACS Nano*, 2011, **5**, 4756–4768.
38. S. M. Danauskas, D. Li, M. Meron, B. Lin and K. Y. C. Lee, *J. Appl. Crystallogr.*, 2008, **41**, 1187–1193.
39. P. A. Staniec, A. J. Parnell, A. D. F. Dunbar, H. Yi, A. J. Pearson, T. Wang, P. E. Hopkinson, C. Kinane, R. M. Dalgliesh, A. M. Donald, A. J. Ryan, A. Iraqi, R. A. L. Jones and D. G. Lidzey, *Adv. Energy Mater.*, 2011, **1**, 499–504.

40. H. Chen, R. Hegde, J. Browning and M. D. Dadmun, *Phys. Chem. Chem. Phys.*, 2012, **14**, 5635–5641.
41. J. W. Kiel, M. E. Mackay, B. J. Kirby, B. B. Maranville and C. F. Majkrzak, *J. Chem. Phys.*, 2010, **133**, 074902.
42. B. W. Batterman and H. Cole, *Rev. Mod. Phys.*, 1964, **36**, 681–717.
43. S. K. Sinha, E. B. Sirota, S. Garoff and H. B. Stanley, *Phys. Rev. B: Condens. Matter Mater. Phys.*, 1988, **38**, 2297–2311.
44. R. Lazzari, *J. Appl. Crystallogr.*, 2002, **35**, 406–421.
45. D. Babonneau, *J. Appl. Crystallogr.*, 2010, **43**, 929–936.
46. C. Durniak, M. Ganeva, G. Pospelov, W. Van Herck and J. Wuttke, Software for simulating and fitting X-ray and neutron small-angle scattering at grazing incidence, http://www.bornagainproject.org.
47. P. Mueller-Buschbaum, *Polym. J.*, 2013, **45**, 34–42.
48. M. A. Ruderer, R. Meier, L. Porcar, R. Cubitt and P. Mueller-Buschbaum, *J. Phys. Chem. Lett.*, 2012, **3**, 683–688.
49. M. Rawolle, K. Sarkar, M. A. Niedermeier, M. Schindler, P. Lellig, J. S. Gutmann, J.-F. Moulin, M. Haese-Seiller, A. S. Wochnik, C. Scheu and P. Mueller-Buschbaum, *ACS Appl. Mater. Interfaces*, 2013, **5**, 719–729.
50. S. Guo, B. Cao, W. Wang, J.-F. Moulin and P. Mueller-Buschbaum, *ACS Appl. Mater. Interfaces*, 2015, **7**, 4641–4649.
51. A. Furrer, J. Mesot and T. Strassle, *Neutron Scattering in Condensed Matter Physics*, 1st edn, 2009.
52. A. Arbe, F. Alvarez and J. Colmenero, *Soft Matter*, 2012, **8**, 8257–8270.
53. J. Colmenero and A. Arbe, *J. Polym. Sci., Part B: Polym. Phys.*, 2013, **51**, 87–113.
54. I. Hoffmann, *Colloid Polym. Sci.*, 2014, **292**, 2053–2069.
55. T. Liu and A. Troisi, *Adv. Funct. Mater.*, 2014, **24**, 925–933.
56. O. D. Bernardinelli, S. M. Cassemiro, L. A. O. Nunes, T. D. Z. Atvars, L. Akcelrud and E. R. deAzevedo, *J. Phys. Chem. B*, 2012, **116**, 5993–6002.
57. D. P. McMahon, D. L. Cheung, L. Goris, J. Dacuna, A. Salleo and A. Troisi, *J. Phys. Chem. C*, 2011, **115**, 19386–19393.
58. J. C. Phillips, *Rep. Prog. Phys.*, 1996, **59**, 1133–1207.
59. J. Obrzut and K. A. Page, *Phys. Rev. B: Condens. Matter Mater. Phys.*, 2009, **80**, 195291.
60. G. Paterno, F. Cacialli and V. Garcia-Sakai, *Chem. Phys.*, 2013, **427**, 142–146.
61. T. Etampawala, D. Ratnaweera, B. Morgan, S. Diallo, E. Mamontov and M. Dadmun, *Polymer*, 2015, **61**, 155–162.
62. D. Djurado, M. Bee, M. Sniechowski, S. Howells, P. Rannou, A. Pron, J. P. Travers and W. Luzny, *Phys. Chem. Chem. Phys.*, 2005, **7**, 1235–1240.
63. M. Sniechowski, D. Djurado, M. Bee, M. A. Gonzalez, M. R. Johnson, P. Rannou, B. Dufour and W. Luzny, *Chem. Phys.*, 2005, **317**, 289–297.
64. D. Djurado, M. Sniechowski, M. Bee, M. Johnson, M. A. Gonzalez, P. Rannou and A. Pron, *Macromol. Symp.*, 2006, **241**, 28–33.
65. M. Moreno, M. Casalegno, G. Raos, S. V. Meille and R. Po, *J. Phys. Chem. B*, 2010, **114**, 1591–1602.
66. R. S. Bhatta, Y. Y. Yimer, D. S. Perry and M. Tsige, *J. Phys. Chem. B*, 2013, **117**, 10035–10045.

CHAPTER 6

Structural Control in Polymeric Semiconductors: Application to the Manipulation of Light-emitting Properties

IOAN BOTIZ,[a,b] COSMIN LEORDEAN[a] AND
NATALIE STINGELIN*[b]

[a] Babes-Bolyai University, Interdisciplinary Research Institute in
Bio-Nano-Sciences, Nanobiophotonics and Laser Microspectroscopy
Centre, Treboniu Laurian Str. 42, Cluj-Napoca 400271, Romania;
[b] Department of Materials and Centre for Plastic Electronics, Imperial
College London, Exhibition Road, London SW7 2AZ, UK
*Email: natalie.stingelin@imperial.ac.uk

6.1 Introduction

In conjugated polymeric materials, the molecular architecture, conformational order, packing and overall microstructure are all known to have a significant influence on much of the optoelectronic landscape, including their emission properties (EPs).[1,2] These structural features therefore have to be controlled and tuned to efficiently make use of the properties of this class of materials. In case of light emission, they are exploited in solution-processable, potentially large area, flexible and lightweight optoelectronic structures such as organic light-emitting diodes (OLEDs)[3,4] or can be integrated into, for example, highly stretchable[5] information displays.[6]

RSC Polymer Chemistry Series No. 21
Semiconducting Polymers: Controlled Synthesis and Microstructure
Edited by Christine Luscombe
© The Royal Society of Chemistry 2017
Published by the Royal Society of Chemistry, www.rsc.org

The emission of light can also be used in organic field-effect transistors,[7] whereas emission quenching (*i.e.* the decrease in the emission intensity of a given system) is another important property that is widely used in organic photovoltaic devices (OPVs) to monitor the efficiency of charge transfer from the donor to the acceptor material.[8,9]

To establish the relevant processing/structure/property interrelationships and to apply them to the EPs of conjugated and/or small molecule semi-conductors, we need first to define the most relevant phenomena. The emission of light refers to the luminescence of a material, *i.e.* the creation of light that is a form of cold body radiation rather than from a process that involves heat. Luminescence can be regarded as a combined effect of various chemical or physical processes. For example, when a polymer material absorbs a photon, the system is excited both electronically and vibrationally. The emission of light from a material after the absorption of photons is called photoluminescence (PL). The emission of light as a result of singlet–singlet electronic relaxation, after the occurrence of vibrational relaxation, is known as fluorescence (FL) (Figure 6.1). The emitted light is always of a

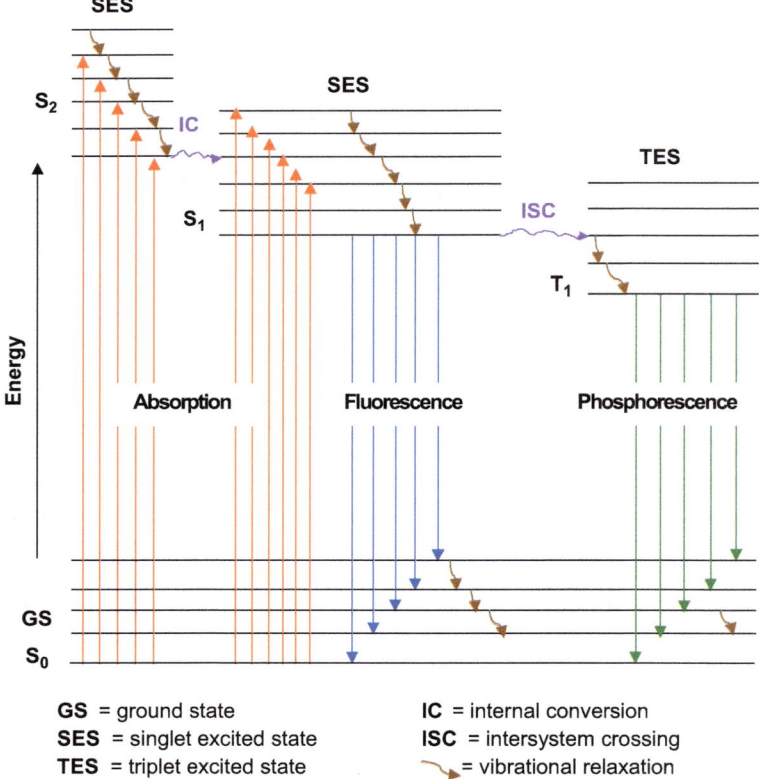

GS = ground state   IC = internal conversion
SES = singlet excited state   ISC = intersystem crossing
TES = triplet excited state   ⤵ = vibrational relaxation

Figure 6.1 Energy diagram illustrating the differences between the main emission phenomena of fluorescence and phosphorescence.

longer wavelength (lower energy) than the wavelength of the absorbed photons. When light emission originates from triplet–singlet electronic relaxation, we refer to this as phosphorescence (Figure 6.1), which generally occurs over longer timescales and at lower intensities than FL. What unifies these different luminescence phenomena is the fact that they can be explained using the well-known Jablonski diagram.[10] They also have found many practical applications, including FL spectrometry and microscopy.[10] Many other types of luminescence phenomena exist, including chemiluminescence (the emission of light as a result of a chemical reaction) or electroluminescence (EL); the latter is important from a technological point of view. EL is defined as the luminescence of a material that responds to either an applied electric current or a strong electric field (leading to a radiative recombination of electrons and holes) and is used in electroluminescent devices such as OLEDs.[3] PL, FL and EL are the main phenomena discussed in this chapter.

Our objective is to review the most effective strategies used by the scientific community to alter the EPs of conjugated polymers (CPs) by structural control using recent examples from the literature. To this end, we classify most of the approaches used into following categories: chemical design (which is strongly correlated with the molecular architecture and electronic structure of polymers and, thus, to their synthesis); physical methods (*i.e.* methodologies that allow the manipulation of the structural conformations and packing of a system *via* physicochemical means); and strategies that either blend CPs with other materials or are based on the use of noble metal structures.

6.2 Approach 1: Chemical Design

In 1990, poly(*p*-phenylenevinylene) (PPV) polymers were shown to possess semiconducting properties because their π-molecular orbitals can be delocalized along the polymer backbone.[3] It was also demonstrated that CPs could be used to fabricate OLEDs capable of emitting light in the green–yellow region.[3] Following this initial report, many other materials with well-controlled molecular architectures were designed and synthesized, all displaying intriguing EPs. We will discuss here some recently reported advanced materials, such as CPs with low energy gaps [*i.e.* low energy between the lowest unoccupied molecular orbital (LUMO) and the highest occupied molecular orbital (HOMO)]; well-investigated, red-emitting systems such as poly[2-methoxy-5-(2-ethylhexyloxy)-1,4-phenylenevinylene] (MEH-PPV)[11] and poly(3-alkylthiophene) (P3AT),[12] as well as CPs with large energy gaps that have a blue emission, such as poly(*p*-phenylene)[13] or polyfluorene (PFO).[14]

These few examples illustrate that the variety of chemical designs available allows the production of CPs exhibiting EPs that cover the entire visible wavelength range (and beyond); hence they, in principle, fulfil the need for colour tunability. However, challenges still exist. For instance, the realization of colour-stable blue OLEDs with a high efficiency has proved to be demanding as a result of difficulties in synthesizing an efficient and stable blue emissive material with a band gap energy larger than ∼3 eV. PFO polymers match these requirements;[15] PFO-based materials, in which there is little aggregation and/or very few inter-chain interactions, also display other attributes that are ideal for the fabrication of OLEDs. However, despite these benefits, also this class of CP can exhibit colour instability.[15]

Synthetic strategies towards CPs with lower HOMO and LUMO energy levels, such as CPs based on thienothiophene vinylene polymers, can provide alternative options.[16] By changing the oxidation state of sulfanyl to sulfonyl substituted thienothiophene monomers, fine-tuning of the energy levels has been demonstrated through random polymerization. The resulting low band gap (1.5 eV) polymers retain the uniformity of their conjugated backbone and show increased FL properties.[16] A series of alternating copolymers composed of gallafluorene with various co-monomers has been synthesized *via* several chemical reactions. Such CPs possess not only good solubility in organic solvents, but are also stable in air and moisture; their optical properties originate from the electronic interaction between gallafluorene and the specific co-monomers and therefore their colour emissions can be tuned to the ultraviolet (UV), blue, green, orange–yellow and red regions (Figure 6.2a).[17]

Additional opportunities are provided by other classes of CP. A good example is the blue, green and red light-emitting CPs based on 9,9-bis(4-(2-ethylhexyloxy)phenyl)fluorene containing a dibenzothiophene-*S,S*-dioxide unit with additional benzothiadiazole or 4,7-di(4-hexylthien-2-yl)-benzothiadiazole moieties.[18] In addition to good thermal stability, these polymers show high FL yields, making them promising candidates for the fabrication of OLEDs. Another interesting advantage of these systems is that mixing them in specific ratios can lead to white emission and therefore to the realization of white OLEDs with good efficiency and colour stability.[18]

Other series of CPs with emission colours ranging from light blue to green, yellow and red that may find applications in OLED fabrication include

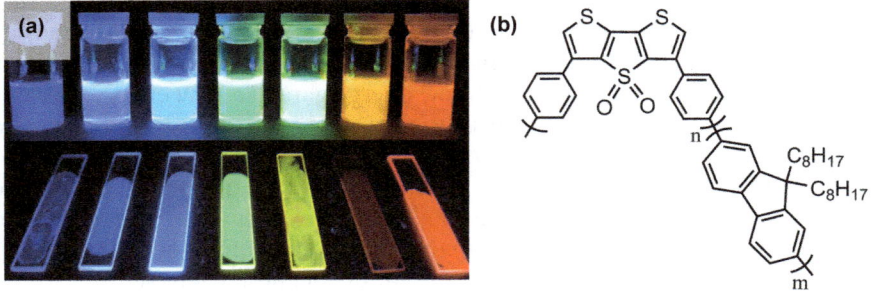

Figure 6.2   (a) Photographs of various CP solutions in chloroform taken under UV light. (b) Chemical structure of a DTT-fluorene-based copolymer. Adapted with permission from ref. 17 (a) and ref. 19 (b). Copyright (2015, 2013) American Chemical Society.

fluorine-based copolymers containing various ratios of dithieno[3,2-*b*;2,3'-*d*]-thiophene-*S,S*-dioxide (DTT-SO) (see Figure 6.2b for an example).[19] Polymers with either the fluorene moiety connected to the peripheral thiophenes of the DTT-SO unit or featuring a thiophene extension between the fluorene unit and DTT-SO were reported to exhibit a red shift in emission (up to 585 and 646 nm, respectively) with increasing DTT-SO content. Conversely, to shift the emission towards blue while also increasing the DTT-SO content, the fluorene moieties can be connected with the DTT-SO unit through the phenyl moiety of DTT.[19] This leads to an emission in the range 400–675 nm. The EPs can be further manipulated by replacing the conjugated di-benzothiophene-*S,S*-dioxide with non-conjugated diphenylsulfone units in the main chain of the poly(spirobifluorene) polymers, which leads to the ex-tinction of charge transfer processes from the pendant 2,3,6,7-tetraoctyloxy-fluorene to the main chain, probably as a result of the weaker electron affinity of diphenylsulfone relative to dibenzothiophene-*S,S*-dioxide. As a consequence, polymers containing diphenylsulfone moieties exhibit pure blue emission, in contrast to poly(spirobifluorene) polymers containing dibenzothiophene-*S,S*-dioxide moieties, which exhibit an undesirable green emission.[20]

A key parameter that dictates the EPs of a CP and which can be controlled *via* chemical design routes is the polymer chain rigidity. Figure 6.3 shows schematic diagrams of polymers of different chain rigidities, from flexible, semi-flexible to rigid macromolecules. One important difference is that more flexible molecules can entangle, whereas rigid chains generally lead to more extended chain structures, which can drastically affect the molecular assembly from the nano- to the macroscale.

One way that the chain rigidity can be manipulated is by varying the ratio of the monomeric components in random copolymers. In *para*-poly(2,5-pyridinediyl) and *meta*-poly(2,6-pyridinediyl), for example, the chain rigidity can be reduced by introducing 'meta' flexible linkages.[21] This reduction leads to both a reduced PL and a reduced conjugation length (a parameter determined by the number of coplanar conjugated rings), resulting in a blue shift in their emission.

The relation between conjugation length and EPs is applicable to many other conjugated systems. Good examples are the oligophenylenevinylene

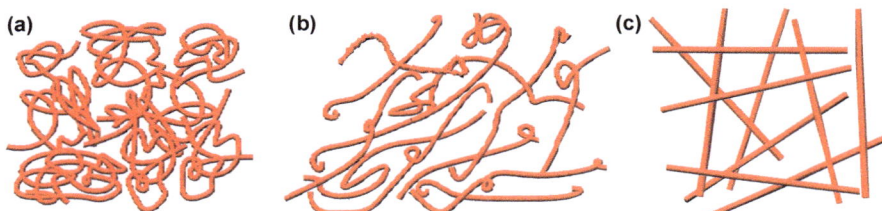

Figure 6.3 Schematic representations of the structures and molecular assemblies formed by (a) flexible, (b) semi-flexible and (c) rigid macromolecules.

derivatives. Disrupting the conjugation length by causing torsional defects on an unsubstituted phenylene ring (to eight, seven and six monomers) has been shown to lead to three types of FL spectral shapes, each corresponding to a distinct conjugation length.[22]

Another example is given by the work of Rathgeber *et al.*,[23] who systematically modified anthracene containing poly(*p*-phenyleneethynylene)-*alt*-PPV polymers by decorating the CP backbone with linear, branched or combinations of linear and branched alkoxy side-chains to manipulate inter-chain interactions *via* changes in the molecular packing. They found that polymers with all-linear side-chains attached close to the anthracenylene–ethynylene unit (more inclined towards ordering) showed red-shifted emission compared with the amorphous polymers with branched side-chains attached at the anthracenylene–ethynylene unit. This is caused by conformational disorder. Theoretical studies using MEH-PPV as a model system have shown that the electronic structure and thus the EPs are largely determined by the conformational disorder of individual macromolecules, even when they are essentially of the same chemical nature.[24]

Another parameter related to chemical design and synthetic methods that influences the EPs of CPs is their chirality (Figure 6.4a). Chirality is a property of a polymer – or, indeed, any organic compound – that makes it distinguishable from its mirror image (the system and its mirror image are not superimposable by means of rotation). This was clearly shown by Resta *et al.*,[25] who synthesized and characterized low molecular weight MEH-PPV in the enantiopure (*R*) form (*R*-MEH-PPV) and then compared this material with its racemic MEH-PPV analogue (Figure 6.4b). Because the aggregates of *R*-MEH-PPV, which form in solution, adopt a helical supramolecular structure, the resulting FL is notably reduced compared with the racemic MEH-PPV. The effect of polymer chirality on the EPs was further illustrated when OLED devices were made using the chiral PPV derivative. These OLEDs emitted light with the right emission intensity being slightly larger than the left intensity, demonstrating the concept of circularly polarized EL.[26]

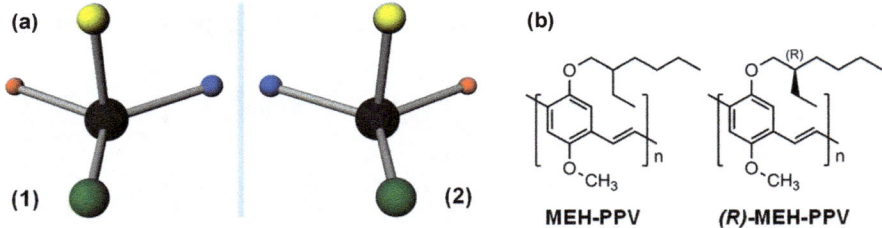

Figure 6.4   (a) Example of a chiral molecule: (1) is the mirror image of (2); (2) cannot be superimposed on (1) by means of rotation. (b) Chemical structure of racemic and enantiopure MEH-PPV. Figure (b) was adapted with permission from ref. 25.
Copyright (2014) American Chemical Society.

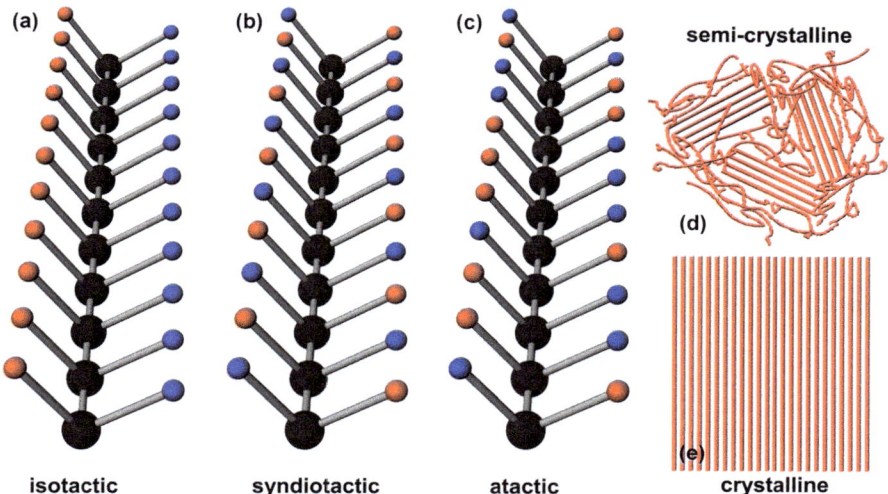

Figure 6.5 (a)–(c) Comparison of macromolecules of different regularities found in CPs. Schematic representation of the molecular packing in (d) semicrystalline and (e) chain-extended structures.

Another parameter that can be controlled *via* synthetic routes to manipulate important features, including the EPs, is the chain regularity of CPs (see Figure 6.5a–c for various regio-regular/-random configurations). A polymer consisting of repeat units that are all derived from the same isomer of the monomer is referred to as regio-regular (*rr*-); otherwise the polymer is regio-random (*rra*-). A prime example is poly(3-hexylthiophene) (P3HT). Although P3HT is not known as a highly emissive polymer, it is frequently used in organic electronics, in field-effect transistors and in bulk heterojunction solar cells. Its chain regio-regularity influences the conjugation length and molecular packing in thin films and therefore the resulting optoelectronic properties, from charge transport and absorption to the EPs. For example, it was shown that in thin films of *rr*-P3HT, most chains can adopt a relatively ordered conformation, in contrast with films consiting of *rra*-P3HT, where the chains are found in a variety of conformations.[27] As a consequence, the *rr*-P3HT system usually adopts a semicrystalline microstructure (Figure 6.5d), although, under specific experimental conditions (when the nucleation and growth parameters are precisely controlled[28]), this system can adopt a paraffinic-like solid-state structure consisting of individual domains of chain-extended crystallites (Figure 6.5e).

The importance of regio-regularity can be further emphasized by the example of *rr*-borane-functionalized poly(3-alkynylthiophene) polymers synthesized from 3-alkynylphenylborane-functionalized thiophene monomers. The thiophene moieties in *rr*-poly(3-alkynylthiophene) polymers adopt a coplanar conformation that favours the planarization of the polymer backbone and therefore leads to changes in the electronic structure. As a consequence, polymers of higher regio-regularity display a red shift in FL (Figure 6.6a).[29]

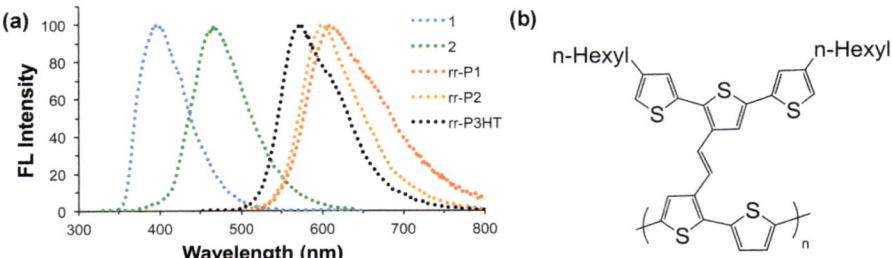

Figure 6.6 (a) FL spectra of model compounds and *rr*-poly(3-alkynylthiophene)s in tetrahydrofuran solutions. (b) Chemical structure of a two-dimensional polythiophene derivative. See references for symbol abbreviations. Adapted with permission from ref. 29 (a) and ref. 31 (b). Copyright (2014, 2013) American Chemical Society.

Odd–even effects need also to be considered: it is not only the way the monomers are coupled that plays a part in the EPs, but also the number of monomer units – especially for oligomers. For example, in thiophene oligomers containing an even number of rings, strong emission is detectable in the region around the pure electronic transition. This is in strong contrast with thiophene-based oligomers containing an odd number of rings, where the electronic emission is suppressed as a result of packing that is different from that of the even-numbered analogue.[30]

Other approaches also exist – for example, the effect of the side-chain architecture on the optoelectronic properties, including emission, has been reported. To illustrate this, we use a recent example of a series of 2-dimensional polythiophene derivatives that comprise a terthiophene side-chain moiety attached to a polymer backbone (Figure 6.6b).[31] These conjugated terthiophene side-chains play a key part in dictating the molecular organization (including the side-chain orientation), which indirectly affects the resulting optoelectronic properties. Polymers consisting of conjugated terthiophene side-chains showed, for instance, not only a very broad absorption range in thin films (from 300 to 700 nm), which was significantly wider than the absorption of neat P3HT, but also featured excitation-independent EPs in solution. The excitation-independent emission is probably caused by energy transfer processes: the emission generated by the terthiophene moiety is absorbed by the polythiophene backbone and leads to another emission phenomenon. These phenomena can be reduced or limited when attaching the terthiophene side-chain such that they are positioned perpendicular to the polythiophene backbone. In this case, excitation-dependent PL properties are found: this means that, for this particular case, different emission peaks are observed when excited at different wavelengths. This suggests that structural inhomogeneities are present and/or or incomplete energy transfer occurs between the side-chains and polythiophene backbone.[31]

Side-chains can also be used to manipulate the molecular arrangement within the resulting thin film structures. Salammal *et al.*[32] have, for instance, shown that in the case of *rr*-P3AT, decreasing the alkyl side-chain length from *n*-octyl to *n*-pentyl can lead to a certain degree of texturing of thin films along the out-of-plane direction, to better π–π stacking and, as a consequence, to the evolution of well-defined vibronic side-bands in the PL spectra.

The complex and yet clear relations between chemical structure and optoelectronic properties can be further emphasized when comparing, for example, the significantly red-shifted FL properties of an insoluble pyrene-conjugated microporous polymer network with those of the corresponding soluble linear, branched polymers. The observed red shift in the network was attributed to the presence of strained closed rings.[33] Further examples relevant to this topic have been reported elsewhere.[34]

6.3 Approach 2: Physical and Physicochemical Methods

The physical and physicochemical methods that are widely used to alter the EPs of CPs often rely on classical polymer science approaches. These include: the selection of solvent quality[35] and polymer concentration;[36] using processes such as solvent vapour annealing,[37,38] the addition of non-solvents[39] and/or post-fabrication treatments such as annealing;[40] and/or exposing the materials to pressure[4,41] or mechanical stretching.[42] Other methods rely on either integrating conjugated molecules into three-dimensional silicon microstructures[43] or forming CP microstructures with a controlled orientation by using microphase separation,[44] ordering,[45] aggregation,[46] electrospinning[47] or chemical cross-linking,[48] to name but a few options. Many other solution-based processes exist.[5,49] Less conventional, yet efficient, methods to enhance the EPs of CPs include processing *via* dewetting[50] or processing by light,[51] or a combination of both approaches.[52] What all these strategies have in common is that they lead to changes in the polymer chain conformation, chain-packing and the overall microstructure and therefore to changes in their EPs.[2]

6.3.1 Solvent Vapour and Solvent Quality

Exposing thin polymer films to a solvent vapour or selecting a suitable solvent quality assists in establishing an approximate control over the resulting solid-state structure and molecular arrangement within CP films over all length scales. Using the latter method, Khan *et al.*[35] have, for instance, controlled the fraction of planarized β-phase within disordered glassy PFO films and have shown that a small fraction of chains adopting a β-phase conformation (a few vol%) generate most of the observed PL due to a higher delocalization of the excited states in these planarized chains compared with that in the disordered glassy chains. Another illustrative example of how the manipulation of chain conformation and short-range molecular packing *via* a solvent vapour annealing method can lead to changes in the EPs is given by single molecule spectroscopy data for MEH-PPV. Exposing thin polymer films to a solvent vapour results in the polymer chains unfolding from a coil-like conformation to more extended conformations, with strong impact on the resulting EPs. The chain-extended conformation has been shown to display pronounced switching between bright and dark states (*i.e.* FL blinking) of the emitting MEH-PPV on continuous photoexcitation.[37]

6.3.2 Pressure

Another efficient way to alter the PL properties of CPs is to vary the surrounding pressure. A higher pressure changes the conformation of PFO macromolecules, for example, and induces a red-shifted PL.[4] This shift has been attributed to a reduced inter-chain distance,[41] but may also partly originate from some backbone planarization. This observation is in agreement with the fact that oriented P3HT 'nanofibres' (whiskers), obtained by self-assembly in solution, were reported to feature a higher *intra*-chain order and structural coherence length, to which an observed red-shifted PL was attributed.[45]

Further evidence that the intra-chain order is important in determining the EPs of CPs has been provided by electro-spun 'nanofibres' obtained from MEH-PPV and PFO, which displayed a controllable polarized PL;[47,53] electrospinning P3HT led to a high intra-chain order and produced 'nanofibres' that exhibited high exciton delocalization and efficient PL quenching on microsecond timescales. This was attributed to the interaction of mobile triplets (from the non-geminate recombination of delocalized polaron species) with emissive singlet excitons.[54]

6.3.3 Dewetting

Another powerful method that leads to a drastic change in the PL in CPs such as MEH-PPV as a result of structural changes is the use of dewetting processes. Dewetting (shown schematically in Figure 6.7a) occurs, for example, when annealing MEH-PPV/polystyrene (PS) thin films above the glass transition temperature of PS. The dewetting process results in droplet-shaped regions of an ultrathin residual polymer monolayer that consists of stretched CP macromolecules (Figure 6.7b). More specifically, during dewetting the molecular conformation of the conjugated component is changed as a result of strong shearing forces. This process can be followed by probing the PL of the system. Before dewetting sets in, the annealing of these thin blend films decreases the PL intensity of the semiconducting component and induces a red shift that is possibly due to chain aggregation and an increase in the conjugation length as a result of aggregation. Once

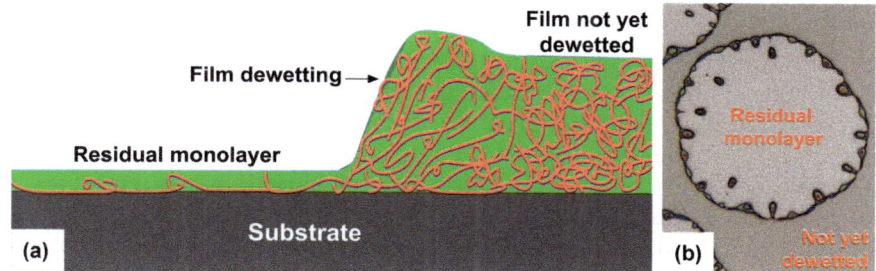

Figure 6.7 (a) Schematic representation of the dewetting phenomenon in thin blend films. (b) Optical micrograph showing the dewetting of a 45 ± 3 nm thick MEH-PPV:PS film ($15:85$) after 25 min. The films were dewetted at 120 °C in an inert nitrogen atmosphere after being deposited on a UV-cleaned silicon wafer that had been previously coated with an ultrathin layer of polydimethylsiloxane. The film was spin-cast from a 6 mg mL^{-1} solution of MEH-PPV/PS in a mixture of tetra-hydrofuran, cyclohexanone and toluene ($1:1:1$ v/v). Size of images is 134×175 μm^2.

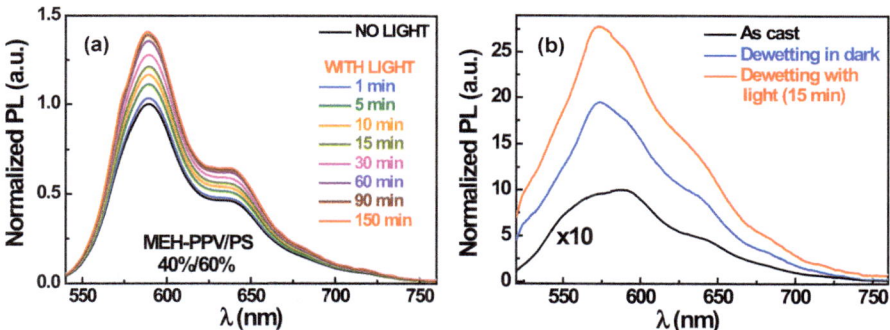

Figure 6.8 PL spectra of MEH-PPV/PS blend films recorded (a) during illumination and (b) inside a dewetting hole after 15 min of dewetting at 100 °C. Adapted with permission from ref. 51 (a) and from ref. 52 (b). Copyright (2014) American Chemical Society and (2014) Elsevier B.V.

dewetting sets in, another PL feature in the blue wavelength regime can be observed, the intensity of which is increased by $\sim$50-fold compared with the non-annealed films.[50]

Importantly, an increased PL intensity was also reported for MEH-PPV/PS blend films when they were exposed to light, indicating that a similar planarization of CP chains can also be induced by photoexcitation (Figure 6.8a).[51] These effects can be combined; recent experiments have shown that the simultaneous exposure of thin conjugated films to light during dewetting leads not only to a blue shift, but also to a further increase in their PL intensity as a result of both dewetting and illumination (Figure 6.8b).[52]

6.3.4 Chemical Cross-linking

Additional strategies that lead to structural changes and therefore changes in the EPs of CPs include chemical cross-linking processes. Using two different families of functionalized poly-(3-hexylthiophene)-*block*-poly(3-methanolthiophene) (P3HT-*b*-P3MT) and poly(3-hexylthiophene)-*block*-poly-(3-hexylthioacetate thiophene) (P3HT-*b*-P3ST) nanofibres as examples, it has been shown that different photophysical properties that depend on the choice of functionalizing moiety and cross-linking strategy can be induced (Figure 6.9).[48] Although the nanofibre families exhibited similar gross structures, the excitonic coupling for the P3ST diblock nanofibres was essentially unchanged during the cross-linking process. By contrast, the cross-linked P3MT nanofibres exhibited a PL similar to unaggregated P3HT molecules, suggesting that the cross-linking process in this case drastically decreased the excitonic coupling because it induced structural perturbations and changes in the chain conformation due to the cross-linking.[48]

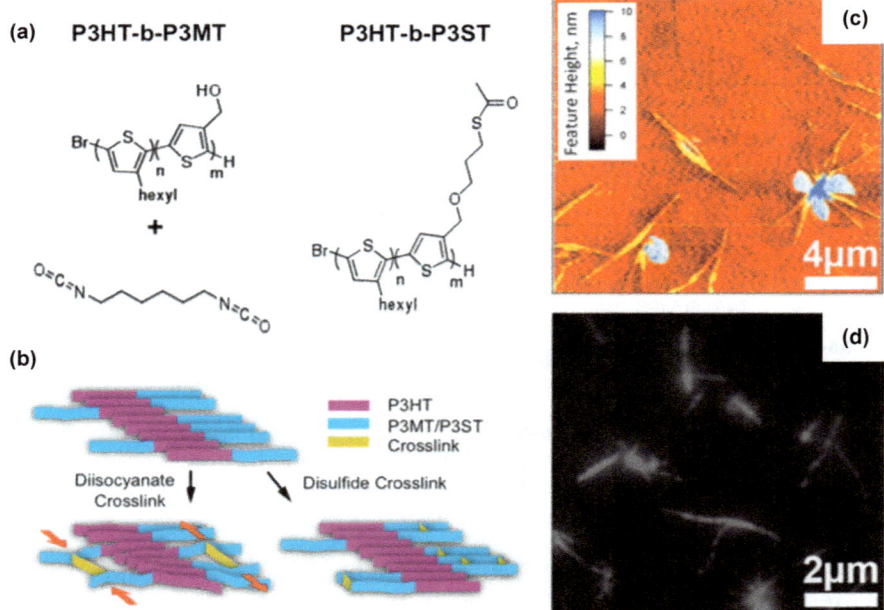

Figure 6.9   (a) Chemical structures of P3HT-*b*-P3MT/P3ST diblock copolymers and hexylmethylene diisocyanate. (b) Proposed molecular arrangement of pre- and post-cross-linked nanofibres. Purple bars indicate lamellar assembly of the P3HT blocks. The blue blocks (P3MT- and P3ST-blocks) are cross-linked (cross-linking bonds indicated in yellow). (c) Atomic force microscopy surface height image of P3MT-containing, cross-linked nanofibres that can display wire, ribbon and nanosheet 'clover leaf'-like structures. (d) PL image of dilute P3MT cross-linked nanofibres cast on glass.
Reproduced from ref. 48. Copyright (2013) American Chemical Society.

6.3.5 Controlled Aggregation

The aggregation of CPs is also known to induce changes in their chain conformation and, in turn, the molecular packing and their overall arrangement. As a consequence, measurable changes in the optoelectronic properties, including the EPs, are observed. Combining a variety of experimental techniques (including time-resolved FL spectroscopy) and theoretical analyses, Panzer *et al.*[46] investigated the aggregation of P3HT in tetrahydrofuran solutions on cooling from 300 to 5 K; they found spectroscopic signatures of two distinct H-aggregate species (characterized by their interchain coupling). More specifically, they identified the existence of two H-type aggregates at 5 K, both with planar polymer backbones, but distinguishable by the order of their side-chains, leading these H-type aggregates to possess clearly different absorption and emission signatures.[46]

6.3.6 Processing in Confined Spaces

Another way to control the molecular arrangement and hence EPs of CPs is through infiltration inside well-defined microstructures. An example of such an infiltration process can be seen in Figure 6.10. Polito *et al.*[43] have infiltrated poly(9,9-di-*n*-octylfluorene-*alt*-benzothiadiazole) (F8*a*BT) into three-dimensional silicon microstructures consisting of holes using deposition methods such as drop-casting, spin- and dip-coating and controlling the hole size, spatial periods and depth-to-width aspect ratios. For both holes with high and low depth-to-width aspect ratios, the FL microscopy data indicated clear confinement effects, which can be deduced to result from light emission in the out-of-plane direction, with each polymer-infiltrated hole acting as a single luminescent light source.[43] Similarly, the *in situ* incorporation of the highly hydrophobic yellow-emitting conjugated polymer poly[9,9-dioctylfluorene-*co*-benzothiadiazole] (F8*c*BT) within mesostructured silica previously synthesized from a tetrahydrofuran-based sol gel in the channels of an anodic alumina membrane using evaporation-induced self-assembly, led to a homogenous distribution of conjugated polymer molecules within the mesostructure. Interestingly, the shape of the PL spectrum

Empty
microstructures   Casting
                       polymer solution   Infiltrated
                                                        microstructures

Figure 6.10 Schematic illustration of the infiltration process of CPs into cylindrical microstructures, leading to more extended chain structures.

corresponding to the infiltrated F8cBT was unchanged when compared with the spectrum of the pristine F8cBT film. However, the PL spectrum corresponding to the infiltrated F8cBT was slightly red-shifted. This shift was attributed to a change in chain conformation, with more extended macromolecules being obtained during the confinement in the cylindrical mesostructured silica channels (leading to an increase in conjugation length) than in the pristine film.[55]

6.3.7 Stretchable Structures

Stretchable structures can be produced, most often from solution, to realize mechanically deformable polymer light-emitting diodes. These can exhibit tunable EPs, show good flexibility and emit light even when exposed to strains as large as 120%.[5] Devices with a luminance >100 cd m^{-2} have been reported.[49] Interestingly, both the EPs and device efficiencies can be manipulated by varying the applied strain, with small stretching ratios significantly enhancing the light-emitting efficiencies of the polymer light-emitting diode.[5] This is in agreement with other experimental observations. For example, when mechanical stretching was used to alter the EPs of MEH-PPV dispersed in an optically inert matrix, the individual molecular strands displayed a several fold enhancement in PL when fully stretched (*i.e.* when the mechanical strain of the matrix polymer was >550%).[42] Going from single-layer devices to triple-layer structures, a luminance as high as 16540 cd m^{-2} was obtained, which is a five-fold increase compared with single-layer devices.[56] Further reading on OLEDs, structuring, CPs and EPs can be found in Romero *et al.*[44]

6.4   Approach 3: Blending

The blending of two or more components can have a drastic effect on molecular packing and the overall microstructure as well as on the optoelectronic landscape. One of the most common examples in which the EPs of a CP are strongly altered by blending is in the case of bulk heterojunctions (an electron donor/electron acceptor blend) used in OPVs.[8] Here an electron-donating polymer is blended with electron acceptors such as fullerenes[57] and electron-accepting polymers,[58] or they can be covalently attached to fullerene moieties[59] or electron-accepting polymer blocks.[60] Scenarios where fullerenes are covalently attached to both ends of an *rr*-P3HT system have also been reported.[61] A strong quenching in emission was observed in all these examples. This is a signature of an efficient charge transfer from the electron-donating to the electron-accepting material. An example where an essentially complete PL quenching can be observed is in rod–coil block copolymers (BCPs), which form highly crystalline domains and make use of a polymer consisting of P3AT and a polyacrylate with perylene bisimide pendant groups.[60]

It often is important to gain a knowledge of the various phases that can be present in OPV blends. Based on a range of photophysical, electrochemical, physicochemical and structural experimental observations (Figure 6.11), Jamieson *et al.*[62] proposed a functional model for the required phase morphology of OPV blends that may have significant implications for the design of alternative acceptor materials to phenyl-C61-butyric acid methyl

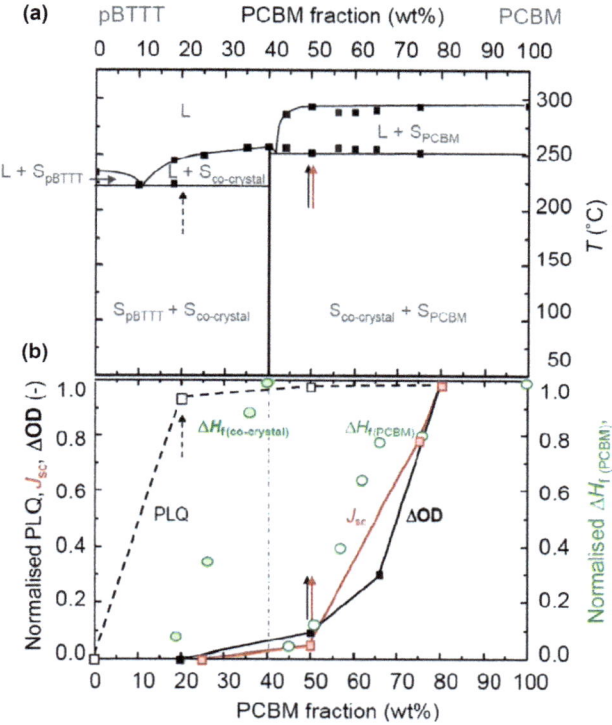

Figure 6.11 Physicochemical, structural and optoelectronic characteristics as a function of the pBTTT/PCBM blend composition. (a) Temperature/composition diagram of the pBTTT/PCBM system showing the formation of a molecular compound (co-crystal) at a PCBM content of ~40 wt%. As a consequence, the pBTTT/PCBM/co-crystal system features two eutectics at ~10 wt% PCBM (pBTTT/co-crystal binary) and 43 wt% PCBM (co-crystal/PCBM binary). Liquid lines were constructed with end-melting and end-dissolution temperatures of the neat components and excess component, respectively, and the peak eutectic temperatures. (b) Normalized PL quenching (PLQ), short circuit photocurrent (J_{SC} measured under simulated AM1.5 irradiation) and transient absorption data (ΔOD, measured at a time delay of 1 ms) plotted for pBTTT/PCBM films as a function of blend composition, overlain with the enthalpies of fusion (ΔH_f) of PCBM crystals and the pBTTT/PCBM co-crystals deduced from the respective first melting endotherms observed in differential scanning calorimetry.
Reproduced from ref. 62 with permission of The Royal Society of Chemistry.

ester (PCBM). The model describes how charge generation in a finely intermixed polymer/fullerene phase followed by the spatial separation of electrons and holes is dependent on the presence of the interface of a mixed phase with the aggregated PCBM domains. More specifically, the formation of such a relatively pure, molecularly ordered phase of PCBM is proposed as the key factor driving the spatial separation of photogenerated electrons and holes in many of these devices because the crystallization of PCBM results in an increase in its electron affinity, providing an energetic driving force for the spatial separation of electrons and holes.[62]

In the case where aggregation dominates blend formation, co-polymerization (especially the formation of block-copolymers) can assist in controlling the phase morphology of certain systems. One example is electron donor–acceptor rod–coil BCPs, synthesized by coupling a PPV block with a flexible coil block consisting of a PS backbone to which acceptor fullerene molecules are grafted.[63,64] This will probably hinder fullerene aggregation, but could increase the interaction between the donor and acceptor moieties, leading to changes in PL quenching. Such interactions may be enhanced by fusing the fullerene into the polymer backbone.[65] Note that in this context, the PL emission intensity of most bulk heterojunctions (using blends or BCPs) is modest. Exceptions include polymers based on 2,6-diethynyltriptycene and aromatic dibromides, which are fluorescent, although the chromophores are discrete and non-conjugated.[66] When blended with electron-accepting fullerenes, these polymers exhibit clear FL quenching (Figure 6.12), which suggests the formation of a polymer–fullerene complex, most probably through π–π interactions between the fullerene and electron donor triptycene.[66] Such quenching is also observable in fluorescent images of the polymer solution when adding fullerene (Figure 6.12, inset).

A blending approach can lead to changes in the inter-chain interactions and hence an alteration in the EPs of CPs. This approach is based on the so-called 'dilution' effect that occurs, for example, when blending conjugated MEH-PPV with non-conjugated PS, the latter system acting as an optically inert matrix (Figure 6.13a). Drastic changes can be achieved (Figure 6.13b). Using, for instance, MEH-PPV : PS (50 : 50) blends as the emitting layers in OLEDs led to a considerable increase in the EL efficiency of the device compared with the EL efficiency observed in OLEDs consisting of neat MEH-PPV.[67] As there is neither energy nor charge transfer occurring in the MEH-PPV : PS blends, the observed enhancement in EL efficiency was attributed to suppressed inter-chain interactions (between the MEH-PPV chains that were 'diluted' in the PS matrix) which determined the PL yield.[67]

Another interesting approach to manipulate the chain conformation of CPs is by dispersing them in a polar insulating macromolecular matrix. Examples of P3HT, poly[2,5-bis(3-tetradecylthiophen-2-yl)thieno[3,2-*b*]-thiophene] (pBTTT), poly[(9,9-dioctylfluorenyl-2,7-diyl)-*alt*-(4,7-bis(3-hexyl-thiophen-5-yl)-2,1,3-benzothiadiazole)-2′,2″-diyl], poly(2-methoxy-5-(3′,7′-dimethyloctyloxy)-1,4-phenylenevinylene or PFO introduced into a matrix

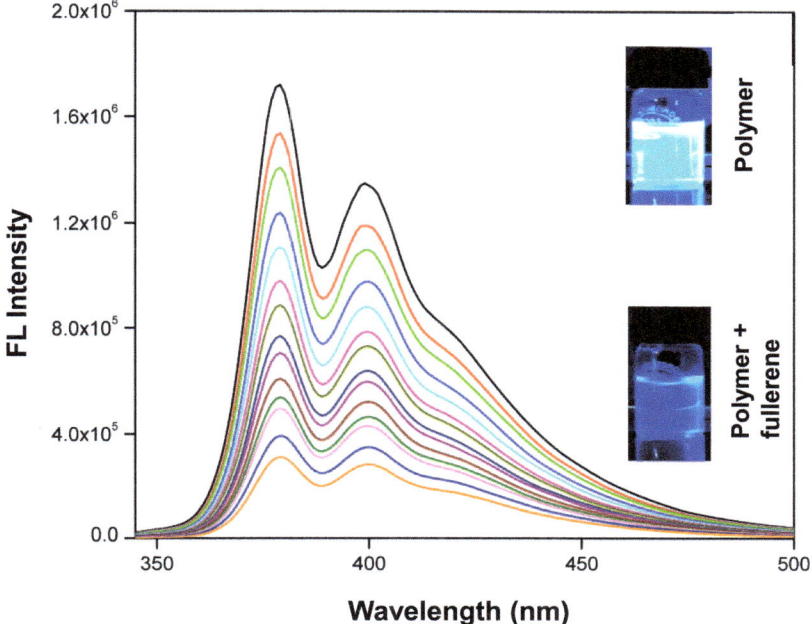

Figure 6.12 FL quenching of polymers based on 2,6-diethynyltriptycene in the presence of fullerenes. Inset: fluorescent images of these polymer solutions without (top) and with (bottom) fullerene under UV light. Adapted with permission from ref. 66. Copyright (2013) American Chemical Society.

such as poly(ethylene oxide) (PEO) have shown that the conjugated backbone of the semiconducting component is planarized on blending. This is deduced from the fact that the electronic coupling of the molecules of the modified CPs is changed, leading to different absorption and EPs in the blends compared with the neat materials[68] (Figure 6.14), similar to the observed transition in PPV-based materials during cooling.

The use of insulating matrices seems to be generally useful for the structural manipulation of CPs. Dispersing PFO in a Zeonex® matrix, for example, led to the majority of the PFO macromolecules adopting an almost ideal β-phase conformation in the single molecule regime. This provided a near-perfect π-electron system in which emission occurs anywhere along the polymer backbone, allowing the material to effectively behave as a molecular quantum wire.[1] Experimental work on similarly isolated β-phase PFO chains further emphasized the relation between changes in the chromophore arrangement (*i.e.* the arrangement of a conjugated region on the polymer backbone in which π-electrons are being delocalized), the conformation of the polymer backbone and the luminescent properties. In one specific case it was shown that bending of the π-system was correlated directly with an increased FL lifetime.[69]

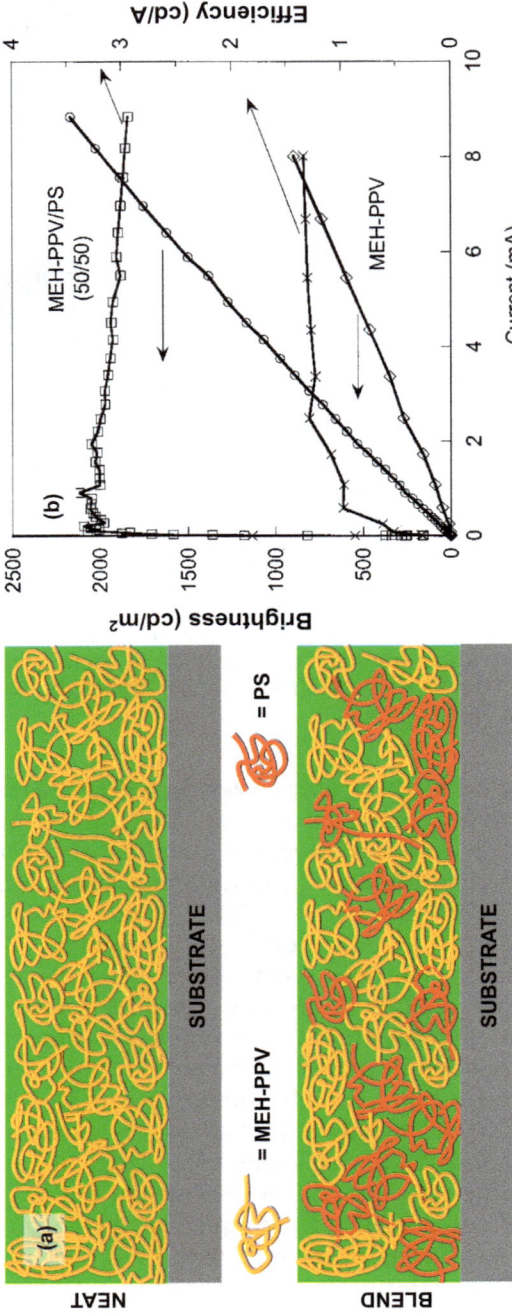

Figure 6.13   (a) Schematic illustration of a neat (top) and blended (bottom) thin film. (b) Brightness–current–efficiency curves of the MEH-PPV/PS (50 : 50) device and a neat MEH-PPV device. Figure (b) was reproduced from ref. 67. Copyright (2002) AIP Publishing LLC.

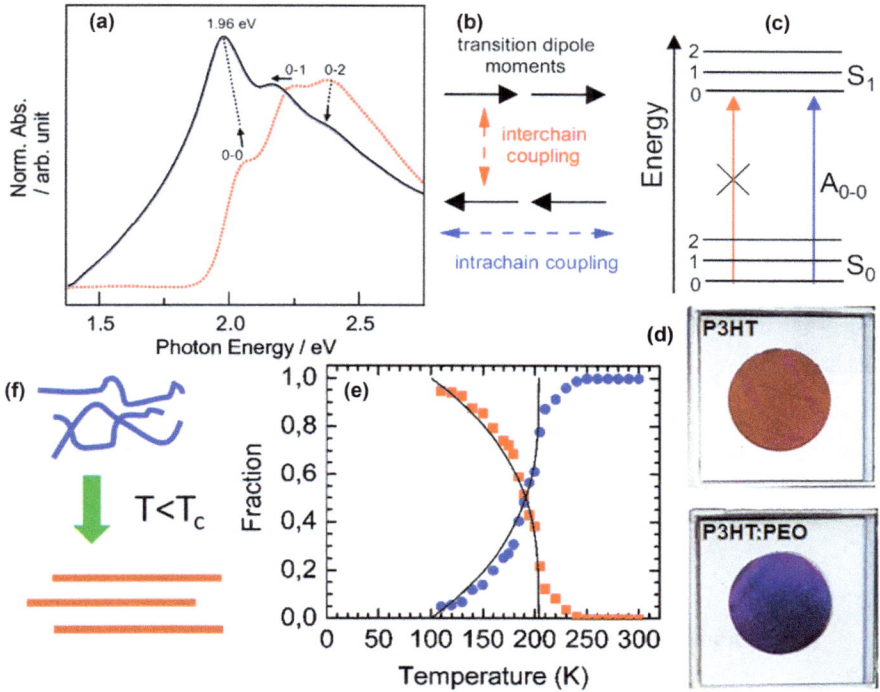

Figure 6.14 (a) Absorption spectra for neat P3HT (dotted) and P3HT:PEO blend (solid line) films and (b, c) how this affects (b) the inter- *versus* intra-chain coupling and (c) the type of aggregation from H-like to J-like. (d) Photographs of a neat P3HT film (top) and a P3HT:PEO blend (bottom). (e) Phase diagram of blue and red phase of MEH-PPV. (f) Schematic diagrams depicting the corresponding conformational arrangements for both the blue and red phase.

Adapted with permission from ref. 68 (a–d) and ref. 40 (e–f). Copyright (2013) WILEY-VCH Verlag GmbH & Co. KGaA, Weinheim and (2012) American Chemical Society.

Another way to manipulate the EPs of an electron donor polymer by structural means, in this case poly(*N*-vinylcarbazole), is to add nanostructured supramolecular acceptors with similar HOMO–LUMO band gaps but with different electron affinities, such as 1-cyano-*trans*-1,2-bis-(4′-methylbiphenyl)-ethylene) (**1**), 1-cyano-*trans*-1,2-bis-(3′,5′-bis-trifluoromethyl biphenyl)ethylene (**2**) and/or 3,3′-(1,4-phenylene)bis(2-(3,5-bis(trifluoromethyl)phenyl)acrylo nitrile) (**3**) (Figure 6.15a–c). As a result, highly fluorescent exciplexes (excited state charge transfer complexes) may form at the interfacial polymer–nanostructured acceptor region and emit blue, green and orange light. By incorporating mixed supramolecular acceptors, each providing independent exciplex emissions, into a poly(*N*-vinylcarbazole) film, a film emitting white light (Figure 6.15d) can be realized.[72]

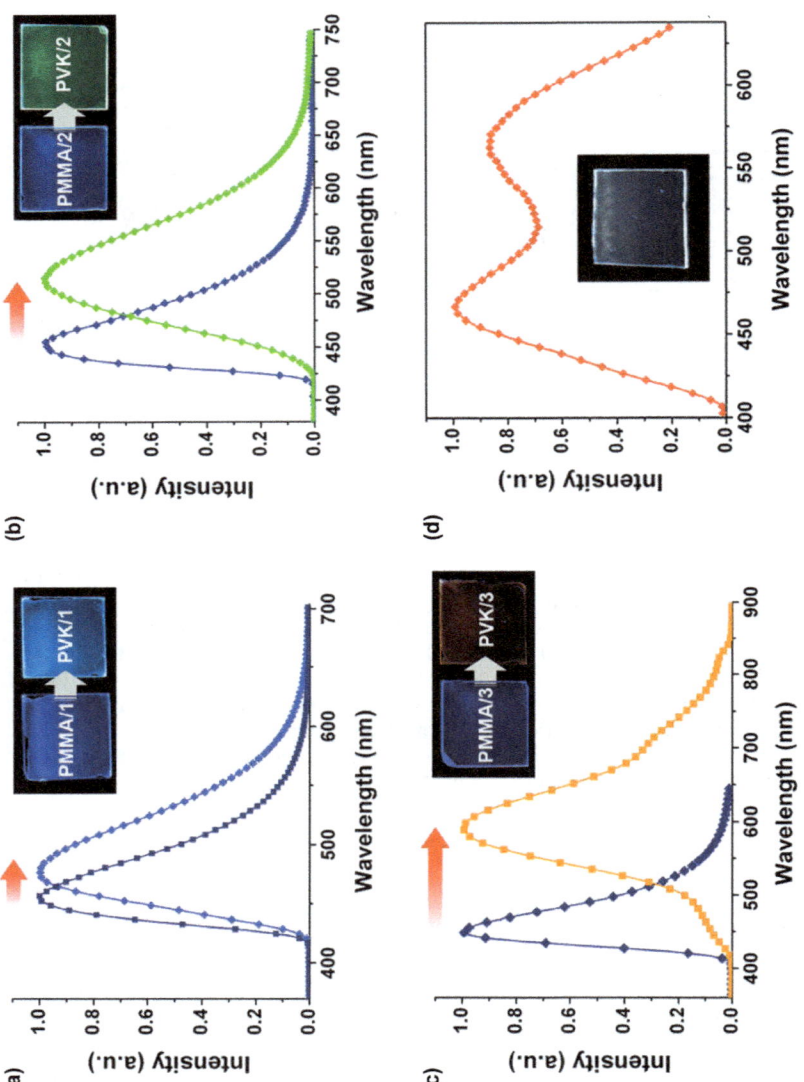

Figure 6.15 Normalized PL emission spectra of (a) poly(methylmethacrylate) PMMA/1 and poly(*N*-vinylcarbazole) PVK/1, (b) PMMA/2 and PVK/2 and (c) PMMA/3 and PVK/3 films. (d) Normalized PL spectrum of white fluorescent film prepared from a blend solution of PVK/1 and 3 (volume ratio of **1 : 3** is **2 : 1**). Insets show changes in emission wavelength of blend films as observed (a–c) by eye and (d) white FL under 365 nm UV light.
Reproduced from ref. 72. Copyright (2014) WILEY-VCH Verlag GmbH & Co. KGaA, Weinheim.

White-emitting films can also be prepared using nano-/microspheres and hybridization. A 2,6-bis(pyrazolyl)pyridine-*co*-octylated phenylethynyl high-molecular weight CP was used to produce thin films and self-assembled spheres, which, in their neat form, emitted a cyan colour under UV light. The emission of a nearly ideal white light was obtained through the further co-ordination of these films and spheres with red phosphor europium(III) metal ions.[73] Such 'hybrid' methods are indeed promising for white (and blue) electrophosphorescent light-emitting devices. Another example relies on designing and synthesizing a high triplet energy host polymer based on car-bazole and tetraphenylsilane units (PCztPSi).[74] The further introduction of bis[2-(4,6-difluorophenyl)pyridinato-N,C2']iridium(III) picolinate into the host polymer was shown to lead to a blue phosphorescent polymer (Figure 6.16a), whereas the further introduction of red emissive bis[2-phenylquinoline-N,C2']iridium(III) picolinate into this blue phosphorescent polymer led to a white phosphorescent polymer (Figure 6.16b). As a result of the efficient en-ergy transfer from the host polymer to the blue and red iridium(III) complexes, the fabricated phosphorescent devices exhibited promising efficiencies, in-cluding a maximum blue luminous efficiency of 3.57 cd A^{-1}.[74]

Combining various types of materials by producing inorganic–organic constructs can generally lead to systems with unique properties. Mixing, for example, PFO-based nanoparticles (NPs) with diameters ranging from 30 to 150 nm with graphene oxide (GO) nanosheets through an emulsification process can result in systems with a unique white PL and a characteristic green emissive band above 500 nm that is distinct from the PL behaviour of the neat PFO NPs.[70] The green band was suggested to be a result of the presence of a GO nanosheet shell surrounding the PFO NPs.[70] PL decay analysis (Figure 6.17a) showed that the PFO NPs wrapped in GO had a longer lumi-nescence lifetime than the neat PFO NPs.[70] Similar observations were made for the EL of other hybrid structures. For instance, bright, full-colour alternating current (AC) EL devices were fabricated from fluorescent polymers blended in solution with multiwalled carbon nanotubes (2 wt%) and self-assembled BCP micelles (20 wt%).[71] This procedure was also shown to lead to extremely bright AC-EL devices with long operating lifetimes (Figure 6.17b) that had lumi-nances of about 2300, 6000 and 5000 cd m^{-2} for blue, green and red emission, respectively (compared with only a few hundred cd m^{-2} for AC-EL devices processed from solutions containing only the polymer). AC-driven FL reson-ance energy transfer was also observed, which, along with the good carrier injection (favoured by the multiwalled carbon nanotubes) and the suppressed inter-chain non-radiative energy quenching (suppressed by the BCP micelles), was exploited in the fabrication of functional AC-EL devices with an improved performance compared with systems without the nanotubes.[71]

6.5   Approach 4: Metal-enhanced Fluorescence

Oriented interfaces between well-engineered polymer–plasmonic NPs have the potential to increase the EPs of the resulting system. It has been shown

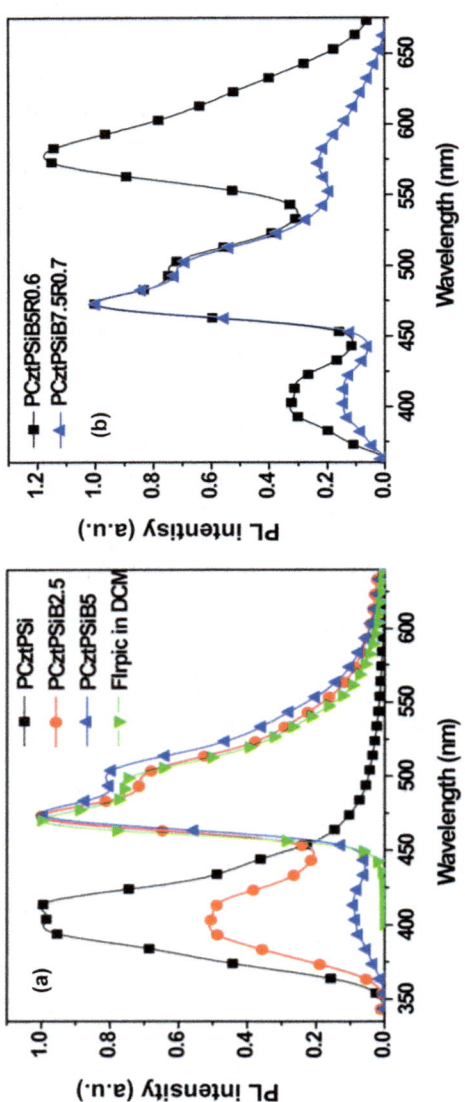

Figure 6.16 (a) Normalized PL spectra of host (PCztPSi) and blue-emitting (PCztPSiB2.5 and PCztPSiB5) polymers in a thin film state. (b) Normalized PL spectra of white-emitting (PCztPSiB5R0.6 and PCztPSiB7.5R0.7) polymers in the film state. Reproduced from ref. 74. Copyright (2014) American Chemical Society.

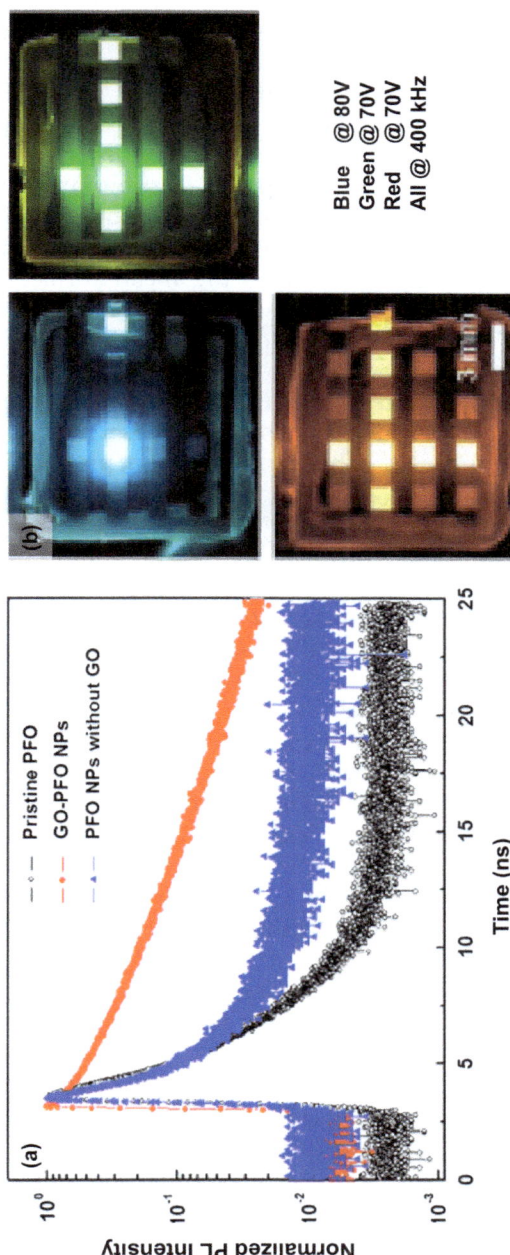

Figure 6.17 (a) PL decay of PFO, PFO NPs wrapped in GO, and PFO NP solutions without GO. (b) Photographs illustrating the blue, green and red emission observed in the AC-EL of the corresponding devices. Adapted with permission from ref. 70 (a) and ref. 71 (b). Copyright (2014, 2013) American Chemical Society.

that the presence of noble metal surfaces can significantly modify the way in which the incident photoexcitation is converted into FL by fluorophores,[75] directly affecting the radiative rates. The enhancement or quenching of FL is thereby determined by the balance between the fluorophore's excitation and the radiative and non-radiative decay rates, which are modified by interactions with the nanostructure of the noble metal.[76] This phenomenon is referred to as metal-enhanced FL (MEF). Metal surfaces in close vicinity to the fluorophore can respond to the oscillating fluorophore dipole and modify the rate of emission and the spatial distribution of the radiated energy. The electric field acting on the fluorophore depends on the interaction of the incident light with the metal surface as well as the interaction of the oscillating fluorophore dipole with the metal surface. It has been shown that such interactions can increase or decrease the field incident on the fluorophore and also increase or decrease the radiative decay rate.[76] Several plasmonic systems have been shown to be effective in enhancing the FL of CPs *via* MEF. These include colloidal NPs that enhance the PL intensity of CPs in solution,[77,78] hybrid nanocomposites that exhibit controllable MEF[79] and the emission of a CP enhanced by different plasmon modes when deposited on an Ag film.[80]

Tang *et al.*[78] produced fluorescent NPs by the self-assembly of cationic conjugated poly[9,90-bis(600-(N,N,N-trimethylammonium)-hexyl)fluorene-2,7-ylenevinylene-*co-alt*-1,4-phenylene dibromide] (PFV) onto core–shell silver–silica NPs (Ag@SiO$_2$). Nanometre-sized Ag NPs were used because they have an intense absorption band in the visible region as a result of surface plasmon resonance; the silica spacer shell was been chosen to increase the inter-particle distance of the Ag/CPs. It was observed that the FL intensity of PFV after assemble on the Ag@SiO2 core–shell NPs was enhanced *via* MEF (by the Ag@SiO$_2$ nanostructure) by 1.3 times compared with the FL intensity of PFV assembled on silica NPs without Ag cores. When replacing the silica shell with a poly(N-isopropylacrylamide) (PNIPAM) shell, Ag@PNIPAM NPs were obtained (Figure 6.18a). The FL intensity of PFV on the Ag@PNIPAM NPs at 50 °C was enhanced *via* MEF by 3.1-fold (Figure 6.18b) compared with the FL intensity of the PFV solution without Ag@PNIPAM NPs.[77]

An interesting way to enhance the FL of conjugated poly[9,99-bis(60-(N,N,N-trimethylammonium)-hexyl)fluorene-*co-alt*-2,5-dimethoxy-1,4-phenylene dibromide] (PFPMO) using MEF is to fabricate a hybrid Ag nanocomposite film. Wang *et al.*[79] loaded Ag NPs onto an agarose matrix structure to form an agarose film containing metallic Ag (Ag@agarose), which was then covered by layers of cationic polyethyleneimine (PEI), anionic sodium alginate and cationic chitosan. A PFPMO layer was drop-cast on top of this assembly. At a specific content of Ag NPs in the Ag@agarose films and an optimum thicknesses of the cationic chitosan/anionic sodium alginate interlayers between the Ag@agarose film and the fluorescent PFPMO, a maximum 8.5-fold increase in the FL of PFPMO was measured (Figure 6.18c). FL photographs of PFPMO with and without the Ag@agarose film illustrate the increase in FL intensity when the Ag@agarose film was used (Figure 6.18c, inset).

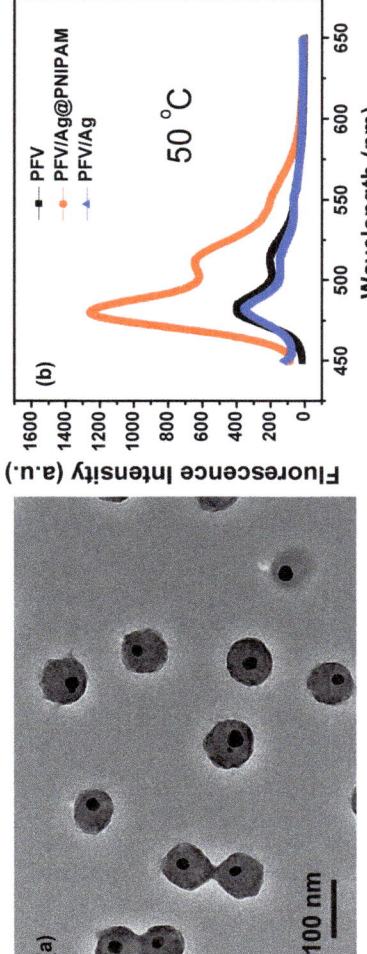

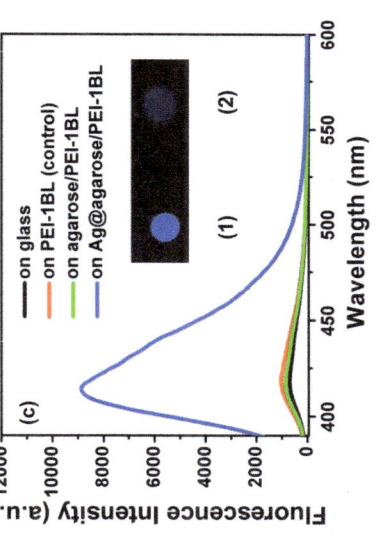

Figure 6.18 (a) Transmission electron micrographs of Ag@PNIPAM NPs. FL spectra of (b) a PFV/Ag@PNIPAM dispersion, a PFV solution and a PFV/Ag dispersion at 50 °C and (c) PFPMO on different substrates (c). Inset in (c) shows FL photographs of PFPMO on Ag@agarose/PEI-1BL (1) and on PEI-1BL (2) following irradiation at 365 nm. PEI-1BL describes a polyethyleneimine layer covered with an interlayer of anionic sodium alginate/cationic chitosan. Adapted from ref. 77 (a–b) with permission of The Royal Society of Chemistry and from ref. 79 (c) by permission from Macmillan Publishers Ltd: [Scientific Reports], Copyright (2014).

Another interesting conjugated system that exhibited enhanced EPs when using metal (surface plasmon)-enhanced emission is the conjugated poly[2,5-bis(3′,7′-dimethyloctyloxy)-1,4-phenylenevinylene] (BDMO-PPV). By comparing the BDMO-PPV deposited on a segregated island-like nano-structured Ag film and on a smooth and continuous Ag film, it was shown that the PL was enhanced in both instances. The enhancement in PL was mainly a result of the coupling of the BDMO-PPV excitons with localized surface plasmons of the Ag NPs in the case of the island-like Ag film, whereas the surface plasmon polaritons dominated the PL enhancement when a continuous Ag film was used.[80]

6.6 Conclusions

The EPs of CPs are currently used in the fabrication of a variety of techno-logical devices that significantly improve our day-to-day life while func-tioning at sufficiently high efficiencies. Promisingly, such devices have the potential to be continuously improved and their efficiency can be further increased using a variety of synthesis and processing methods to alter and control their molecular arrangement and packing over many length scales. This will allow the manipulation of their EPs. The development of plasmonic methods based on MEF is expanding the possibilities for the enhancement of the EPs of CPs and is broadening their applicability to novel devices still to be imagined and designed.

References

1. E. D. Como, N. J. Borys, P. Strohriegl, M. J. Walter and J. M. Lupton, Formation of a defect-free p-electron system in single b-phase poly-fluorene chains, *J. Am. Chem. Soc.*, 2011, **133**, 3690–3692.
2. I. Botiz and N. Stingelin, Influence of Molecular Conformations and Microstructure on the Optoelectronic Properties of Conjugated Poly-mers, *Materials*, 2014, **7**, 2273–2300.
3. J. H. Burroughes, D. D. C. Bradley, A. R. Brown, R. N. Marks, K. Mackay, R. H. Friend, P. L. Burns and A. B. Holmes, Light-emitting diodes based on conjugated polymers, *Nature*, 1990, **347**, 539–541.
4. S. Guha, M. Chandrasekhar, U. Scherf and M. Knaapila, Tuning struc-tural and optical properties of blue-emitting polymeric semiconductors, *Phys. Status Solidi B*, 2011, **248**, 1083–1090.
5. J. Liang, L. Li, X. Niu, Z. Yu and Q. Pei, Elastomeric polymer light-emitting devices and displays, *Nat. Photonics*, 2013, **7**, 817–824.
6. Y. Sungryul, P. Suntak, P. Bongjae, P. Seung Koo, H. Prahlad, P. von Guggenberg and K. Ki-Uk, Polymer-Based Flexible Visuo-Haptic Display, *IEEE/ASME Trans Mechatronics*, 2014, **19**, 1463–1469.
7. W. S. C. Roelofs, W. H. Adriaans, R. A. J. Janssen, M. Kemerink and D. M. de Leeuw, Light Emission in the Unipolar Regime of Ambipolar Organic Field-Effect Transistors, *Adv. Funct. Mater.*, 2013, **23**, 4133–4139.

8. W. Li, K. H. Hendriks, A. Furlan, W. Roelofs, S. C. Meskers, M. M. Wienk and R. A. Janssen, Effect of the Fibrillar Microstructure on the Efficiency of High Molecular Weight Diketopyrrolopyrrole-Based Polymer Solar Cells, *Adv. Mater.*, 2014, **26**, 1565–1570.

9. I. Botiz and S. B. Darling, Self-assembly of poly(3-hexylthiophene)-block-polylactide rod-coil block copolymer and subsequent incorporation of electron acceptor material, *Macromolecules*, 2009, **42**, 8211–8217.

10. J. W. Lichtman and J.-A. Conchello, Fluorescence microscopy, *Nat. Methods*, 2005, **2**, 910–919.

11. D. Braun and A. J. Heeger, Visible light emission from semiconducting polymer diodes, *Appl. Phys. Lett.*, 1991, **58**, 1982–1984.

12. O. Yutaka, U. Masao, M. Keiro and Y. Katsumi, Effects of alkyl chain length and carrier confinement layer on characteristics of poly(3-alkylthiophene) electroluminescent diodes, *Solid State Commun.*, 1991, **80**, 605–608.

13. G. Grem, G. Leditzky, B. Ullrich and G. Leising, Realization of a blue-light-emitting device using poly(p-phenylene), *Adv. Mater.*, 1992, **4**, 36–37.

14. D. Neher, Polyfluorene Homopolymers: Conjugated Liquid-Crystalline Polymers for Bright Blue Emission and Polarized Electroluminescence, *Macromol. Rapid Commun.*, 2001, **22**, 1365–1385.

15. C. L. Chochos, J. K. Kallitsis and V. G. Gregoriou, Rod–Coil Block Copolymers Incorporating Terfluorene Segments for Stable Blue Light Emission, *J. Phys. Chem. B*, 2005, **109**, 8755–8760.

16. J. A. Schneider, A. Dadvand, W. Wen and D. F. Perepichka, Tuning the Electronic Properties of Poly (thienothiophene vinylene) s via Alkylsulfanyl and Alkylsulfonyl Substituents, *Macromolecules*, 2013, **46**, 9231–9239.

17. T. Matsumoto, K. Tanaka and Y. Chujo, Synthesis and Characterization of Gallafluorene-Containing Conjugated Polymers: Control of Emission Colors and Electronic Effects of Gallafluorene Units on π-Conjugation System, *Macromolecules*, 2015, **48**, 1343–1351.

18. L. Yu, J. Liu, S. Hu, R. He, W. Yang, H. Wu, J. Peng, R. Xia and D. D. Bradley, Red, Green, and Blue Light-Emitting Polyfluorenes Containing a Dibenzothiophene-S,S-Dioxide Unit and Efficient High-Color-Rendering-Index White-Light-Emitting Diodes Made Therefrom, *Adv. Funct. Mater.*, 2013, **23**, 4366–4376.

19. I. Osken, A. S. Gundogan, E. Tekin, M. S. Eroglu and T. Ozturk, Fluorene–dithienothiophene-S, S-dioxide copolymers. Fine-tuning for OLED applications, *Macromolecules*, 2013, **46**, 9202–9210.

20. X. Wang, L. Zhao, S. Shao, J. Ding, L. Wang, X. Jing and F. Wang, Poly(spirobifluorene)s Containing Nonconjugated Diphenylsulfone Moiety: Toward Blue Emission Through a Weak Charge Transfer Effect, *Macromolecules*, 2014, **47**, 2907–2914.

21. M. Tammer, L. Horsburgh, A. P. Monkman, W. Brown and H. D. Burrows, Effect of Chain Rigidity and Effective Conjugation Length

on the Structural and Photophysical Properties of Pyridine-Based Luminescent Polymers, *Adv. Funct. Mater.*, 2002, **12**, 447–454.

22. H. Kobayashi, K. Tsuchiya, K. Ogino and M. Vacha, Spectral multitude and spectral dynamics reflect changing conjugation length in single molecules of oligophenylenevinylenes, *Phys. Chem. Chem. Phys.*, 2012, **14**, 10114–10118.

23. S. Rathgeber, D. Bastos de Toledo, E. Birckner, H. Hoppe and D. A. Egbe, Intercorrelation between structural ordering and emission properties in photoconducting polymers, *Macromolecules*, 2009, **43**, 306–315.

24. T. Qin and A. Troisi, Relation between structure and electronic properties of amorphous MEH-PPV polymers, *J. Am. Chem. Soc.*, 2013, **135**, 11247–11256.

25. C. Resta, S. Di Pietro, M. Majerić Elenkov, Z. Hameršak, G. Pescitelli and L. Di Bari, Consequences of Chirality on the Aggregation Behavior of Poly [2-methoxy-5-(2'-ethylhexyloxy)-p-phenylenevinylene](MEH-PPV), *Macromolecules*, 2014, **47**, 4847–4850.

26. E. Peeters, M. P. T. Christiaans, R. A. J. Janssen, H. F. M. Schoo, H. P. J. M. Dekkers and E. W. Meijer, Circularly Polarized Electroluminescence from a Polymer Light-Emitting Diode, *J. Am. Chem. Soc.*, 1997, **119**, 9909–9910.

27. T. Adachi, J. Brazard, R. J. Ono, B. Hanson, M. C. Traub, Z.-Q. Wu, Z. Li, J. C. Bolinger, V. Ganesan and C. W. Bielawski, Regioregularity and single polythiophene chain conformation, *J. Phys. Chem. Lett.*, 2011, **2**, 1400–1404.

28. K. Rahimi, I. Botiz, N. Stingelin, N. Kayunkid, M. Sommer, F. P. V. Koch, H. Nguyen, O. Coulembier, P. Dubois, M. Brinkmann and G. Reiter, Controllable Processes for Generating Large Single Crystals of Poly(3-hexylthiophene), *Angew. Chem., Int. Ed.*, 2012, **51**, 11131–11135.

29. F. Guo, X. Yin, F. Pammer, F. Cheng, D. Fernandez, R. A. Lalancette and F. Jäkle, Regioregular Organoborane-Functionalized Poly(3-alkynylthiophene)s, *Macromolecules*, 2014, **47**, 7831–7841.

30. F. Meinardi, M. Cerminara, A. Sassella, A. Borghesi, P. Spearman, G. Bongiovanni, A. Mura and R. Tubino, Intrinsic Excitonic Luminescence in Odd and Even Numbered Oligothiophenes, *Phys. Rev. Lett.*, 2002, **89**, 157403.

31. C.-Y. Kuo, Y.-C. Huang, C.-Y. Hsiow, Y.-W. Yang, C.-I. Huang, S.-P. Rwei, H.-L. Wang and L. Wang, Effect of side-chain architecture on the optical and crystalline properties of two-dimensional polythiophenes, *Macromolecules*, 2013, **46**, 5985–5997.

32. S. T. Salammal, S. Dai, U. Pietsch, S. Grigorian, N. Koenen, U. Scherf, N. Kayunkid and M. Brinkmann, Influence of Alkyl Side Chain length on the In-plane Stacking of Room Temperature and Low Temperature Cast Poly(3-alkylthiophene) Thin Films, *Eur. Polym. J.*, 2015, **67**, 199–212.

33. M. A. Zwijnenburg, G. Cheng, T. O. McDonald, K. E. Jelfs, J.-X. Jiang, S. Ren, T. Hasell, F. Blanc, A. I. Cooper and D. J. Adams, Shedding Light on Structure–Property Relationships for Conjugated Microporous

Polymers: The Importance of Rings and Strain, *Macromolecules*, 2013, **46**, 7696–7704.

34. I. Botiz and S. B. Darling, Optoelectronics Using Block Copolymers, *Mater. Today*, 2010, **13**, 42–51.

35. A. L. T. Khan, P. Sreearunothai, L. M. Herz, M. J. Banach and A. Köhler, Morphology-dependent energy transfer within polyfluorene thin films, *Phys. Rev. B: Condens. Matter Mater. Phys.*, 2004, **69**, 085201.

36. C. D. Danesh, N. S. Starkweather and S. Zhang, In Situ Study of Dynamic Conformational Transitions of a Water-Soluble Poly (3-hexylthiophene) Derivative by Surfactant Complexation, *J. Phys. Chem. B*, 2012, **116**, 12887–12894.

37. J. Vogelsang, T. Adachi, J. Brazard, D. A. Vanden Bout and P. F. Barbara, Self-assembly of highly ordered conjugated polymer aggregates with long-range energy transfer, *Nat. Mater.*, 2011, **10**, 942–946.

38. Z. Hu, T. Adachi, R. Haws, B. Shuang, R. J. Ono, C. W. Bielawski, C. F. Landes, P. J. Rossky and D. A. Vanden Bout, Excitonic Energy Migration in Conjugated Polymers: The Critical Role of Interchain Morphology, *J. Am. Chem. Soc.*, 2014, **136**, 16023–16031.

39. T. Adachi, L. Tong, J. Kuwabara, T. Kanbara, A. Saeki, S. Seki and Y. Yamamoto, Spherical Assemblies from π-Conjugated Alternating Copolymers: Toward Optoelectronic Colloidal Crystals, *J. Am. Chem. Soc.*, 2013, **135**, 870–876.

40. A. Köhler, S. T. Hoffmann and H. Bässler, An Order-Disorder Transition in the Conjugated Polymer MEH-PPV, *J. Am. Chem. Soc.*, 2012, **134**, 11594–11601.

41. Y.-S. Huang, J. Gierschner, J. P. Schmidtke, R. H. Friend and D. Beljonne, Tuning interchain and intrachain interactions in polyfluorene copolymers, *Phys. Rev. B: Condens. Matter Mater. Phys.*, 2011, **84**, 205311.

42. K.-P. Tung, C.-C. Chen, P. Lee, Y.-W. Liu, T.-M. Hong, K. C. Hwang, J. H. Hsu, J. D. White and A. C.-M. Yang, Large Enhancements in Optoelectronic Efficiencies of Nano-plastically Stressed Conjugated Polymer Strands, *ACS Nano*, 2011, **5**, 7296–7302.

43. G. Polito, S. Surdo, V. Robbiano, G. Tregnago, F. Cacialli and G. Barillaro, Two-Dimensional Array of Photoluminescent Light Sources by Selective Integration of Conjugated Luminescent Polymers into Three-Dimensional Silicon Microstructures, *Adv. Opt. Mater.*, 2013, **1**, 894–898.

44. D. B. Romero, M. Schaer, L. Zuppiroli, B. Cesar, G. Widawski and B. Francois, Light-emitting diodes based on copolymer organic semiconductors, *Opt. Eng.*, 1995, **34**, 1987–1992.

45. M. Baghgar, J. Labastide, F. Bokel, I. Dujovne, A. McKenna, A. M. Barnes, E. Pentzer, T. Emrick, R. Hayward and M. D. Barnes, Probing Inter- and Intrachain Exciton Coupling in Isolated Poly(3-hexylthiophene) Nanofibers: Effect of Solvation and Regioregularity, *J. Phys. Chem. Lett.*, 2012, **3**, 1674–1679.

46. F. Panzer, M. Sommer, H. Bässler, M. Thelakkat and A. Köhler, Spectroscopic Signature of Two Distinct H-Aggregate Species in Poly(3-hexylthiophene), *Macromolecules*, 2015, **48**, 1543–1553.
47. A. Camposeo, I. Greenfeld, F. Tantussi, M. Moffa, F. Fuso, M. Allegrini, E. Zussman and D. Pisignano, Conformational Evolution of Elongated Polymer Solutions Tailors the Polarization of Light-Emission from Organic Nanofibers, *Macromolecules*, 2014, **47**, 4704–4710.
48. M. Baghgar, E. Pentzer, A. J. Wise, J. A. Labastide, T. Emrick and M. D. Barnes, Cross-linked functionalized poly (3-hexylthiophene) nano-fibers with tunable excitonic coupling, *ACS Nano*, 2013, **7**, 8917–8923.
49. M. S. White, M. Kaltenbrunner, E. D. Głowacki, K. Gutnichenko, G. Kettlgruber, I. Graz, S. Aazou, C. Ulbricht, D. A. Egbe, M. C. Miron, Z. Major, M. C. Scharber, T. Sekitani, T. Someya, S. Bauer and N. S. Sariciftci, Ultrathin, highly flexible and stretchable PLEDs, *Nat. Photonics*, 2013, **7**, 811–816.
50. P. W. Lee, W.-C. Li, B.-J. Chen, C.-W. Yang, C.-C. Chang, I. Botiz, G. Reiter, Y. T. Chen, T. L. Lin, J. Tang, J.-H. Jou and A. C.-M. Yang, Massive Enhancement of Photoluminescence through Nanofilm Dewetting, *ACS Nano*, 2013, **7**, 6658–6666.
51. I. Botiz, P. Freyberg, C. Leordean, A.-M. Gabudean, S. Astilean, A. C.-M. Yang and N. Stingelin, Enhancing the Photoluminescence Emission of Conjugated MEH-PPV by Light Processing, *ACS Appl. Mater. Interfaces*, 2014, **6**, 4974–4979.
52. I. Botiz, P. Freyberg, C. Leordean, A.-M. Gabudean, S. Astilean, A. C.-M. Yang and N. Stingelin, Emission properties of MEH-PPV in thin films simultaneously illuminated and annealed at different temperatures, *Synth. Met.*, 2015, **199**, 33–36.
53. V. Fasano, A. Polini, G. Morello, M. Moffa, A. Camposeo and D. Pisignano, Bright Light Emission and Waveguiding in Conjugated Polymer Nanofibers Electrospun from Organic Salt Added Solutions, *Macromolecules*, 2013, **46**, 5935–5942.
54. A. K. Thomas, J. A. Garcia, J. Ulibarri-Sanchez, J. Gao and J. K. Grey, High Intrachain Order Promotes Triplet Formation from Recombination of Long-Lived Polarons in Poly (3-hexylthiophene) J-Aggregate Nanofibers, *ACS Nano*, 2014, **8**, 10559–10568.
55. A. Keller, S. Kirmayer, T. Segal-Peretz and G. L. Frey, Mesostructured Silica Containing Conjugated Polymers Formed within the Channels of Anodic Alumina Membranes from Tetrahydrofuran-Based Solution, *Langmuir*, 2012, **28**, 1506–1514.
56. R. Trattnig, L. Pevzner, M. Jäger, R. Schlesinger, M. V. Nardi, G. Ligorio, C. Christodoulou, N. Koch, M. Baumgarten, K. Müllen and J. W. E. List, Bright Blue Solution Processed Triple-Layer Polymer Light-Emitting Diodes Realized by Thermal Layer Stabilization and Orthogonal Solvents, *Adv. Funct. Mater.*, 2013, **23**, 4897–4905.
57. M. Scarongella, J. De Jonghe-Risse, E. Buchaca-Domingo, M. Causa', Z. Fei, M. Heeney, J.-E. Moser, N. Stingelin and N. Banerji, A Close Look

at Charge Generation in Polymer:Fullerene Blends with Microstructure Control, *J. Am. Chem. Soc.*, 2015, **137**, 2908–2918.

58. C. Mu, P. Liu, W. Ma, K. Jiang, J. Zhao, K. Zhang, Z. Chen, Z. Wei, Y. Yi, J. Wang, S. Yang, F. Huang, A. Facchetti, H. Ade and H. Yan, High-Efficiency All-Polymer Solar Cells Based on a Pair of Crystalline Low-Bandgap Polymers, *Adv. Mater.*, 2014, **26**, 7224–7230.

59. A. Cravino and N. S. Sariciftci, Organic electronics: Molecules as bipolar conductors, *Nat. Mater.*, 2003, **2**, 360–361.

60. M. Sommer, A. S. Lang and M. Thelakkat, Crystalline-crystalline donor-acceptor block copolymers, *Angew. Chem., Int. Ed.*, 2008, **47**, 7901–7904.

61. A. T. Healy, B. W. Boudouris, C. D. Frisbie, M. A. Hillmyer and D. A. Blank, Intramolecular Exciton Diffusion in Poly(3-hexylthiophene), *J. Phys. Chem. Lett.*, 2013, **4**, 3445–3449.

62. F. C. Jamieson, E. B. Domingo, T. McCarthy-Ward, M. Heeney, N. Stingelin and J. R. Durrant, Fullerene crystallisation as a key driver of charge separation in polymer/fullerene bulk heterojunction solar cells, *Chem. Sci.*, 2012, **3**, 485–492.

63. S. Barrau, T. Heiser, F. Richard, C. Brochon, C. Ngov, K. van de Wetering, G. Hadziioannou, D. V. Anokhin and D. A. Ivanov, Self-assembling of novel fullerene-grafted donor-acceptor rod-coil block copolymers, *Macromolecules*, 2008, **41**, 2701–2710.

64. U. Stalmach, B. de Boer, C. Videlot, P. F. van Hutten and G. Hadziioannou, Semiconducting diblock copolymers synthesized by means of controlled radical polymerization techniques, *J. Am. Chem. Soc.*, 2000, **122**, 5464–5472.

65. R. C. Hiorns, P. Iratcabal, D. Begue, A. Khoukh, R. de Bettignies, J. Leroy, M. Firon, C. Sentein, H. Martinez, H. Preud'homme and C. Dagron-Lartigau, Alternatively linking fullerene and conjugated polymers, *J. Polym. Sci., Part A: Polym. Chem.*, 2009, **47**, 2304–2317.

66. S. Mondal, S. Chakraborty, S. Bhowmick and N. Das, Synthesis of Triptycene-Based Organosoluble, Thermally Stable, and Fluorescent Polymers: Efficient Host–Guest Complexation with Fullerene, *Macromolecules*, 2013, **46**, 6824–6831.

67. G. He, Y. Li, J. Liu and Y. Yang, Enhanced electroluminescence using polystyrene as a matrix, *Appl. Phys. Lett.*, 2002, **80**, 4247–4249.

68. C. Hellmann, F. Paquin, N. D. Treat, A. Bruno, L. X. Reynolds, S. A. Haque, P. N. Stavrinou, C. Silva and N. Stingelin, Controlling the interaction of light with polymer semiconductors, *Adv. Mater.*, 2013, **25**, 4906–4911.

69. T. Adachi, J. Vogelsang and J. M. Lupton, Chromophore Bending Controls Fluorescence Lifetime in Single Conjugated Polymer Chains, *J. Phys. Chem. Lett.*, 2014, **5**, 2165–2170.

70. D. Y. Yoo, N. D. K. Tu, S. J. Lee, E. Lee, S.-R. Jeon, S. Hwang, H. S. Lim, J. K. Kim, B. K. Ju, H. Kim and J. A. Lim, Graphene Oxide Nanosheet Wrapped White-Emissive Conjugated Polymer Nanoparticles, *ACS Nano*, 2014, **8**, 4248–4256.

71. S. H. Cho, S. S. Jo, I. Hwang, J. Sung, J. Seo, S.-H. Jung, I. Bae, J. R. Choi, H. Cho, T. Lee, J. K. Lee, T.-W. Lee and C. Park, Extremely Bright Full Color Alternating Current Electroluminescence of Solution-Blended Fluorescent Polymers with Self-Assembled Block Copolymer Micelles, *ACS Nano*, 2013, **7**, 10809–10817.
72. J. H. Kim, B. K. An, S. J. Yoon, S. K. Park, J. E. Kwon, C. K. Lim and S. Y. Park, Highly Fluorescent and Color-Tunable Exciplex Emission from Poly (N-vinylcarbazole) Film Containing Nanostructured Supramolecular Acceptors, *Adv. Funct. Mater.*, 2014, **24**, 2746–2753.
73. Y. S. Narayana, S. Basak, M. Baumgarten, K. Müllen and R. Chandrasekar, White-Emitting Conjugated Polymer/Inorganic Hybrid Spheres: Phenylethynyl and 2, 6-Bis(pyrazolyl)pyridine Copolymer Coordinated to Eu(tta)3, *Adv. Funct. Mater.*, 2013, **23**, 5875–5880.
74. F. Xu, J.-H. Kim, H. U. Kim, J.-H. Jang, K. S. Yook, J. Y. Lee and D.-H. Hwang, Synthesis of High-Triplet-Energy Host Polymer for Blue and White Electrophosphorescent Light-Emitting Diodes, *Macromolecules*, 2014, **47**, 7397–7406.
75. J. Seelig, K. Leslie, A. Renn, S. Kuhn, V. Jacobsen, M. van de Corput, C. Wyman and V. Sandoghdar, Nanoparticle-Induced Fluorescence Lifetime Modification as Nanoscopic Ruler: Demonstration at the Single Molecule Level, *Nano Lett.*, 2007, **7**, 685–689.
76. R. M. Bakker, H.-K. Yuan, Z. Liu, V. P. Drachev, A. V. Kildishev, V. M. Shalaev, R. H. Pedersen, S. Gresillon and A. Boltasseva, Enhanced localized fluorescence in plasmonic nanoantennae, *Appl. Phys. Lett.*, 2008, **92**, 043101.
77. F. Tang, N. Ma, X. Wang, F. He and L. Li, Hybrid conjugated polymer-Ag@PNIPAM fluorescent nanoparticles with metal-enhanced fluorescence, *J. Mater. Chem.*, 2011, **21**, 16943–16948.
78. F. Tang, F. He, H. Cheng and L. Li, Self-Assembly of Conjugated Polymer-Ag@SiO2 Hybrid Fluorescent Nanoparticles for Application to Cellular Imaging, *Langmuir*, 2010, **26**, 11774–11778.
79. X. Wang, F. He, X. Zhu, F. Tang and L. Li, Hybrid silver nanoparticle/conjugated polyelectrolyte nanocomposites exhibiting controllable metal-enhanced fluorescence, *Sci. Rep.*, 2014, **4**, 4406.
80. Y. Zhang, W. Hong, M. Li, Z. Zhao, L. Gan, J. Ou, X. Chen and M. Zhang, Fluorescence emission of BDMO-PPV enhanced by different plasmon Modes, *J. Phys. Chem. Solids*, 2015, **85**, 75–80.

CHAPTER 7

Structure and Order in Organic Semiconductors[†]

CHAD R. SNYDER,*[a] DEAN M. DELONGCHAMP,[a]
RYAN C. NIEUWENDAAL[a] AND ANDREW A. HERZING[b]

[a] Materials Science and Engineering Division, National Institute of
Standards and Technology(NIST), 100 Bureau Drive, Stop 8541,
Gaithersburg, MD 20899-8541, USA; [b] Materials Measurement Science
Division, National Institute of Standards and Technology(NIST),
100 Bureau Drive, Stop 8371, Gaithersburg, MD 20899-8371, USA
*Email: chad.snyder@nist.gov

7.1 Introduction

Across the spectrum of the current and potential applications for semi-conducting polymers, one of the most critical parameters is the nature and level of structure and order in the polymer. The functionality of a semi-conducting polymer depends on the phases or distributions of phases, which can range from disordered amorphous to liquid crystalline to semi-crystalline (with a distribution of crystalline and amorphous domains).[1] Broadly speaking, charge transport occurs most rapidly along the main chain axis in more highly planar chains, so increasing the level of order can improve charge transport; however, the subtleties of the morphology[2,3] also need to be accounted for. Figure 7.1 illustrates this through the differences

---

[†]Official contribution of the National Institute of Standards and Technology; The U.S. Government is authorized to reproduce and distribute reprints for Government purposes notwithstanding any copyright notation hereon.

RSC Polymer Chemistry Series No. 21
Semiconducting Polymers: Controlled Synthesis and Microstructure
Edited by Christine Luscombe
© The Royal Society of Chemistry 2017
Published by the Royal Society of Chemistry, www.rsc.org

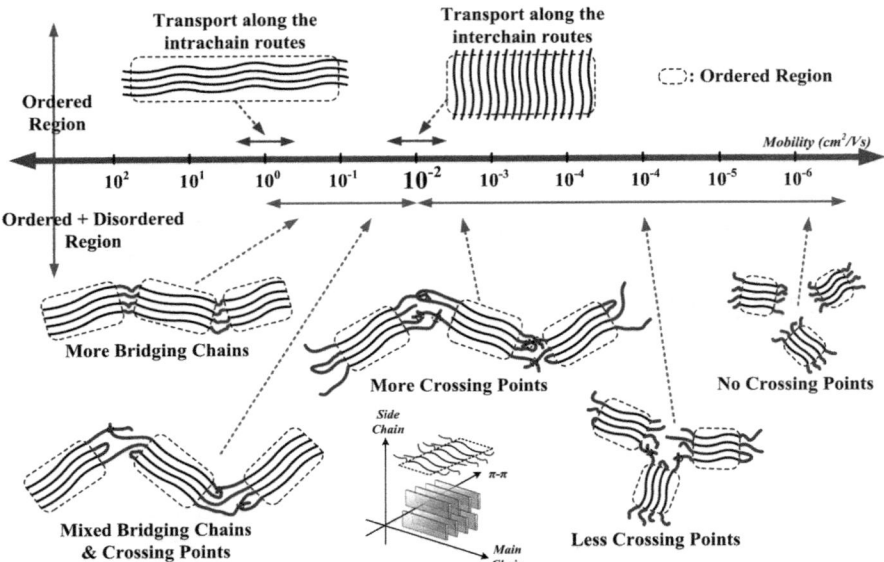

Figure 7.1 Typical region of the calculated charge mobility associated with the molecular conformation and chain structure of poly(3-hexylthiophene-2,5-diyl) (P3HT) in the ordered and disordered domains.
Reprinted with permission from Y.-K. Lan and C.-I. Huang, *J. Phys. Chem. B*, 2009, **113**, 14555–14564. Copyright 2009 American Chemical Society.

in the calculated charge mobility for different regions in semicrystalline poly(3-hexylthiophene-2,5-diyl) (P3HT).

For these reasons, it is important to characterize semiconducting polymers using methods that probe different aspects of the polymer structure, including both the degree and nature of the order. This chapter presents several different techniques for characterizing semiconducting polymer order on different length scales. As the behavior of highly conjugated polymers can differ distinctly from the more classic, highly saturated polymers for which some of these techniques were originally developed (*e.g.* polyethylene), the hazards and pitfalls associated with applying these techniques to semiconducting polymers are discussed where appropriate. It will be assumed that the reader is familiar with the techniques discussed; however, references will also be given for more fundamental introductions to the techniques.

7.2 Differential Scanning Calorimetry

7.2.1 Introduction to Differential Scanning Calorimetry

Differential scanning calorimetry (DSC) is often used for characterizing semiconducting semicrystalline polymers because of its ease of use and widespread access to the instrumentation. DSC instruments can largely be broken down into two distinct types: power compensation and heat flux designs.[4,5] In heat flux designs, there is a single furnace and the temperature difference between the sample and a reference, and potentially other locations in the furnace, is measured and the heat flow in the system is then modeled mathematically. By contrast, in power compensation DSC there are two furnaces, one for the sample and one for the reference. In addition to the sample temperature, the power necessary (constant, added or removed) to maintain the same temperature between the sample and reference is measured directly. At the simplest level, the data obtained from both types of DSC measurements are interpreted in terms of heat flow into (endothermic) and out of (exothermic) the sample as a function of temperature and heating rate. With proper calibration, this heat flow can be transformed into the sample's apparent heat capacity. Typical features sought in the DSC heating traces of semiconducting polymers are the glass transition (T_g) (step in heat capacity), cold crystallization (T_{cc}) (exotherm), and melting (T_m) or isotropization (T_i) (endotherm) temperatures, and, on cooling from the melt, the temperature of maximum crystallization rate (T_x) (exotherm) (often referred to simply as the crystallization temperature). Figure 7.2 shows a schematic diagram of some of these transitions.

7.2.2 Qualifying and Quantifying Crystallinity in Semiconducting Polymers

Many excellent papers related to instrument (temperature and heat flow) calibration,[6–10] sample measurement,[11] and the proper estimation of baselines[12–16] have been published. The focus herein is on methods for quantifying order in semiconducting polymers and their particular quirks that make traditional methods of quantifying polymer order more challenging. One of the first steps that needs to be taken to *quantify* the order is to *qualify* the order. In this context, qualifying order means determining whether the polymer is amorphous, liquid crystalline, or semicrystalline. The difference between the latter two, liquid crystalline *versus* semicrystalline, is critical as the calorimetric measures of order for these two states are very different. The simplest method of distinguishing between a liquid crystalline polymer and a semicrystalline polymer is through a comparison of the difference (undercooling) between the highest temperature exothermic transition on cooling and the subsequent highest temperature endothermic isotropization transition on heating – that is, the temperature at which the sample becomes an isotropic liquid. For liquid crystals this is known as the clearing (or isotropization) temperature and for crystals as the melting temperature. For a semicrystalline polymer there will be a difference >10 °C between the endotherm and exotherm due to the need for a nucleation event, whereas for a liquid crystal the difference will be either 0 °C or very small because it is a nearly equilibrium process. This can be seen by comparing the highest

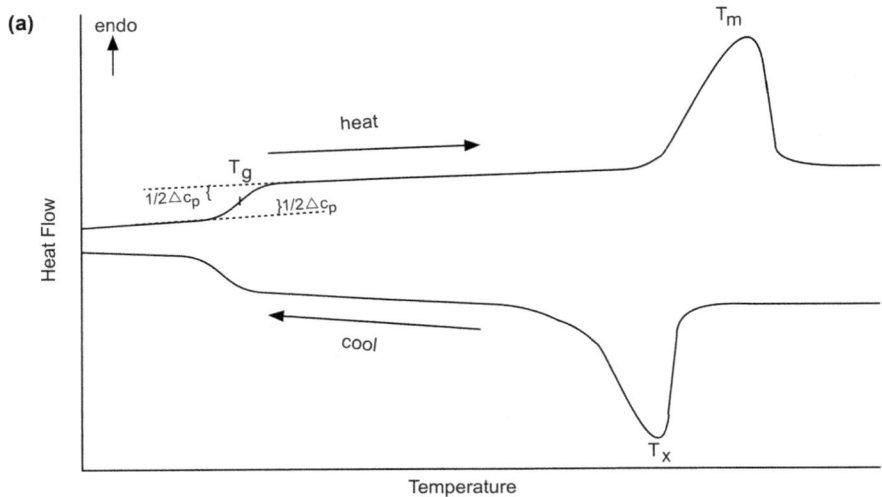

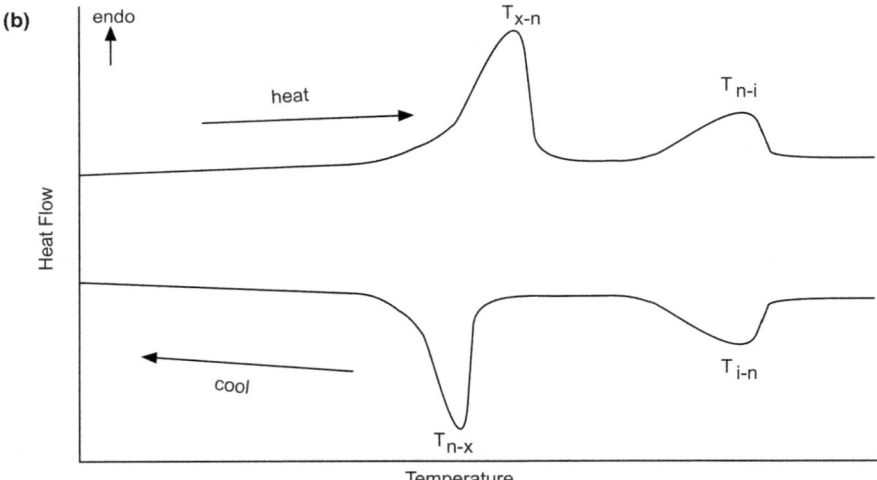

Figure 7.2 Schematic diagram of differential scanning calorimetry heating/cooling curves for (a) a typical semicrystalline polymer and (b) a liquid crystalline polymer exhibiting crystalline (x), nematic (n), and isotropic (i) phases. In part (a), the glass transition temperature has been defined as the point midway ($1/2\Delta c_p$) between the extrapolated glassy and melt heat capacities. The direction of endothermic transitions is indicated on the plot.

temperature endotherms and exotherms in Figure 7.2a and 7.2b.[17] Caution should be exercised, however, because superheating effects in liquid crystals[18,19] can lead to the erroneous conclusion that there is a large difference in undercooling. Superheating refers to the condition wherein crystals are in a metastable state above their melting temperature, as characterized by an increase in the observed melting temperature $T_{m,obs}$ with an increase in the heating rate β. This phenomenon typically obeys the following functional form:

$$T_{m,obs} = T_m + A\beta^z, \tag{7.1}$$

where T_m is the melting point at zero heating rate, A is a constant, and z is a constant with values between 0 and 0.5.[20] The difference between semi-crystalline and liquid crystalline states is critical because the enthalpy (latent heat) of fusion of a semicrystalline polymer can be used to determine the mass fraction of crystalline material in the sample, whereas for a liquid crystalline polymer the enthalpy of isotropization will correspond to the isotropization of almost 100% of the material.

The lack of an endotherm in the baseline is often taken as an indicator of an amorphous material; however, in the case of infusible semicrystalline materials such as poly(thiophene),[21] the materials will undergo a degradation process prior to melting. Similarly, the lack of a clear glass transition temperature (the temperature below which a supercooled liquid or mesophase behaves like a solid)[22] is often taken as an indication that the material falls into the crystalline infusible class, although there can be other reasons for the lack of a clear glass transition temperature. For example, it has been shown that the step in heat capacity at the glass transition temperature in an amorphous material $\Delta c_{p,am}$ is often about 11 J K^{-1} mol^{-1} of mobile units and that rigid aromatic groups typically only count as two to three of these mobile units.[5] Because heat flow in DSC is measured on a per unit mass basis, not on a per mole basis, the step in heat capacity is often fairly small when the molar mass per repeat unit is taken into account. In addition, because side-chains are typically added to semiconducting polymers to improve their solubility, there is a possibility of changes in the relaxation processes being reflected in the change in heat capacity at the glass transition temperature, as has been seen in poly(n-alkyl methacrylates).[23] Scattering methods such as those discussed in Section 7.5 are probably the best way to determine whether the material is completely amorphous through the presence of only an amorphous halo, or possibly a liquid crystalline or semicrystalline polymer through diffraction peaks suggestive of ordering.

In the light of this discussion of the glass transition temperature, it is appropriate to mention that an upper bound estimate can be made on the mass fraction of crystallinity if the step in heat capacity at the glass transition temperature of a perfectly amorphous reference sample $\Delta c_{p,am}$ can be measured. This is based on the work of Wunderlich and coworkers.[24] For a semicrystalline polymer, the system is most correctly treated calorimetrically as a three-phase system consisting of a crystalline fraction x_c, a mobile amorphous fraction x_{maf}, and, possibly, a rigid amorphous fraction x_{raf}.[25] The rigid amorphous fraction is most often attributed to the crystal–amorphous interfacial region;[26] however, it is only rigorously defined through the following equation:[24]

$$x_c = 1 - x_{maf} - x_{raf} = 1 - \frac{\Delta c_{p,sc}}{\Delta c_{p,am}} - x_{raf} \leq 1 - \frac{\Delta c_{p,sc}}{\Delta c_{p,am}}. \tag{7.2}$$

The inequality on the right-hand side of eqn (7.2) demonstrates the method for determining the upper bound on crystallinity. By combining a value for $\Delta c_{p,am}$ with the measured step in heat capacity at the glass transition of the semicrystalline sample of interest, $\Delta c_{p,sc}$, the upper bound on x_c, is obtained.

Although fairly useful in establishing bounds, eqn (7.2) is not typically used to determine the mass fraction of crystallinity x_c by DSC. Instead, the following relationship is typically employed:

$$x_c \cong \frac{\Delta H_f{}^*}{\Delta H_u}, \tag{7.3}$$

where $\Delta H_f{}^*$ is the measured enthalpy (integrated heat flow under the melting peak) and ΔH_u is the enthalpy of fusion per crystalline repeat unit, or the enthalpy of a 100% infinitely thick crystalline sample. However, this equation is an approximation that can underestimate x_c. A better approximation is given by:[27,28]

$$x_c \cong \left(\frac{\Delta H_f{}^*}{\Delta H_u}\right)\left(\frac{T_m^0}{T_m}\right), \tag{7.4}$$

where T_m^0 is the equilibrium melting temperature, which is the melting temperature of an infinitely thick crystal, and T_m is the melting temperature of the sample being evaluated. This equation corrects for the effect of smaller imperfect crystals melting at T_m. The best way to use this equation is to multiply the measured heat flow as a function of temperature T by (T_m^0/T) prior to integration of the melting peak to obtain $\Delta H_f{}^*$, then the previous equation can be used because each T slice of the heat flow corresponds to different crystal sizes and this procedure accounts for them accordingly. Thus to correctly determine the crystallinity by DSC, two parameters must be known in advance: T_m^0 and ΔH_u. For regio-defective materials and co-polymers with non-crystallizable co-monomers, T_m^0 is replaced in eqn (7.4) by the copolymer equilibrium melting temperature T_m^c, which can be readily calculated for the case where the co-monomer is completely excluded from the crystal via:[29,30]

$$\frac{1}{T_m^c} - \frac{1}{T_m^0} = -\left(\frac{R}{\Delta H_u}\right)\ln p, \tag{7.5}$$

where R is the gas constant and p is the mole fraction of the crystallizable monomer. It has been shown that this seems to be a reasonable assumption for regio-defects in P3HT.[31] A more generalized expression has been derived by Sanchez and Eby[32] that includes an enthalpic penalty for the partial inclusion of the co-monomer in the crystal; however, a far more extensive analysis beyond the scope of the present chapter is required to calculate all of the associated parameters necessary to use this expression. When determining the mass fraction of crystallinity by the enthalpy of fusion, caution should always be exercised, particularly with respect to semi-rigid semiconducting polymers, because it is possible for some of the enthalpy to be attributable to an oriented amorphous fraction.[33]

7.2.3 Determination of the Equilibrium Melting Temperature

There are several ways of determining T_m^0 that have been applied to semiconducting polymers: the Gibbs–Thomson equation,[34] the Broadhurst equation,[35] and the Hoffman–Weeks equation.[36] Each method has its own associated precautions that must be exercised. The Gibbs–Thomson relationship is derived from thermodynamic considerations and the resulting equation is:

$$T_m = T_m^0 \left[1 - \frac{2m_0\sigma_e}{\rho_c l\Delta H_u} \right], \tag{7.6}$$

where m_0 is the relative molar mass per repeat unity, ρ_c is the crystal unit cell density, σ_e is the crystal/melt fold surface interfacial free energy, and l is the crystal thickness in the chain axis direction. It should be noted that this form of the Gibbs–Thomson equation is based on the assumption that the chain axis thickness is much less than the lateral extent of the crystal,[34,37] although this is an assumption that may be questionable for semiconducting polymers. It should also be noted that eqn (7.6) is more complicated than typical forms because the enthalpy in the denominator is normally in units of energy per unit volume, but, to remain consistent with other equations involving ΔH_u in this chapter, it is in units of energy per mole of repeat unit, requiring the addition of m_0 and ρ_c to the equation. The classical implementations of this method are based on the characterization of l by small-angle X-ray scattering (SAXS)[38] or cross-sectional transmission electron microscopy (TEM).[39] This information, combined with the melting temperature T_m obtained from DSC, allows the determination of T_m^0 from a plot of T_m *versus* l^{-1}. An example of the Gibbs–Thomson analysis applied to poly(9,9-di-*n*-octyl-2,7-fluorene) is shown in Figure 7.3.[40] To the best of our knowledge, the only other semiconducting polymer for which Gibbs–Thomson analyses have been performed is P3HT.[31,41,42] For regio-defective materials (see supporting information to ref. 31) and copolymers,[43] it has been suggested that, rather than using the endotherm peak maximum as T_m, the temperature at which the melting process is complete should be used. The reason for this is that there is an additional entropy of mixing as a result of the heterogeneity of the material and therefore only the last crystals melt into the most representative melt. It should also be realized that the melting temperature from DSC might not

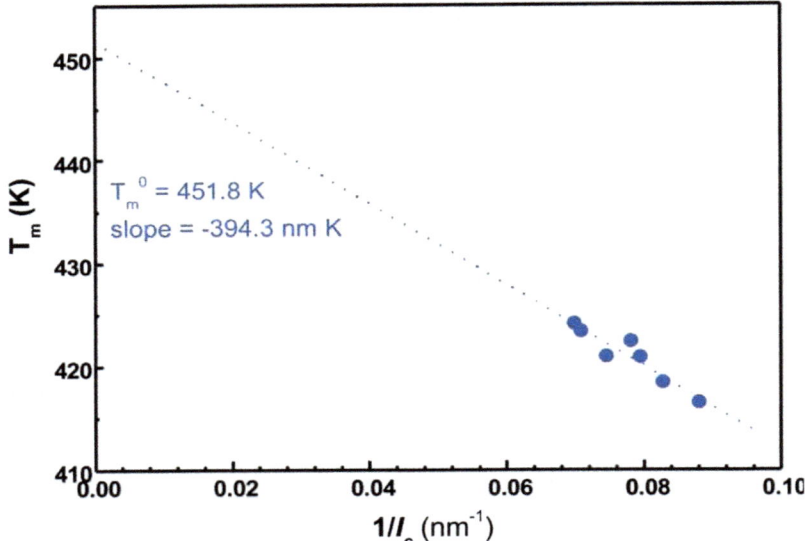

$T_m^0 = 451.8$ K

slope = -394.3 nm K

Figure 7.3 Gibbs–Thomson plot for poly(9,9-di-*n*-octyl-2,7-fluorene). Reproduced with permission of the International Union of Crystallography from *J. Appl. Crystallogr.*, 2007, **40**, s573–s576.

correspond to the melting of crystals of size *l* for several reasons, including the possibility of lamellar thickening during heating. For this reason, it has been argued that *in situ* SAXS measurements should be performed to determine T_m at the disappearance of l.[43]

The second method that has been successfully used for *n*-alkanes[35] and for monodispersed oligomers of 3-hexylthiophene[44] and P3HT[28] is through the Broadhurst relationship between the number of oligomer/polymer repeat units *n* and the observed melting temperature:

$$T_m = T_m^0 \left(\frac{n+a}{n+b} \right), \tag{7.7}$$

where *a* and *b* are treated as fitting constants, which, in the simplest version of the model, are given by $\Delta H_e / \Delta H_u$ and $\Delta S_e / \Delta S_u$, respectively, with ΔH_e the enthalpic and ΔS_e the entropic penalties due to the fold or chain end interfaces, and ΔS_u the entropy of fusion per crystalline repeat unit. It is important to note that this relationship is only valid for low molecular weight fractions that have not undergone chain folding. SAXS can be used to verify this; however, in the absence of SAXS measurements a thermal technique known as self-nucleation and successive annealing (SSA) has been demonstrated to be able to evaluate the onset of chain folding in P3HT fractions[28] in the absence of defects or co-monomers in the backbone (a brief discussion of SSA is given later in this chapter). An example of the application of the Broadhurst relationship is given in Figure 7.4.

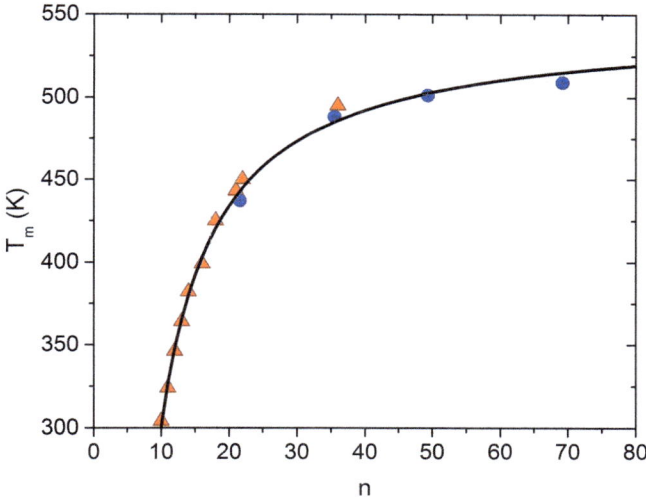

Figure 7.4 Example of observed melting temperature *versus* number (or number average) of repeat units for 100% regio-regular 3-hexylthiophene oligomers (triangles) and poly(3-hexylthiophene) narrow fractions (circles). The solid line is a fit to eqn (7.7).
Adapted with permission from *Macromolecules*, 2014, **47**, 3942–3950. Copyright 2014 American Chemical Society.

Experimental observations[45–47] of a dependence of T_m on the isothermal crystallization temperature T_x led to the development of the final method for determining T_m^0, the simplified Hoffman–Weeks[36] relationship based on kinetic arguments from the Lauritzen–Hoffman[34] secondary surface nucleation theory and eqn (7.6).

$$T_m = T_m^0 \left[1 - \frac{1}{\gamma} \right] + \frac{T_x}{\gamma}, \tag{7.8}$$

where γ is a factor that is supposed to account for crystal lamellar thickening above the initial kinetically determined thickness; in this simple form it is often treated as a constant. Great care should be employed in using this relationship beyond a simple estimation because Marand and co-workers[48–50] have shown that there are problems associated with the linear extrapolation procedure. To address these concerns, they recommended a more involved non-linear fitting process[50] based on first extrapolating the observed melting temperature at a given crystallization temperature back to the initially formed crystal nucleus *via* annealing studies. Strobl[51] has also argued against any validity of the Hoffman–Weeks method (outside the concerns of Marand and coworkers), so the previous two methods for characterizing T_m^0 might be preferred as a result of the associated uncertainties in the Hoffman–Weeks method. Although one article[31] has made a brief mention of an attempt to use the Marand and co-workers[48–50] approach on P3HT, this was unsuccessful and, to the best of our knowledge, no other study has tried this approach for semiconducting polymers and hence only the simple linear approach has been used. Some examples of semiconducting polymers to which the simple linear Hoffman–Weeks extrapolation method has been applied include poly(3-butylthiophene),[52] poly(3-hexylthiophene),[53] poly(3-(2'-ethyl)hexylthiophene),[31,54] poly(3-octyl-thiophene),[52,53] poly(3-dodecylthiophene),[52,53] and poly(9,9-di-*n*-octyl-2,7-fluorene).[40]

An additional important benefit arises from the application of the Gibbs–Thomson and Broadhurst approaches to semiconducting polymers. Once the parameters for the equations have been determined, they then provide a means of quickly characterizing the crystal thickness in the chain axis direction from the observed melting temperature of the sample. For semiconducting polymers, this information is often absent from the X-ray diffraction data and can often be difficult to derive from small-angle scattering data. The crystal thickness, in concert with the mass fraction of crystallinity, can then be used to help understand the electron transport properties of the aggregate system.

7.2.4 Determination of the Enthalpy of Fusion Per Repeat Unit

There are several different methods for determining the heat of fusion, including enthalpy/crystal size extrapolation,[55] the diluent method,[30,56,57] enthalpy/specific volume extrapolation,[55] and point-wise methods combining a separate measure of order such as X-ray diffraction,[58] solid-state nuclear magnetic resonance (NMR) spectroscopy, or, potentially, spectroscopic methods. Most of these methods have been used for semiconducting polymers.

The crystal size, ζ, extrapolation is based on simple thermodynamics and is predicated on the assumption that the crystal extent in the plane of the crystal is greater than the crystal thickness in the chain axis dimension. The equation typically used is:[55]

$$\frac{\Delta H_f^*}{x_c} \cong \Delta H_u + \frac{\Delta H_e}{\zeta}. \tag{7.9}$$

For bookkeeping purposes, the ΔH_e in eqn (7.9) is often written as $2\Delta H_e$ to indicate the two opposite crystal surfaces; however, for simplicity, we are folding the factor of two into the definition of ΔH_e and follow this nomenclature throughout the chapter. Note that the presence of x_c in this equation indicates the requirement that the crystallinity of each sample is known. If monodisperse, pure oligomers with the same crystal polymorph as the polymer of interest can be prepared and characterized, then the assumption of 100% crystallinity should be reasonable;[28,44] however, for polymer fractions this assumption cannot be made and a separate method is required to

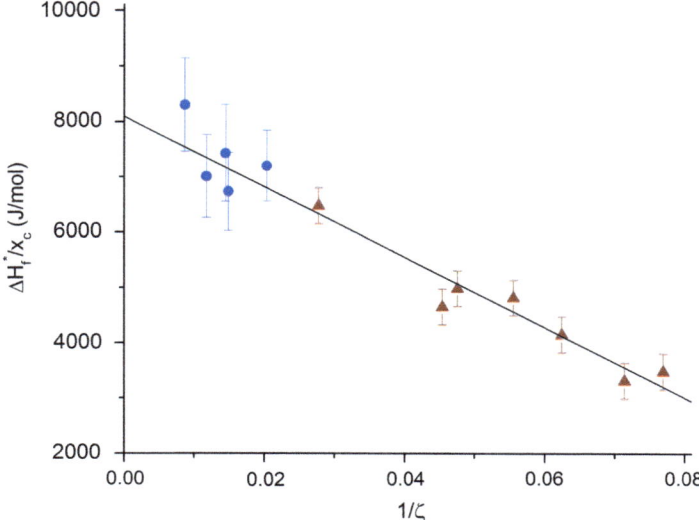

Figure 7.5 Enthalpy of fusion (measured enthalpy ΔH_f^* divided by the mass fraction of crystallinity x_c) as a function of inverse crystal size (in units of repeat units) for monodispersed 3-hexylthiophene oligomers (triangles) and 100% regio-regular P3HT fractions (circles). For the 3-hexythiophene oligomers it was assumed that $x_c = 1$ and for the P3HT fractions x_c was determined from solid-state NMR.
Adapted with permission from *Macromolecules*, 2014, **47**, 3942–3950. Copyright 2014 American Chemical Society.

estimate x_c for each sample.[28] In the absence of chain folding, $\zeta = 1/n$, where n is the number of repeats in the monodisperse oligomer, or n_n, the number average number of repeats in polydisperse samples. This equation demonstrates why the crystal size correction in eqn (7.4) is necessary, as ΔH_e is negative and, for fixed crystal sizes, the measured enthalpy will be less than ΔH_u, as can be seen in Figure 7.5 for P3HT. It also shows why the point-wise measurements discussed previously should only be used if T_m^0 is known and hence a crystal size correction can be used, or if they are used in concert with this extrapolation method.

The diluent method is based on the classic equation derived by Flory and coworkers:[30,56,57]

$$\frac{1}{T_m} - \frac{1}{T_m^0} = \frac{R}{\Delta H_u}\left(\frac{v_u}{v_1}\right)\left[(1 - \phi_p) - \chi(1 - \phi_p)^2\right] \qquad (7.10)$$

for the observed depression of the melting point T_m below T_m^0 due to the addition of a second dilute component. In this equation, R is the gas constant, χ is the Flory–Huggins interaction parameter, v_u is the molar volume per repeat unit of the polymer, v_1 is the molar volume of the diluent, and ϕ_p is the volume fraction of the polymer in the system. It should be noted first

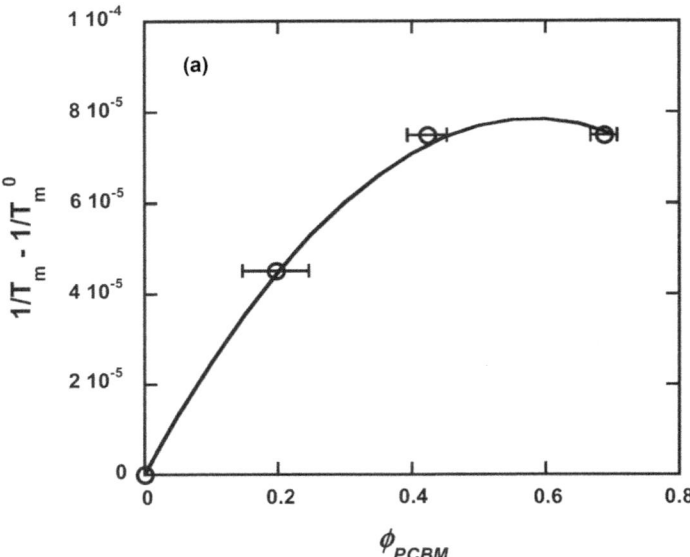

Figure 7.6 Depression of the melting temperature of P3HT due to the presence of
PC$_{61}$BM. ϕ_{PCBM} is the volume fraction of PC$_{61}$BM in the blend. The solid
line is a fit to eqn (7.10).
Reprinted with permission from *Macromolecules*, 2011, **44**, 5722–5726.
Copyright 2011 American Chemical Society.

that T_m^0 is rarely, if ever, used in the application of this equation. Instead, the
melting point for the specific polymeric sample in the absence of diluent is
used. The first challenge behind the diluent method, which has long been
recognized, is ensuring that the crystal size does not change on the addition of
the diluent because changes in the crystal size will also affect T_m. To cir-
cumvent this, samples are typically annealed for long times at elevated temperatures to maximize the crystal thickness, or are slow-cooled from the melt.
An example of a melting point depression plot for P3HT is shown in Figure 7.6,
where the diluent was [6,6]-phenyl C$_{61}$ butyric acid methyl ester (PC$_{61}$BM).[59]
An analysis of the data yielded an enthalpy of fusion for P3HT of about
7800 J mol^{-1}, which is below the extrapolated value from Figure 7.5, but within
its estimated uncertainty; this lower value could also be a possible consequence of finite crystal size effects. A more appropriate form for eqn (7.10) is:

$$\frac{1}{T_m} - \frac{1}{T_\zeta} \cong \frac{T_m^0}{\Delta H_u T_\zeta} \left(\frac{v_u}{v_1}\right) \left[(1 - \phi_p) - \chi(1 - \phi_p)^2 \right], \tag{7.11}$$

where T_ζ is the equilibrium melting temperature for a crystal of thickness ζ.
This equation takes into account the initial melting point depression due to
the finite size of the starting crystal[60,61] and then uses an approximation[27,28]
to enable the substitution of the enthalpy of fusion for a crystal of finite size
ΔH_ζ with that of the infinite crystal limit $\Delta H_\zeta = (T_\zeta/T_m^0)\Delta H_u$.

In the traditional approach to the specific volume extrapolation, the challenge is ensuring the crystal size ζ is the same for all samples or that an estimate of the enthalpic penalty due to chain ends ΔH_e is known.[55] It is to be stressed that, in the specific volume extrapolation, it is the specific volume V (not the density $\rho = 1/V$) that must be used to obtain the appropriate values from the extrapolation. Furthermore, in the traditional approach, the extrapolated enthalpy will correspond to the crystal thickness of the samples on which the specific volume was measured.

$$\Delta H_f^* = \frac{V_{amorph} - V}{V_{amorph} - V_{cryst}}(\Delta H_u + B), \qquad (7.12)$$

where V_{amorph} is the specific volume of the pure amorphous phase, V_{cryst} is the specific volume of the pure crystalline phase, and $B = \Delta H_e/\zeta$. From this equation, it should be apparent that when $\Delta H_f^* = 0$, $V = V_a$, and at $V = V_c$, $\Delta H_f^* = (\Delta H_u + B)$. Therefore only V_a can be determined from the plot with no additional information; the extrapolated value of $(\Delta H_u + B)$ requires *a priori* knowledge of V_c and hence the crystal repeat unit density ρ_c. When uncertainties exist in ρ_c, only ranges for ΔH_u can be provided, as was in Lee and Dadmun.[62] Some of the uncertainties in this method can be reduced if the same crystal size correction discussed previously in this chapter is applied to the specific volume equation. In this case, the following approximation can be obtained that eliminates B and the requirement of identical crystal sizes:

$$\Delta H_f^* \left(\frac{T_m^0}{T_m} \right) \cong \frac{V_{amorph} - V}{V_{amorph} - V_{cryst}} \Delta H_u. \qquad (7.13)$$

Thus the new plot will be the crystal size corrected enthalpy as described with eqn (7.4) as a function of specific volume. Figure 7.7 shows an example of a plot of enthalpy *versus* specific volume for P3HT.

7.2.5 Self-nucleation and Successive Annealing

As mentioned previously, SSA can be used to determine whether a polymer is of sufficient chain length to undergo chain folding in the crystal.[28] The procedure for SSA is described in detail elsewhere[63,64] and is shown schematically in Figure 7.8. A few clarifying points about the SSA process shown in the figure are as follows. The sample is initially melted at a temperature sufficiently high to destroy existing nuclei (typically at least 25 °C above the melting point of the polymer), so that a "standard state" can be generated. The cooling steps generate nuclei and crystals that are annealed or melted on subsequent heating steps. The first annealing temperature $T_{s,ideal}$ is the optimum self-nucleation temperature determined from separate DSC measurements.[63] The final heating scan "reads" out the distribution of crystal sizes generated.

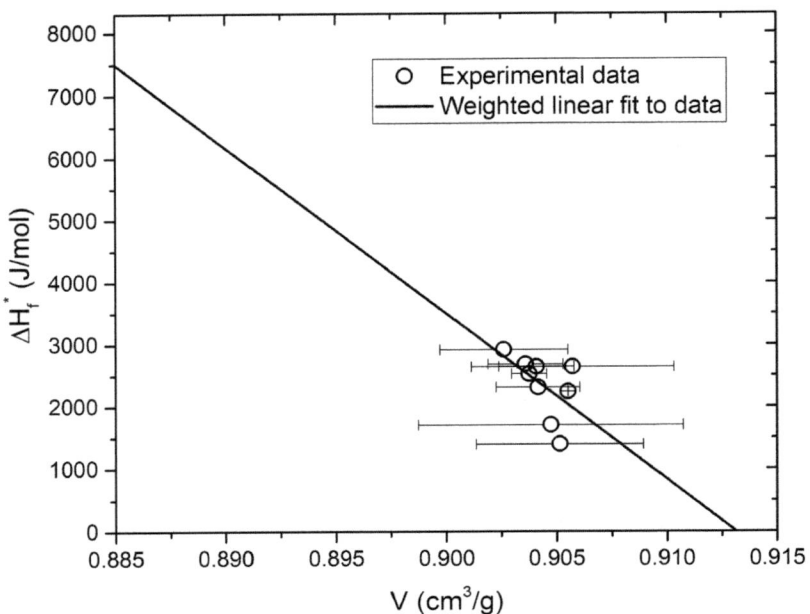

Figure 7.7   Enthalpy/density data of Lee and Dadmun[62] converted to enthalpy/
specific volume and analyzed using a weighted linear least-squares fit
to eqn (7.12). Because the melting temperatures of all the samples were
close to each other, the linear form assuming *B* as a constant was used.
The error bars correspond to one standard deviation in the experimental
uncertainty based on the reported values.

It is important to note that, for the results of the SSA analysis to be used
to determine the onset of chain folding, the polymer must not consist of
non-crystallizable units such as regio-defects or co-monomers. The reason
is that SSA effectively "bins" the polymer by its distribution of crystallizable
sequence lengths. Non-crystallizable units define the ends of the crystallizable sequence lengths; however, in their absence the SSA will reflect the
polydispersity of the material until the onset of chain folding, at which
point chains of different lengths will be incorporated into the chain-folded
crystal in a continuous, rather than discrete, distribution. This transition
can be clearly seen when the results of the SSA analysis transition from a
series of discrete peaks corresponding to specific crystallizable sequence
lengths to a continuous curve; the key signature of this transition is shown
in Figure 7.9.

7.2.6   Methods for Characterizing Thin Film Samples

One additional complication related to the analysis of semiconducting
polymers is due to the limited sample mass available. There are two sources
of this limitation: (1) syntheses tend to produce limited amounts of sample;

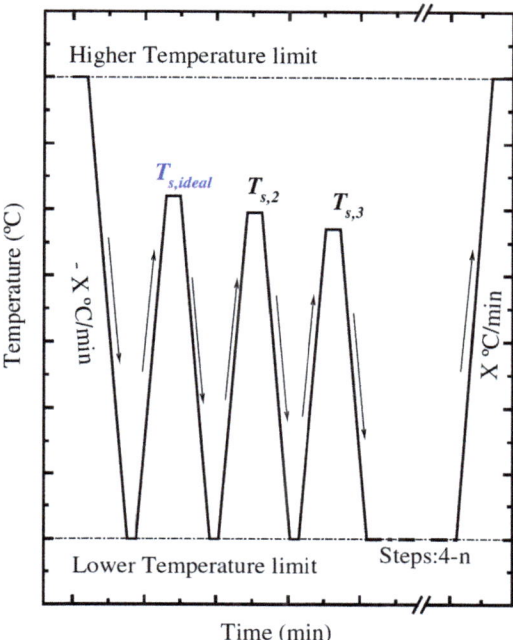

Figure 7.8   Thermal scheme for self-nucleation and successive annealing. The heating and cooling steps are performed at a constant rate *X*.
Reprinted from A. J. Müller, R. M. Michell, R. A. Pérez and A. T. Lorenzo, Successive Self-nucleation and Annealing (SSA): Correct design of thermal protocol and applications, *Eur. Polym. J.*, **65**, 132–154, Copyright (2015), with permission from Elsevier.

and (2) for semiconducting polymers the device-relevant samples for characterization are in film form, with thicknesses typically of the order of 200 nm or less. If such a film is prepared by spin-coating on a 50.8 mm silicon wafer and if a density of 1.1 $g\,cm^{-3}$ is assumed, then this corresponds to a sample mass of <0.4 mg. This is far below the 2–15 mg used for typical DSC measurements, depending on the phenomena being examined and the weakness of the transition. One common method used for small sample masses (thin films) is the stacking of a number of these samples until a sufficient mass has been reached; however, a number of alternative, less involved, procedures also exist. For current DSC technologies, the measurements capable of quantifying such small sample masses can be broken into traditional and chip-based DSC techniques. Two broad approaches exist for the traditional DSC measurements. The first is achieved in heat flux DSC through ultra-linearization of the baseline, such that even minor deviations from the baseline are discernable, *e.g.* recent advances based on the original method of Danley.[65] The second method is achieved in both heat flux and power compensation DSC and is based on the simple observation that the DSC signal is proportional to the heating rate and therefore faster heating

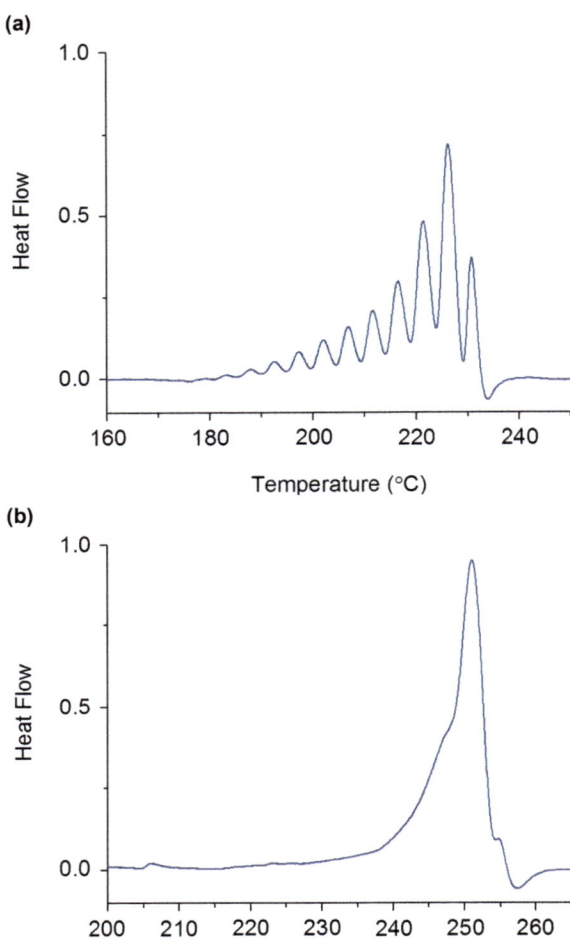

Figure 7.9   Self-nucleation and successive annealing analysis of 100% regio-regular
P3HT fractions with number-averaged relative molar masses $\langle M_n \rangle$ of (a)
8.2 kg mol^{-1} (below the onset) and (b) 15.9 kg mol^{-1} (above the onset).
The chain folding onset was demonstrated to be at about 11.5 kg mol^{-1}
via this analysis.[28]
Adapted with permission from *Macromolecules*, 2014, **47**, 3942–3950.
Copyright 2014 American Chemical Society.

rates can be used to compensate for smaller sample masses. Different ver-
sions have been developed for heat flux[66] and power compensation[10,67,68]
DSC methods. For the sample masses used in these bulk measurements,
analytical ultra-microbalances are typically needed to provide the resolution
in sample mass to properly compute the measured enthalpy of fusion.
For chip-based DSC measurements performed at high speeds, *e.g.* up to
10^6 K s^{-1}, the sample mass must be determined by characterization of the

melt heat capacity of the sample and through comparison with measurements of the heat capacity of a bulk sample to enable back-calibration to the mass of the sample of interest.[69,70]

7.3 Solid-state NMR Spectrometry

7.3.1 Introduction

NMR is a classic spectroscopic technique for studying soft matter. As long-range order is not a requirement, NMR can be utilized on essentially any diamagnetic system to understand the local packing structures of molecules. NMR utilizes radiofrequency excitation to induce nuclear spin transitions at the Larmor frequency. Spectral information is related to the atomic level structure because the magnetic field environments local to the nuclei perturb the spin-flip transitions. For organic chemists synthesizing the complex macromolecules used in organic electronics, solution NMR is the gold standard analytical technique for assigning molecular structures to synthetic products. However, this section will not include the impacts of solution NMR on the organic synthetic products used for semiconductors; instead, it will deal with solid-state NMR and how it can be utilized to determine structural and dynamic information in materials in end-use applications. It should also be noted that solid-state NMR has been used since the beginning of the field of organic electronics.[71–74] Solid-state NMR has been utilized for studying the electronic structure in organic conductors since the genesis of the organic conductor; a review of solid-state NMR as applied to conducting polymers was written as early as 1983.[75] Initially, interest surrounded the impact of conduction electron spin density on the Knight shift,[76–78] dynamic nuclear polarization,[79] and NMR relaxation[80,81] to understand the electronic spin phases. NMR was more focused on electronic structure and phase and was not so heavily used to disentangle local structures in the as-processed heterogeneous media, which is the focus of this section.

To begin, we should acknowledge the stark contrast in solid-state NMR methods from those of the more familiar solution-state NMR methods. The resonance lines in solid-state spectra are generally much broader than those of solutions and this lack of resolution can translate into a loss of molecular structure information. The persistence of anisotropic chemical shifts, dipole–dipole interactions, and (when applicable) quadrupole interactions can translate into broader resonances in solid samples. For organic semiconductors, of course, there may be multiple nuclei to study, such as ^{1}H, ^{13}C, ^{14}N, ^{19}F, and ^{33}S. Each nuclear spin system has its own prevailing set of interactions, each with its own set of opportunities and drawbacks. For instance, higher spin ($>1/2$) nuclei such as ^{14}N ($I=1$) and ^{33}S ($I=3/2$) have strong anisotropic quadrupole interactions and broad resonance lines (>100 kHz), which, in principle, could be used to measure molecular orientations, but the spectral widths make the determination of absolute crystallinity or order *via* the resonance position difficult because the differences in chemical shift between the crystalline and non-crystalline sites tend to be <1 ppm (10–100 Hz). However, with common, modern day NMR spectrometers, we can turn many of these challenges into benefits. These interactions can be manipulated, or "selectively averaged," *via* magic angle spinning (MAS) and application-specific pulse sequences to garner atomic level structural information such as inter-nuclear proximities, molecular orientations and conformations, domain sizes, and molecular dynamics. Under MAS, solid-state NMR can yield high-resolution spectra capable of chemical shift contrasts high enough to distinguish chemical functional groups, such as methylene sites from methyl sites. Hence MAS NMR is a quantitative, atom-specific identification tool. Nucleus-to-nucleus proximities can be determined from dipole–dipole mediated spin exchange, which depends on spatial proximity ($\propto r^{-3}$, where r is the inter-nuclear distance). Molecular dynamics are garnered *via* spin relaxation, exchange, and line narrowing methods. We discuss aspects of these measurements as applied to organic semiconductors in the following sections.

7.3.2 Crystallinity and Order

Unlike X-ray diffraction, NMR is not sensitive to long-range periodicity or order. However, local order that exists in semicrystalline polymers can be measured because the resonance position can be sensitive to the molecular packing arrangement. The chemical shift is inherently sensitive to molecular conformation because magnetic shielding depends on the electronic geometry. Local magnetic fields can vary from site to site because of aromatic ring currents and variations in the magnetic susceptibility. The magnitudes of these effects on ^{13}C NMR resonances have been documented.[82]

As organic semiconductor thin films are generally heterogeneous, the NMR resonances will tend to be convolutions of spins in multiple kinds of environments. For instance, a thiophene ^{13}C nucleus in the crystal could resonate at a slightly different frequency than those outside the crystal. It is often simply assumed that these peaks have a certain functional form, such as Gaussian, Lorentzian, or Voigt. In some cases this assumption is valid and there is enough spectral separation to warrant such an assumption. However, in many cases NMR peaks can be asymmetrical and, in cases of spectral overlap, we stress that it is beholden on the scientist to experimentally determine the shape of these resonances *via* spectral editing or multidimensional techniques. Simply fitting to a standard line shape is not a good idea because it may lead to conclusions that are not physically sound. In many ways, diagnosing the shape, breadth, and position of NMR resonances is the whole point of the spectroscopic investigation and will lead to structural insights.

High-resolution ^{1}H NMR can be used to study order. In general ^{1}H–^{1}H dipole interactions are strong and the nuclei can become strongly coupled, so dipolar ("homogeneous" or "lifetime") broadening can exist in magnitudes of 10–100 kHz. However, by using line narrowing techniques such as combined rotation and multiple pulse spectroscopy (CRAMPS)[83] or very fast MAS, it is possible to achieve sufficient resolution to probe order. Our group used ^{1}H CRAMPS for the qualitative measurement of order in thin film blends of P3HT/PCBM.[84] Fast MAS (25 kHz) ^{1}H NMR has been used to acquire high-resolution ^{1}H NMR spectra in bulk P3HT.[85]

As shown in Figure 7.10, the peak at 5.5–7.5 ppm (thiophene ^{1}H NMR resonances) was least-squares fitted to two Gaussian curves; one centered at 6 ppm (Figure 7.10, orange peak) and another at 6.6 ppm (Figure 7.10, blue peak) and were ascribed to crystalline and non-crystalline P3HT, respectively. The crystallinity was defined as the relative integrated area of the crystalline peak at 6 ppm relative to the entire peak (black line, 5.5–7.5 ppm). The crystallinity, however, showed significant disagreement with the crystallinities measured by X-ray diffraction and DSC. Hence, although the assignments of the peak fits ($\delta_{XTAL} = 6.0$ ppm, $\delta_{NONXTAL} = 6.6$ ppm) were supported by [1]H–[1]H and [13]C–[1]H correlation experiments and quantum chemical calculations (not shown), the lack of support from other techniques suggested that this assignment was uncertain. In general, ^{1}H NMR cannot exhibit a higher resolution than about 1 ppm, which is not enough to distinguish between similar chemical shifts (methylene and methyl). New homonuclear dipolar coupling methods such as decoupling using mind-boggling optimization (DUMBO)[86] are promising candidates for higher resolution, but generally suffer the same resolution (about 1 ppm) as classic homonuclear dipolar decoupling schemes such as MREV-8.[87]

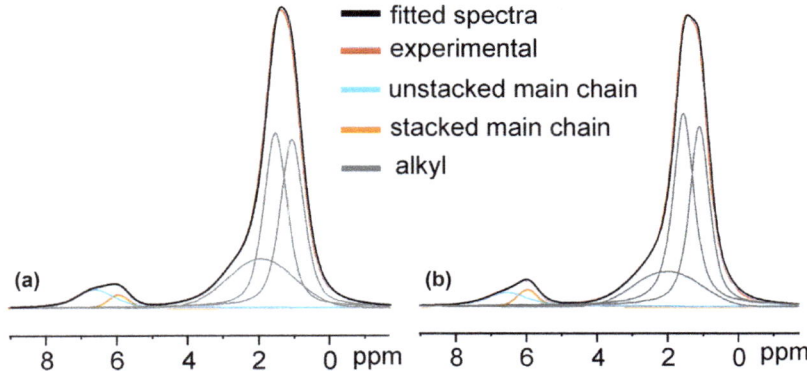

Figure 7.10 Deconvolution of fast MAS ^{1}H NMR spectra for two P3HT samples of varied molar mass and regio-regularity. Sample (a): mass-averaged relative molar mass $M_w = 60.0$ kg mol^{-1}, dispersity $Đ = M_w/M_n = 2.2$, regio-regularity = 94%. Sample (b): $M_w = 25.0$ kg mol^{-1}, $Đ = 1.6$, regio-regularity >98%. Fits were made to five resonances with variable Gaussian/Lorentzian contributions.
From D. Dudenko, A. Kiersnowski, J. Shu, W. Pisula, D. Sebastiani, H. W. Spiess and M. R. Hansen, *Angew. Chem., Int. Ed.*, 2012, **51**, 11068–11072. Copyright © 2012 by John Wiley & Sons, Inc. Reprinted by permission of John Wiley & Sons, Inc.

^{13}C is the most useful nucleus for studying local order in organic semi-conductors. Although ^{13}C NMR is less intense than ^{1}H, a popular signal enhancement technique called cross-polarization MAS (CPMAS) can be used.[88] ^{13}C CPMAS has been used to study absolute order, mesophase order, and the dynamics therein. ^{13}C CPMAS was used to study the mesophases in bulk polyalkylthiophenes.[89] In a later study on P3HT,[90] ^{13}C NMR was acquired from 173 to 363 K and a phase transition involving twisting of the main chain was shown around 0 °C. No crystallinity value was determined from the CPMAS spectra because it is widely known that the CPMAS signal intensity is dependent on the cross-polarization time and matching conditions.[88] Single-pulse excitation can be more reliably quantitative, but suffers from significantly worse signal-to-noise ratios and requires longer experimental times. Single-pulse excitation was used to quantitatively measure the crystallinity in bulk powders of Form II P3HT.[91]

As shown in Figure 7.11, the assignment was based on the spectral contrast of the hexyl side-chain methyl resonance at approximately 15 ppm, which, in earlier studies of polyolefins,[92,93] was shown to be correlated with trans and anti-molecular conformation, *i.e.* the γ-gauche effect.[92] The resolution was clear because the ratio of crystalline to non-crystalline was temperature-dependent. Later studies on Form I P3HT from our group[94] showed that the crystallinity could not be measured *via* the side-chain conformation because both crystals and non-crystals alike contain gauche conformers at room temperature. However, we demonstrated that ^{13}C CPMAS could be utilized to quantitatively measure main chain order to within 7% in P3HT using $T_{1\rho H}$-based spectral editing and thiophene resonance spectral contrast (Figure 7.12). As more order was observed *via* NMR than the DSC melt enthalpy, our early hypothesis was that local order must persist within the non-crystalline regions of P3HT. However, our follow-up study[28] of P3HT of varied molar mass (and crystal size) showed that once the melt enthalpy has been corrected for crystal size, the DSC-determined crystallinity matches the NMR-determined ordered fraction fairly well (Figure 7.13). Hence it was concluded that ^{13}C CPMAS NMR could be utilized for determining "absolute crystallinity."

7.3.3 Relaxation and Dynamics

Solid-state NMR can also be utilized to investigate molecular dynamics. There are multiple ways that dynamics can be studied *via* NMR, such as simple relaxation time experiments, exchange based experiments such as the center band only detection of exchange (CODEX),[95] or experiments based on ^{1}H NMR motional narrowing such as dipolar chemical shift correlation (DIPSHIFT), which was originally named the dipolar rotational spin echo (DSRE) technique.[96] These experiments vary in their sensitivity to the various aspects of molecular dynamics, such as site selectivity, the motional amplitudes, and the timescale.

NMR relaxation is an important tool for studying molecular dynamics. One particularly powerful relaxation time is the "spin lattice" relaxation, or

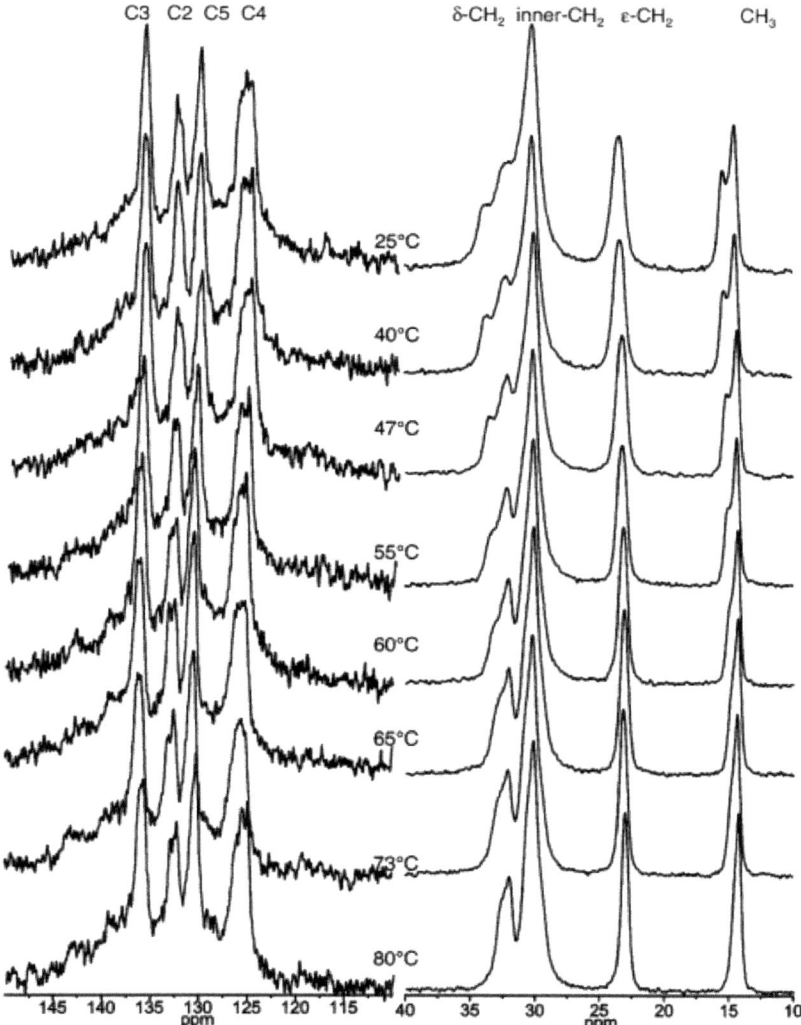

Figure 7.11 ^{13}C Single-pulse excitation MAS spectra of P3HT as a function of temperature. The P3HT sample had size-exclusion chromatography values of $M_n = 5.2$ kg mol^{-1}, $M_w = 6.0$ kg mol^{-1}, and $D = 1.15$, and a MALDI value of $M_n = 3.2$ kg mol^{-1}.
Reprinted with permission from O. F. Pascui, R. Lohwasser, M. Sommer, M. Thelakkat, T. Thurn-Albrecht and K. Saalwächter, *Macromolecules*, 2010, **43**, 9401–9410. Copyright 2010 American Chemical Society.

T_1, because it can elucidate between different modes of molecular dynamics based on variations in the spectral density of ''lattice fluctuations'' at the Larmor frequency. NMR people refer to ''the lattice'' as all other motions that are not the spin, such as vibrations, rotations, and translations. The spin lattice relaxation time, T_1, is controlled by the spectral density of lattice

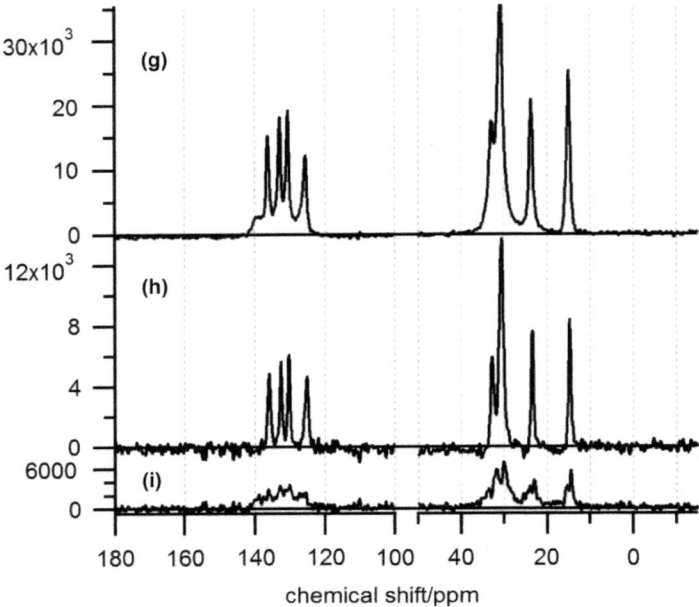

Figure 7.12 ^{13}C CPMAS spectra of (g) P3HT and its crystalline (h) and non-crystalline (i) components. The P3HT sample was $M_n = 64.5$ kg mol^{-1}, $Đ < 2.5$, regio-regularity (head-to-tail) >98%.
Reprinted with permission from R. C. Nieuwendaal, C. R. Snyder and D. M. DeLongchamp, *ACS Macro Lett.*, 2014, **3**, 130–135. Copyright 2014 American Chemical Society.

motions at the Larmor frequency (10^8–10^9 Hz) and the spin lattice relaxation time in the rotating frame, $T_{1\rho}$, is controlled by lattice motions at the nutation frequency (10^4 Hz). Phenomenological relations were first established by the classical theory of Bloembergen *et al.*[97] Later, a quantum mechanical based theory of spin relaxation was formalized by Bloch,[98,99] Wangsness,[100] and Redfield.[101,102] Comprehensive texts on relaxation are given elsewhere.[103–106] We merely state that all molecular dynamics information that can be garnered from NMR relaxation depends on the functional form of the spectral density function, which, in principle, can be related to the motions *via* time-correlation functions.[107–109] Relaxation is sensitive to: (1) whether the motion is on a slower or faster timescale than the observation frequency (fast or slow fluctuation limit); (2) the amplitude of the motion or the strength of the local field (*i.e.* the second moment); and (3) the relaxation mechanism (dipolar fields, chemical shift anisotropy, quadrupolar). As so many parameters govern the relaxation, measurements are typically performed as a function of temperature or magnetic field[106] because a single relaxation time measurement for determining molecular dynamics is ambiguous. Extensive modeling or analytical calculations should be performed, but, despite these, the interpretations from such fits can be cloudy because

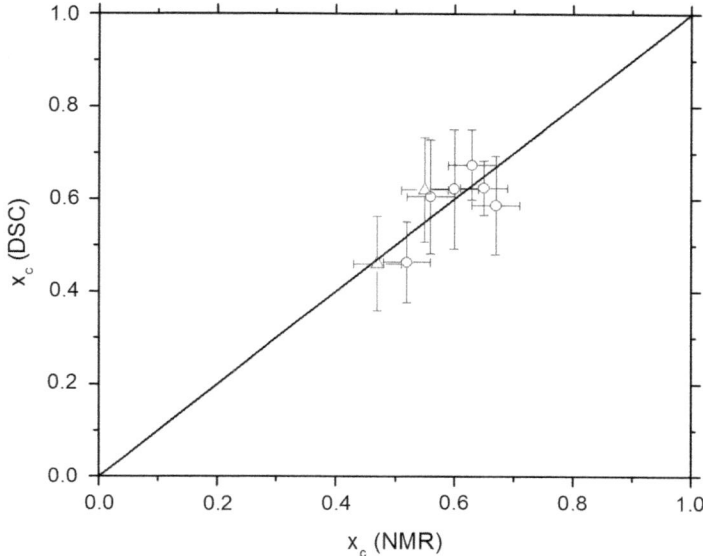

Figure 7.13 Mass fraction crystallinity for P3HT estimated from ^{13}C CPMAS NMR *versus* that estimated from DSC applying the finite crystal size correction of eqn (7.3) for samples of varying M_n from 8 to 49 kg mol^{-1}. The line indicates the location of perfect agreement between the two techniques.
Reprinted with permission from *Macromolecules*, 2014, **47**, 3942–3950. Copyright 2014 American Chemical Society.

very different models of motion can lead to similar spectral density functions.[110] Extensive analytical work has been performed by Geppi and coworkers[111] to discern molecular dynamics in P3HT/PCBM from NMR relaxation data by way of fitting T_1 *versus* temperature curves to analytical models. This may be promising for T_{1C}, but the issue regarding similarities of spectral density functions should be addressed. For T_{1H}, we should not ignore the impact of ^{1}H–^{1}H spin exchange, which serves to spatially average or "homogenize" the relaxation time over multiple spins, despite differences in dynamics from nucleus to nucleus.

   When extracting molecular dynamics information from T_1 *versus* temperature plots, we should also be careful to recognize which "time" is related to which physical process. For instance, when plotting relaxation time (*i.e.* $T_{1\rho}$) *versus* temperature, it may be tempting to extract some Arrhenius activation energy. However, this is non-physical because the nuclear spin relaxation is not governed by transition state theory and, in fact, over a broad enough temperature range can display both negative and positive slopes depending on whether the motion is in the slow or fast fluctuation limit ($t_c \ll 1/\omega_0$ or $t_c \gg 1/\omega_0$) (see Figure 7.14). If we are interested in obtaining the activation energy of a given type of lattice motion, such as a two-site hop or rotation-on-a-cone, then we have to plot the correlation time as a function of

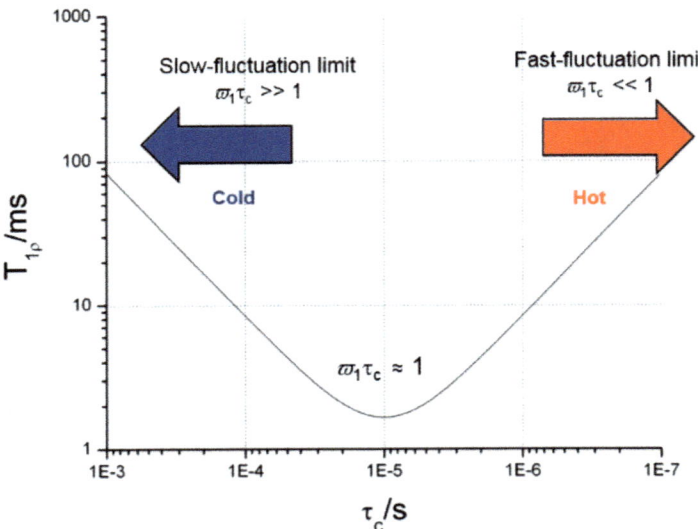

Figure 7.14   Calculated plot of $T_{1\rho}$ as a function of the logarithm of the correlation time of motion. Note that $T \propto t_c^{-1}$.

temperature, not the spin lattice relaxation time. The Arrhenius equations do not describe NMR relaxation. They only influence NMR relaxation indirectly by way of the lattice motions. The slope of the correlation time *versus* temperature plot (or $\ln k$ *versus* $1/T$) should always be negative and hence could theoretically be used to extract an activation energy. Recall from transition state theory that:

$$k = k_\infty \exp\left[-\frac{E_a}{k_B T}\right], \tag{7.14}$$

where k is the rate, k_∞ is the rate in the $T = \infty$ limit, E_a is the activation energy, k_B is Boltzmann's constant, and T is the absolute temperature.

CODEX can also be utilized for probing slow reorientation motions in organic semiconductors.[89] CODEX is an experiment that is based on the exchange of the principal components of the chemical shift tensor.[95,112] In CODEX, refocusing of the chemical shift anisotropy occurs in the absence of motion and slow (10^{-2}–1 s) reorientations cause destructive interference, and hence decay, of the refocused signal. By adjusting the preparation time and the mixing time, we can probe the amplitude (*e.g.* angle of reorientation) and timescale of the fluctuations, respectively. As shown in Figure 7.15, deAzevedo and coworkers[113] demonstrated that CODEX could be used to measure the tilt angles and timescales of main chain motions in MEH-PPV. In CODEX, two spectra are acquired: an exchange spectrum (denoted S) and a reference spectrum (denoted S_0), which eliminates the effects of other relaxation processes on the measured value of E ($= (S_0 - S)/S_0$).

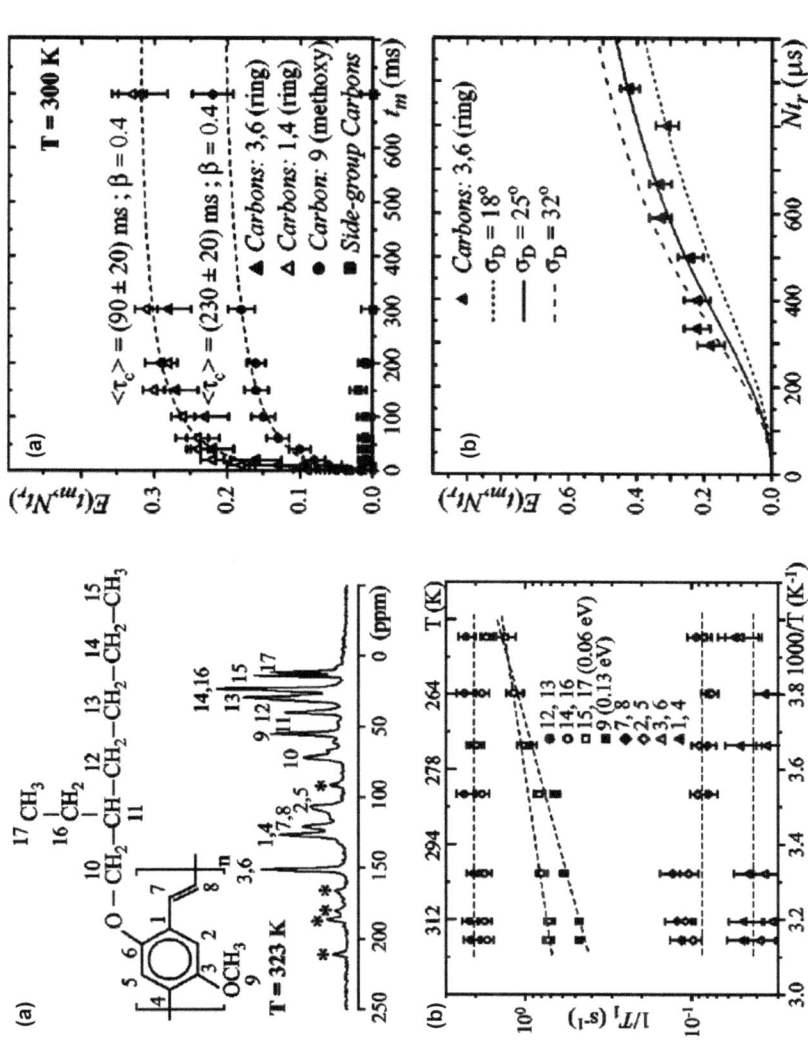

Figure 7.15 Left-hand panels: (a) structure of MEH-PPV; and (b) ^{13}C spectra of the CODEX exchange (S), the CODEX reference (S_0), and the difference ($S_0 - S$), which is referred to as $E(t_m, Nt_r)$ on the y-axes of the plots in the right-hand panels. Right-hand panels: (a) CODEX intensity (E) plotted as a function of the mixing time for finding the timescale of the reorientations via fits (dotted lines); and (b) E plotted as a function of the preparation time (Nt_r, where N is the number of rotor cycles and t_r is the rotor period) to find the tilt angle via fits (dotted lines). Reprinted figure with permission from A. C. Bloise, E. R. deAzevedo, R. F. Cossiello, R. F. Bianchi, D. T. Balogh, R. M. Faria, T. D. Z. Atvars and T. J. Bonagamba, *Phys. Rev. B: Condens. Matter Mater. Phys*, 71, 2005, 174202. Copyright 2005 by The American Physical Society.

As shown in Figure 7.15a (right-hand panel), when performed as a function of the mixing time, t_m, we can obtain the correlation time (*i.e.* hopping rate) for each site that re-orients, such as the main chain carbons 1,4 and 3,6 of MEH-PPV. Furthermore, when performed as a function of the total preparation time, we can deduce the tilt angle (out of the plane of the main chain) of the reorientation (Figure 7.15b, right-hand panel).

Yazawa *et al.*[89] showed *via* CODEX that slow twisting motions occur in the main chain of P3BT at 353 K, which they claimed is above a critical transition. These motions were also observed at 303 K, but were much less pronounced, indicating either longer timescales, smaller amplitudes, or fewer sites involved in the motion.

NMR line narrowing is another way to study molecular dynamics. NMR lines can be homogeneously ("lifetime") broadened by dipolar interactions. In general, ^{1}H NMR resonances are homogeneously broadened, so although a narrow distribution of chemical shifts exists (about 10 ppm), resonances can be broadened >50 kHz because the time domain signal loses coherence by way of ^{1}H–^{1}H mixing. Inhomogeneously broadened lines, on the other hand, are a result of the distribution of sites. Hence when motional narrowing occurs, it depends on the type of broadening mechanism in play. Motional narrowing of a homogeneous line is a result of increased coherence (or longer T_2) due to weaker dipolar interactions. Motional narrowing of an inhomogeneous line is a result of the dynamic averaging of "static" disorder. As NMR is a fixed timescale technique, if motions occur much faster than the acquisition time (milliseconds), then sites will appear as a "dynamic" average, but if motions occur much slower than the acquisition time, then they will appear as "static." Dynamic averaging of conformational disorder has been reported in crystalline P3HT[94] and observed elsewhere.[90] In P3HT, the motions involved in the dynamic disorder are faster than about 3 ms at temperatures >0 °C and occur in both the hexyl side-chain and the thiophene main chain (Figure 7.16).

Motional narrowing of homogeneously broadened systems has also been observed in organic semiconductors.[91,114] Experiments can be as simple as measuring the ^{1}H line width as a function of temperature. One of the drawbacks, however, is that it is difficult to relate a single number (*i.e.* line width or $1/T_2^*$) to what can be fairly complex molecular dynamics. If the molecule is complicated or the system is heterogeneous, the number of pertinent parameters that govern the dynamics can be too great as a result of the number of sites and the distributions of couplings. More sophisticated experiments rely on ^{1}H motional narrowing with ^{13}C detection, such as two-dimensional ^{13}C–^{1}H wide line spectroscopy (2D WISE) or separated local field experiments such as DIPSHIFT. However, they suffer from poorer signal-to-noise ratios and longer experiment times than ^{1}H NMR experiments, so are difficult or impossible to implement on small sample volumes such as thin films. If enough sample is available for measurement, DIPSHIFT experiments have an advantage over 2D WISE in that DIPSHIFT is sensitive to the motional narrowing of only ^{1}H–^{13}C couplings, whereas 2D

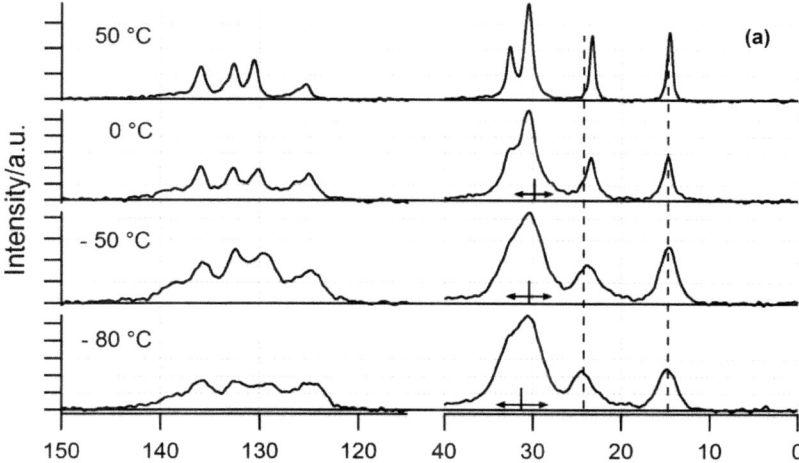

Figure 7.16 ^{13}C CPMAS spectra of P3HT acquired at various temperatures. Reprinted with permission from R. C. Nieuwendaal, C. R. Snyder and D. M. DeLongchamp, *ACS Macro Lett.*, 2014, **3**, 130–135. Copyright 2014 American Chemical Society.

WISE spectroscopy reports all couplings (^{1}H–^{1}H and ^{1}H–^{13}C) and so can be less selective.

The DIPSHIFT pulse sequence has been utilized to study the side-chain dynamics of poly(9,9-dioctylfluorene-*alt*-benzothiadazole) (F8BT),[113] poly(phenylene vinylene) (PPV) derivatives,[113,115,116] and Form II P3HT.[91] One of the drawbacks of this experiment is that, when performed as a function of temperature, the relative effects of the rates and amplitudes of intermediate regime (10^{-3}–10^{-7} s)] motions on the DIPSHIFT curve can be difficult to disentangle. Although possible ways to do so have been laid out,[116] similar curves can be interpreted very differently. Assuming a fixed-angle, temperature-independent geometry of motion (diffusion on a cone), deAzevedo and coworkers[114] showed that by fitting to DIPSHIFT curves (Figure 7.17b and 7.17c) as a function of temperature, the fluctuation rates of side-chain methylene units in F8BT (Figure 7.17a) could be ascertained. Assuming Arrhenius behavior, the outermost methylene on the hexyl side-chain of F8BT exhibited an activation energy of 0.23 ± 0.04 eV. The four interior methylene groups, which spectrally overlap in the ^{13}C spectrum, were found to exhibit an average activation energy of 0.30 ± 0.06 eV.

Saalwächter and coworkers[91] utilized DIPSHIFT experiments to measure P3HT hexyl side-chain dynamics (Figure 7.18). Based on strong intensity changes in the DIPSHIFT pattern at $t_r/2$ with temperature (and weaker changes at t_r) for the methyl groups and four outermost methylene groups (Figure 7.18b–7.18d), it was concluded that the amplitudes of motions (*i.e.* tilt angles) changed with temperature. Fluctuation rates also increased with temperature, but the authors noted that the effect on the DIPSHIFT curves was less pronounced because the motions were in the fast limit regime

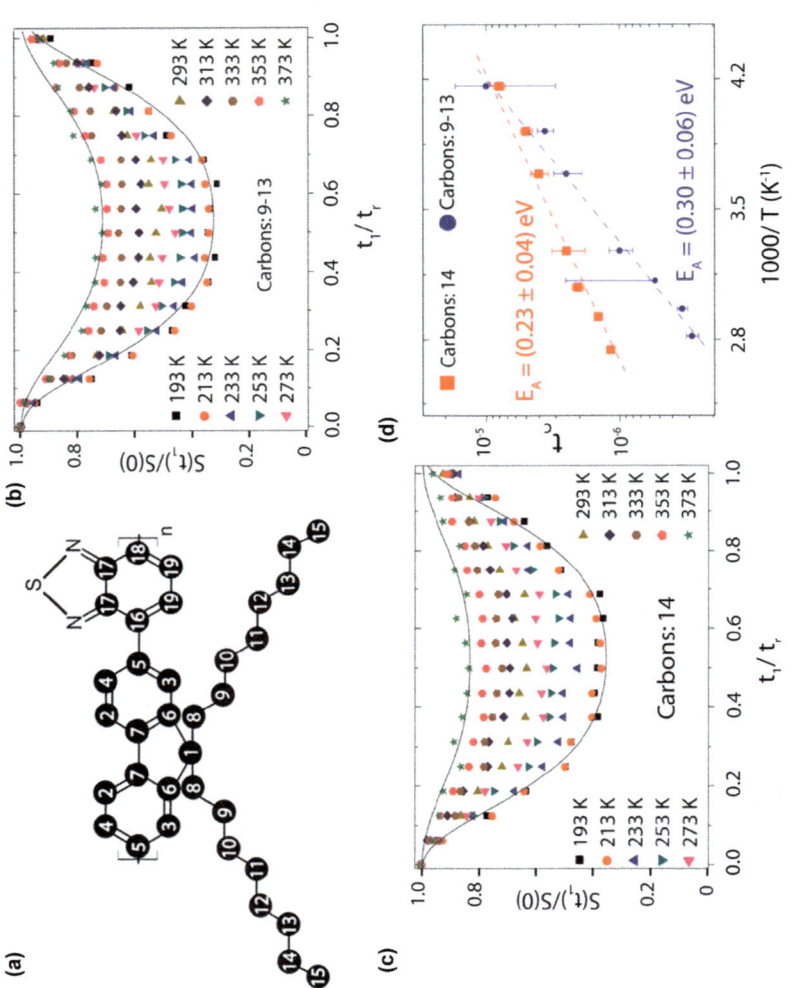

Figure 7.17 (a) Chemical structure of F8BT. DIPSHIFT curves for side-chain carbons (b) 9–13 and (c) 14, where t_r refers to the rotor period. (d) Arrhenius plots for carbons 9–14. Adapted with permission from G. C. Faria, E. R. deAzevedo and H. von Seggern, *Macromolecules*, 2013, **46**, 7865–7873. Copyright 2013 American Chemical Society.

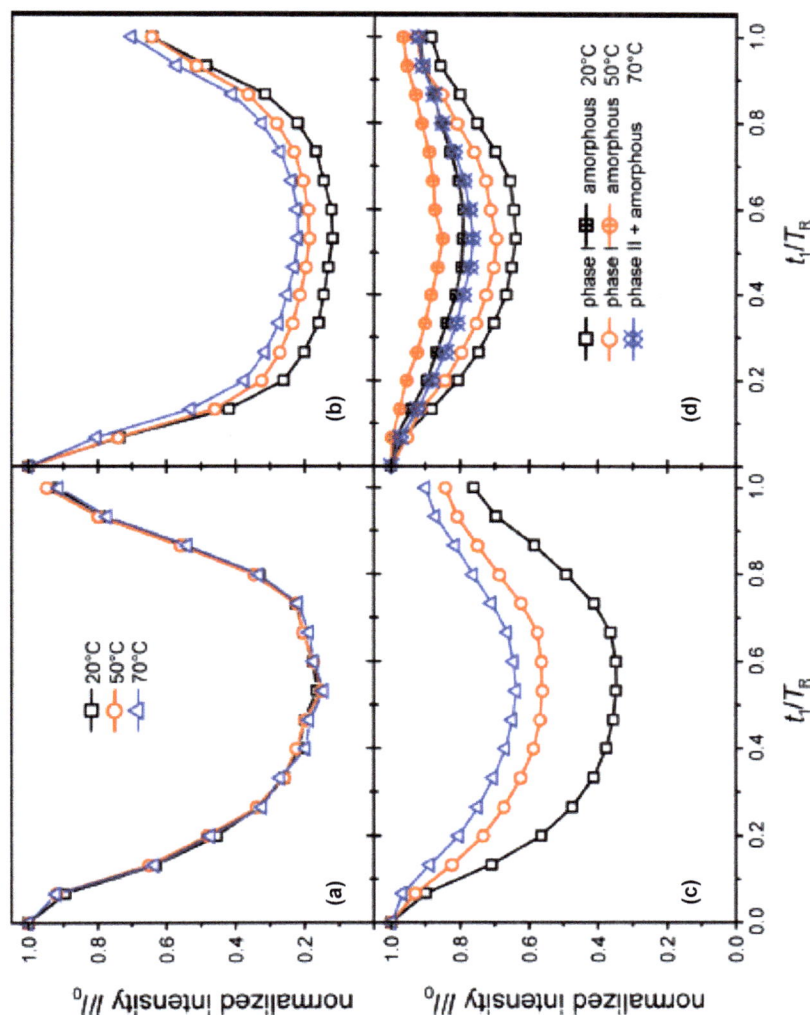

Figure 7.18 DIPSHIFT curves for various P3HT carbons: (a) protonated thiophene carbon; (b) four interior side-chain methylene groups; (c) ε-methylene; and (d) methyl. Reprinted with permission from O. F. Pascui, R. Lohwasser, M. Sommer, M. Thelakkat, T. Thurn-Albrecht and K. Saalwächter, *Macromolecules*, 2010, **43**, 9401–9410. Copyright 2010 American Chemical Society.

(>100 kHz). Comparing the DIPSHIFT patterns in Figures 7.17b and 7.18c, we can see that similar patterns can be interpreted in different ways from the authors claim of rate changes in Figure 7.17b and changes in amplitude in Figure 7.18c. The possibility of distinguishing between amplitude and timescale changes *via* DIPSHIFT curves is laid out in the seminal work on the subject by deAzevedo *et al.*[116]

7.3.4 Domain Sizes and Interfacial Structures in Donor/ Acceptor Blends

One of the most useful applications of solid-state NMR in organic electronics is the ability to measure degrees of mixing and interfacial structures of donor and acceptor molecules in organic photovoltaics (OPVs). Most microscopic methods are unable to easily distinguish between the donor and acceptor phases in a blend as a result of their similar electron densities and chemical structures, and even those that do cannot achieve the sub-Ångström level of spatial resolution that NMR can achieve. NMR has an inherent contrast between the donor and acceptor molecules because of the chemical shift and (potentially, although never observed) dynamics. The size scales of mixing can be directly probed *via* dipole–dipole mediated spin exchange. The rate of exchange between sites can be directly related to the length scale of separation.

If we are interested in measuring the domain size, then the gold standard is ^{1}H spin diffusion NMR. The experimental and analytical details on spin diffusion are given elsewhere.[117,118] Although the spins do not physically move or "diffuse," they are modeled as such in a mean-field sense. Domain sizes are not extracted from spin diffusion data by modeling individual exchange events among the various sites because this would take too much computational time. Rather, domains are extracted by way of Fick's laws and assuming a continuous polarization intensity profile of the two components. Assuming some morphology model, a domain size can be extracted *via* Fick's laws using the interfacial surface area to volume ratio relationships. Of course, the continuous nature of polarization transfer is an assumption, but it is a good one: it has been shown to be generally valid for ^{1}H NMR because of its strong dipolar couplings and high spin densities (>0.1 mol cm^{-3}). Fick's laws are less useful in weakly coupled systems, such as ^{13}C, and may require more extensive modeling to explain the exchange data.

There are several ways to measure domain sizes. One simple way is to just measure T_{1H} (or $T_{1\rho H}$) relaxation time differences among the different components. If the two components exhibit the same T_{1H}, then the two components are mixed on size scales finer than $\langle x \rangle = \sqrt{4DT_{1H}}$, where $\langle x \rangle$ is the domain size diameter, D is the spin diffusion coefficient (typically 0.1–0.8 nm^2 ms^{-1}), and T_{1H} is the relaxation time. If there are T_{1H} differences, the components must be separated on distances greater than $\sqrt{4DT_{1H}^{slow}}$, where T_{1H}^{slow} is the T_{1H} of the slower relaxing component. The same arguments apply for $T_{1\rho H}$ differences. Using $T_{1\rho H}$ and T_{1H} differences, maximum size scale limits were placed on the PCBM domain sizes in blends with P3HT[119] and poly(2-methoxy-5-(3′,7′-dimethyloctyloxy)-1,4-phenylene-vinylene) (MDMO-PPV).[120] The advantage of these experiments is that they can be applied to spun-cast thin film samples (about 0.1 mg) and the analysis is straightforward. In these experiments, we assume that the molecular dynamics of the neat components are similar to the blend, leading to similar values of about 10 kHz $(T_{1\rho H})$ and 100 MHz (T_{1H}) of the spectral densities and ^{1}H spin diffusion coefficients. These assumptions are typically valid, but there have been exceptions.[121] Comparing the ^{1}H line width of the blend with the neat components is a fast diagnostic for the spin diffusion coefficient and is based on established, empirical relations.[122]

Domain sizes can be ascertained directly from one-dimesional chemical shift based (CSB) ^{1}H spin diffusion NMR.[122] The advantages are that: (1) experiments can be straightforwardly run on thin film samples (about 0.1 mg); and (2) no assumption needs be made regarding relaxation times, although we still assume that the spin diffusivity is calculable *via* the same, established empirical relations. One-dimensional CSB ^{1}H spin diffusion NMR (Figure 7.19) is a one-dimensional variant of the original two-dimensional experiment and relies on the establishment of a polarization gradient after a fixed number of homonuclear dipolar coupling cycles (MREV, FSLB, eDUMBO), a variable mixing time, and high-resolution ^{1}H detection under CRAMPS. One-dimensional CSB ^{1}H spin diffusion was used to follow the degree of mixing during thermally induced PCBM crystallization (>150 °C), which was correlated with decreased photoluminescence quenching.[123]

The slope of the ^{1}H spin diffusion plot relates directly to the interfacial area, which can then be related to a domain size *via* Fick's laws. Steep slopes indicate intimate mixing and more gradual slopes suggest coarsening. In the infinitely coarsened limit (*i.e.* a physical mixture), there is no decay on a ^{1}H spin diffusion plot. In the pure domain limit, the number of spins that exhibit a given slope is directly relatable to the amount of material in a given "domain." From ^{1}H spin diffusion NMR, we can directly and unambiguously calculate the absolute volume fraction of the mixed phase in a semicrystalline blend, even in a thin OPV film in which the domain shape is not known.[84] For instance, as shown from Figure 7.19 (lower panel, open circles), the volume fraction of the mixed phase in a thin film blend of P3HT–PCBM that has been spun-cast (from dichlorobenzene) at 2000 rpm (209 rad s^{-1}) and not annealed is about 0.8 because $\Delta M = 0.2$. Quantitative estimates of the size scales of larger domains in most bulk heterojunctions (BHJs) are, unfortunately, more difficult. As shown in Figure 7.19 (lower panel), there are multiple slopes in each BHJ, which indicates that there is composition heterogeneity on multiple length scales (from a few nanometers to almost 100 nm). As it is not known whether the domains are "pure," it is difficult to ascertain a "domain size distribution" from the limited number of data points on a spin diffusion plot (typically 10–20) because the domain shape is not known. For coarse estimates of the domain sizes, it is possible, by way of fitting spin diffusion curves, to bin the number

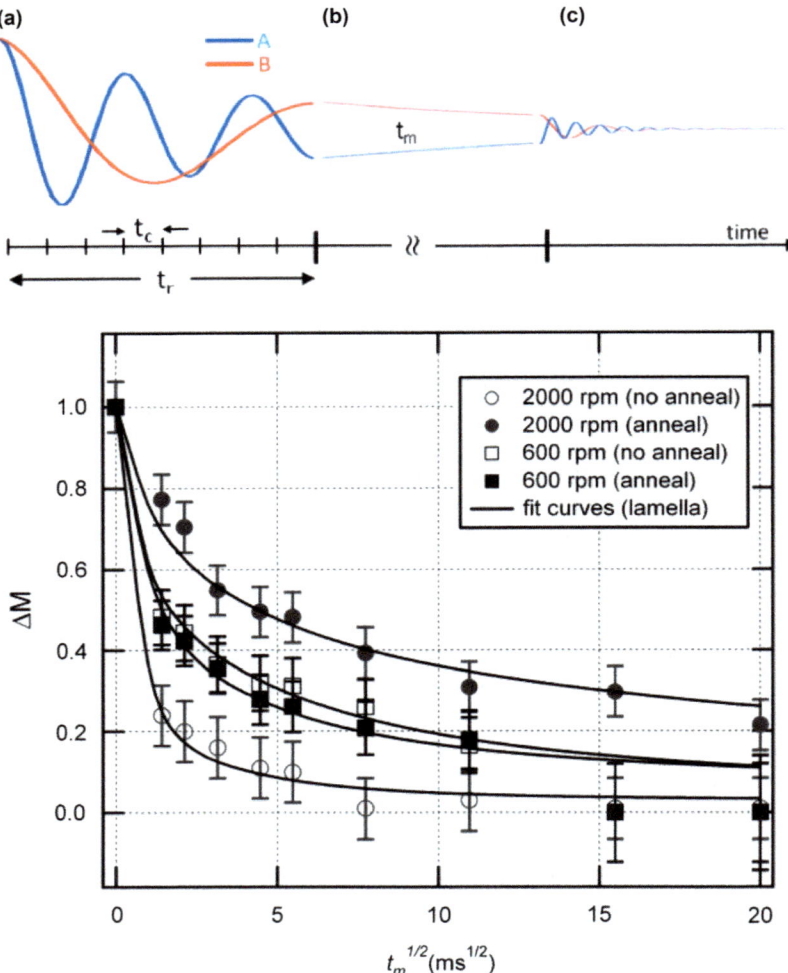

Figure 7.19   Upper panel: schematic diagram of the one-dimensional CSB spin
diffusion experiment which utilizes (a) a preparation of some fixed
number ($n \times t_r/t_c$) of decoupling cycles (t_c = cycle time, t_r = rotor period,
n = integer), (b) a variable mixing time, t_m; and (c) CRAMPS detection.
Lower panel: ^{1}H spin diffusion curves of thin films of P3HT–PCBM
blends for various spin-casting rates and annealing protocols, where
ΔM refers to the normalized change in polarization.

Part (b) is from R. C. Nieuwendaal, H. W. Ro, D. S. Germack, R. J. Kline,
M. F. Toney, C. K. Chan, A. Agrawal, D. Gundlach, D. L. VanderHart and
D. M. DeLongchamp, *Adv. Funct. Mater.*, 2012, **22**, 1255–1266. Copyright
© 2012 by John Wiley & Sons, Inc. Reprinted by permission of John
Wiley & Sons, Inc.

of spins that belong to a given domain to achieve a "domain size distribution" (although we use that term with caution), with the caveat that ^{1}H spin diffusion cannot by itself distinguish pure domains from impure domains because the "domain" from the point of view of spin diffusion simply means "size scale of composition heterogeneity."[84] For instance, in the case of large domains, which exhibit significant spin polarization gradients ($\Delta M > 0.1$) at long mixing times (>10 ms$^{1/2}$), equations have been established[123] that relate the persistent spin polarization value (ΔM) to the composition and volume fraction of the domain in question. Unfortunately, there are typically ranges of composition and volume fraction that fit a given spin diffusion curve. Perhaps these issues could be alleviated with more sophisticated modeling and comparison with other techniques such as scattering.

The donor/acceptor interface in a BHJ can also be characterized *via* two-dimensional ^{13}C–^{1}H heteronuclear correlation (HETCOR) spectroscopy.[124–126] In HETCOR, relative proximities at the interface are garnered by observing correlations that arise from the ^{1}H spins of one molecule (acceptor) with the ^{13}C spins of the other (donor). The analysis is similar in spirit to the first two-dimensional CSD ^{1}H spin diffusion NMR experiment,[122] but with ^{13}C CPMAS detection instead of direct ^{1}H detection. Although a promising technique as a result of the better spectral resolution of ^{13}C, HETCOR suffers from a poorer signal-to-noise ratio than ^{1}H NMR and much longer experiment times as a result of the multidimensional nature of the experiment. Hence it is rare to achieve the multiple mixing times needed for precise distances or relative proximities and it is unlikely/impossible to be achievable with small sample sizes such as thin films. Correlation peaks in two-dimensional HETCOR arise indirectly *via* ^{1}H–^{1}H spin diffusion then ^{1}H–^{13}C cross-polarization, not through direct ^{13}C–^{1}H couplings as in separated local field techniques such as DIPSHIFT. When HETCOR is performed as a function of contact time (with no other mixing time), correlation peaks arise by way of ^{13}C–^{1}H cross-polarization kinetics (which is typically due to direct or nearest neighbor ^{13}C–^{1}H couplings) and ^{1}H–^{1}H spin mixing, so establishing proximity relations based on correlated peak intensities can be ambiguous. Preferably a ^{1}H–^{1}H homonuclear dipolar decoupling scheme such as frequency-switched Lee-Goldburg or phase-switched Lee-Goldburg is implemented on the ^{1}H channel during the cross-polarization to minimize ^{1}H–^{1}H mixing and proximities are established *via* direct ^{1}H spin diffusion.[121,127] In this way, domain sizes can be unambiguously measured by monitoring the growth of a correlation peak as a function of mixing time similar to ^{1}H spin diffusion (Figure 7.19, lower panel). Alternatively, relative proximities could be measured at a single mixing time by comparing the intensity of a correlation peak relative to some other correlated peak of a known coupling, such as a directly bonded ^{13}C–^{1}H pair. Trying to interpret a single correlation peak from HETCOR for a single mixing time can be ambiguous. Variations such as the MEdium and LOng DI (MELODI) HETCOR experiment,[128] which attenuates direct-bond ^{1}H–^{13}C correlations, should be performed to alleviate these ambiguities.

7.3.5   Future Prospects

The future impact of solid-state NMR in the field of organic electronics looks bright. There are multiple reasons for optimism. First, there seems to be a shift away from laboratory-tested OPV cells using spin-casting in favor of more sample-preserving and (potentially) large area deposition methods such as doctor blading, flow-coating, and roll-to-roll coating.[129] As the technology continues to grow from laboratory test bed, to demonstration modules, to industry-level large area film deposition, this will only mean more sample for NMR spectroscopists to load into their rotors to achieve the always-desired higher signal-to-noise ratios. The available sample mass has been the single largest barrier in the field, with common test films containing so little (about 0.1 mg) and solid-state NMR spectroscopists needing so much (>10 mg). Second, there is a trend in the field of NMR to gain more and more signal from less and less sample. With the advent of hyperpolarization NMR techniques such as dynamic nuclear polarization,[130] which is now commercially available for MAS experiments, it may be that achieving ^{13}C NMR spectra in small sample volumes (<1 mg) will soon be possible.

7.4   Transmission Electron Microscopy of Organic Semiconductors

7.4.1   Introduction to Transmission Electron Microscopy

Transmission electron microscopy is a powerful method for characterizing the nano- and atomic-scale structure and chemistry of materials based on their electron-scattering behavior. In TEM, a high-energy beam (typically 60–300 keV) is directed *via* a series of electromagnetic lenses onto a thin section of material and the forward-scattered electrons are collected by a post-specimen detector. The main advantage of the technique is the high spatial resolution that can be achieved by virtue of the much shorter wavelength of high-energy electrons relative to photons and X-rays. Although residual lens aberrations and instabilities prevent the theoretical limit of spatial resolution being reached, atomic-scale imaging and spectroscopy can be readily achieved.

There are two major modes of TEM analysis: conventional TEM (CTEM) and scanning TEM (STEM), both of which are available in most modern instruments. In CTEM, a broad, parallel beam is used to simultaneously illuminate a relatively large area of the specimen and the desired image is collected in parallel. By contrast, in the STEM mode a highly focused probe is scanned over the specimen and the signal is collected in series on a pixel-by-pixel basis.

Many different image signals can be collected in TEM. The most common of these is the bright field signal, which is formed by limiting the maximum scattering angle allowed to contribute to the image. In CTEM, this is accomplished by inserting an aperture in the back focal plane of the microscope's objective lens centered around the transmitted beam, whereas in STEM the post-specimen optics are configured so that any higher angle scattering intensity is excluded from an on-axis detector. In this way, areas with a greater thickness and/or density will appear darker than those that are thinner and/or less dense. If the specimen contains crystalline domains, then the objective aperture or post-specimen optics can be chosen in such a way as to exclude any electrons that have undergone Bragg diffraction, thus revealing structural and orientational differences within the specimen.

A variety of dark field (DF) signals are also available by TEM. For CTEM, the most commonly used DF imaging technique involves centering the objective aperture on a specific Bragg reflection or family of reflections and forming the image exclusively with this signal. In STEM, the most common approach is to use annular detectors that integrate the scattered intensity over a given angular range. Low- and high-angle annular DF imaging (LAADF and HAADF, respectively) can be very useful for generating images with high contrast that can often be directly interpreted in terms of differences in the mass thickness or composition of the specimen at any given point.

7.4.2 Challenges for Characterizing Polymeric Materials in TEM

The spatial resolution that can be achieved in TEM, combined with the wide variety of signals and collection geometries that can be used, make the technique especially powerful. However, the robust characterization of polymeric materials remains challenging, largely because they are often extremely sensitive to irradiation by the electron beam.[131–135] Like all organic matter, polymers are radiation-sensitive and the high-energy electrons used in TEM can cause very rapid chemical and structural alterations, to the point where a crystalline specimen can be quickly rendered amorphous by the beam before any meaningful image can be formed. This is illustrated dramatically in the selected-area electron diffraction (SAED) patterns shown in Figure 7.20. The pattern in the left-hand panel was acquired from a P3HT:PCBM blend on initial exposure to the electron beam, whereas the pattern in the right-hand panel was acquired 60 s later from the same area. The intense, narrow ring due to the {010} family of reflections in P3HT is readily visible in the initial pattern. This ring continually broadens and weakens as the crystal structure breaks down under irradiation until, after 60 s, the ring is no longer visible, indicating that the short-range order of the crystalline regions has been completely disrupted. Although these patterns were generated using a much higher beam current than typically used when analyzing these materials, the rapidity with which the crystalline order was completely destroyed serves as a cautionary example.

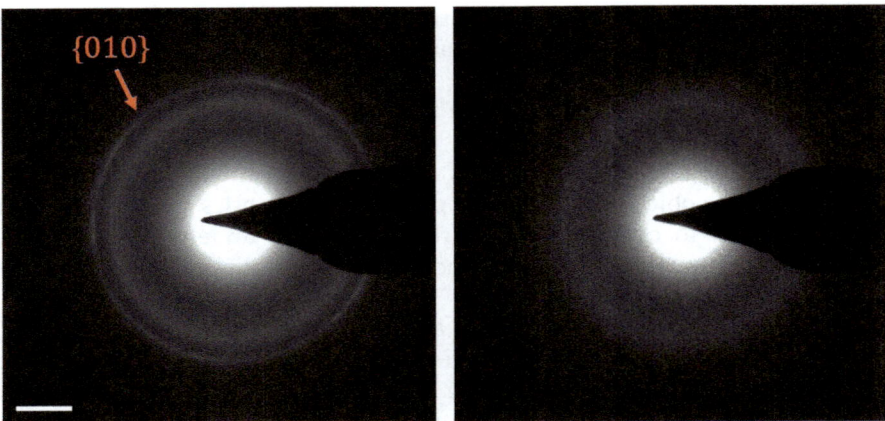

Figure 7.20   Selected-area electron diffraction patterns acquired from a 50:50 P3HT:PCBM blend (1:1 *m/m*) showing the effects of electron beam irradiation. The pattern on the left was acquired immediately after initial exposure to a 300 keV beam in the transmission electron microscope. The sharp ring due to the P3HT {010} family of lattice spacings has completely disappeared in the pattern on the right, which was acquired after 60 s of exposure to the beam.

The degree of specimen alteration can be significantly reduced by the use of lower beam currents, low-dose image acquisition approaches, and by cryogenically cooling the specimen using a specialized holder. Rather counter-intuitively, damage can also be reduced by using higher primary beam energies because the inelastic scattering cross-sections are far lower at higher energies.

In addition to the challenges associated with beam irradiation, polymers are also inherently difficult to analyze by TEM because they are composed mostly of light elements that only weakly scatter the incident electrons. This produces images with little to no contrast, especially at high primary beam energies. This often necessitates increased exposure times to collect sufficient signal to make the required measurement, a requirement that obviously works against the previous discussion of beam damage in which the need for a minimum electron dose was emphasized.

7.4.3   Methods for Characterizing Order and Morphology in Polymer Materials

Despite these challenges, TEM remains an indispensable tool for studying the local structure and morphology of polymeric materials and has played a key part in elucidating the structure of new materials and their relationship with processing parameters. The following sections provide an overview of the various approaches to characterizing the structure and morphology of polymers that are available to the TEM analyst.

7.4.3.1 Electron Diffraction of Polymeric Materials

Although X-ray diffraction techniques are preferred for the characterization of polymeric single crystals as a result of their greater precision, electron diffraction is often the only technique available for the determination of the structure of single crystals when crystals large enough to be suitable for X-ray analysis cannot be produced.[136,137] In CTEM, the spatial resolution of the resulting SAED patterns is limited by an area-selecting aperture located at the image plane of the objective lens, which is typically a few hundred nanometers in diameter. Once the area of interest has been located at low magnification, a pattern must be acquired very quickly to limit the alteration of the crystal structure.

In STEM mode analysis, the spatial resolution is limited by the size of the electron beam at the specimen (≤ 1 nm), plus any broadening of the beam due to interactions with the specimen. Analysis in this mode has further advantages when the electron dose is considered. As the diffraction data are collected serially, only the small areas of the specimen from which data are being collected are exposed to the beam at any given time, in contrast with CTEM SAED, in which the entire crystal is exposed at all times. This fact was utilized by Geiss *et al.*[137] to determine the full structure of poorly crystallized polypyrrole and sub-micron crystals of poly(*p*-hydroxybenzoic acid) (PHBA), both of which were inaccessible *via* X-ray diffraction. For a full structural determination, electron diffraction patterns had to be acquired from several different crystallographic orientations. This was accomplished by simply tilting the crystal between acquisitions, re-positioning the beam onto a fresh point on the crystal, and then acquiring the next pattern. One such pattern, collected along the basal plane of PHBA, is shown in Figure 7.21.

7.4.3.2 Dark Field Techniques for Imaging Ordered Polymer Materials

The reciprocal space information present in the electron diffraction patterns discussed in the preceding section can also be used to produce real space images that show the spatial distribution of specific crystalline orientations. Such DF images are formed by tilting the incident electron beam so that a specific diffraction beam is centered in the microscope's objective aperture, rather than the transmitted beam used in bright field imaging. In this way, only the specific diffracted beam plus any diffuse background intensity that may be present are allowed to contribute to the intensity in the image. Alternatively, in a technique known as hollow cone DF, the beam can then be dynamically tilted about this azimuth and an image integrated over the entire family of reflections at this orientation. This process is shown schematically in Figure 7.22.

A particularly elegant use of these techniques was reported by Zhang *et al.*,[138] who performed a textural analysis of poly(2,5-bis(3-alkylthiophen-2-yl)thieno[3,2-*b*]thiophene (pBTTT) thin films using DF TEM techniques. The

Figure 7.21 STEM electron diffraction pattern from a sub-micrometer crystal of
PHBA oriented along the basal plane.
Reprinted with permission from R. Geiss, G. Street, W. Volksen and
J. Economy, *IBM J. Res. Dev.*, 1983, **27**, 321–329.

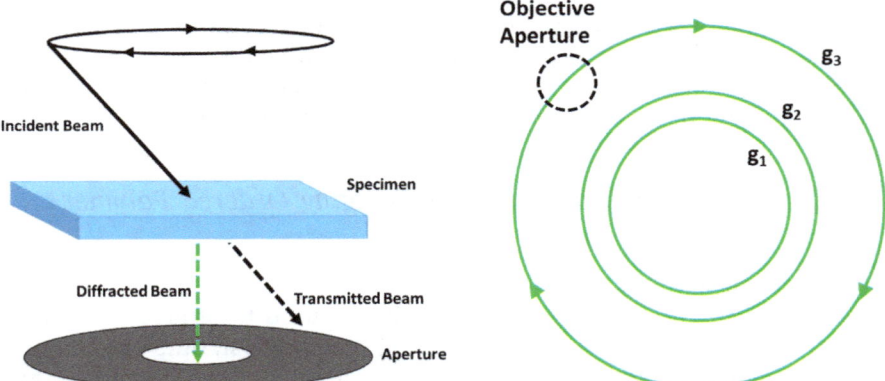

Figure 7.22 Schematic illustration of hollow cone DF imaging technique. The
incident beam is tilted and rotated azimuthally (left) in such a way
that the desired reciprocal lattice distance remains centered in the
objective aperture (right).

SAED patterns and images resulting from DF analysis are shown in
Figure 7.23. Close inspection of the images in Figure 7.23d–g shows that the
intensity in the domains changes dramatically with the orientation of the
diffracted beam used to form the image. By collecting the full series of azimuthal DF images and carefully classifying each domain, a full crystal
orientation map was produced. This showed that pBTTT exhibits an in-plane

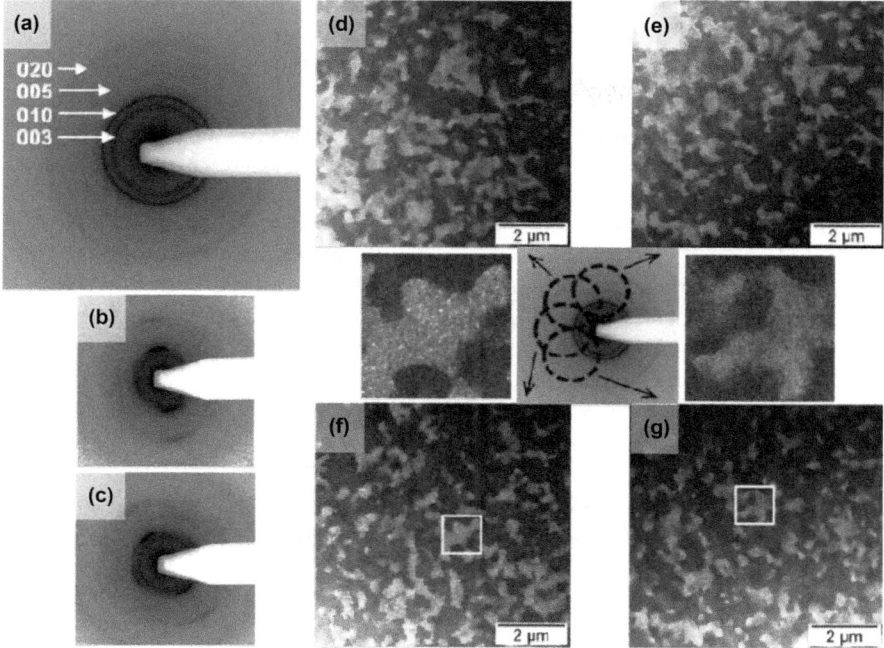

Figure 7.23 Hollow cone DF imaging analysis of a pBTTT-C_{14} film. (a) Diffraction pattern from an area about 8 μm in diameter; (b, c) diffraction patterns from areas about 250 nm in diameter; (d–g) DF images with different beam-tilt configurations as illustrated on the diffraction pattern (center inset). The position of objective aperture relative to the diffraction pattern is shown. White squares in (f) and (g) correspond to regions that are shown as magnified images immediately above each of the images.
Reprinted with permission from X. Zhang, S. D. Hudson, D. M. DeLongchamp, D. J. Gundlach, M. Heeney and I. McCulloch, In-Plane Liquid Crystalline Texture of High-Performance Thienothiophene Co-polymer Thin Films, *Adv. Funct. Mater.*, **20**, 4098–4106. Copyright (c) 2010 WILEY-VCH Verlag GmbH & Co. KGaA, Weinheim.

liquid crystalline structure characterized by low-angle domain boundaries, in contrast with small molecule semiconductors or P3HT-type polymers that exhibit high-angle domain boundaries.

7.4.3.3 Low-dose High-resolution Imaging

In TEM instruments equipped with highly coherent electron sources, such as a cold-field or Schottky-assisted field-emission gun, the phase contrast produced by interference of the wave function of the incident electron beam with the periodic structure within a crystalline specimen can be used to inspect the atomic ultrastructure.[139] These so-called high-resolution TEM (HR-TEM) images are highly dependent on the periodic structure within the specimen, making the technique one of the most sensitive to electron beam irradiation damage, which has limited its application to polymer research. However, when suitable precautions are taken during sample preparation and data acquisition, HR-TEM is capable of providing essential information about the local structure of crystalline polymers (see Martin and Thomas[140] and references cited therein).

To successfully image the lattice structure of polymeric materials, the electron dose delivered to the specimen area under investigation must remain below the threshold for significant alteration prior to the image being collected. Many modern TEM instruments are equipped with software and hardware accessories that aid in this task. The most important of these is a beam blanker, which deflects the beam away from the specimen and ensures that the area of interest is only a receiving dose when data are being collected. In addition to this, the system will typically be equipped to allow the beam to be deflected a short distance away from the area to be analyzed so that focusing and the correction of astigmatism can be carried out on a different region of the specimen. Low-dose imaging also necessitates the use of very low beam currents, short exposure times, and a broad beam that spreads the dose over a large area. Although this process is by no means simple, it can yield great rewards when carried out carefully. An example is shown in Figure 7.24, which shows the low-dose HR-TEM analysis of a thin

Figure 7.24 Low-dose high-resolution TEM analysis of 100 nm thick P3HT film showing the (100) fringes of the lamellar crystals. The inset fast Fourier transform of the image shows the peaks at 1.6 nm spacing.
Reprinted with permission from L. F. Drummy, R. J. Davis, D. L. Moore, M. Durstock, R. A. Vaia and J. W. P. Hsu, *Chem. Mater.*, 2011, **23**, 907–912. Copyright 2011 American Chemical Society.

film specimen of P3HT reported by Drummy *et al.*[141] The crystalline lattice spacing at 1.6 nm, corresponding to the (100) interlamellar spacings, is clearly visible in the image and even more so in the inset fast Fourier transform image.

7.4.3.4 Characterization of Organic Material via Electron Energy Loss Spectroscopy and Energy-filtered Imaging

In addition to the imaging signals already discussed, most TEM instruments are also capable of assessing the specimen *via* electron energy loss spectroscopy (EELS). In this technique, the energy lost by electrons as a result of a variety of inelastic scattering processes can be analyzed with the aid of an energy-dispersive spectrometer to determine the chemical and dielectric properties of the specimen. Alternatively, the spectrometer can be configured in such a way that a spatially resolved image can be formed using only those electrons that have lost a specific amount of energy. This latter approach is known as energy-filtered imaging and is especially powerful for polymeric materials as the dose is distributed over a large field of view.

A thorough discussion of this topic is beyond the scope of this chapter and the reader is referred to the standard textbook by Egerton[142] for an in-depth treatment. For our purposes, we will limit the discussion of the EELS signal to two phenomena: inner shell ionization and plasmon excitation. In the former, an incident electron ejects an inner shell electron from an atom in the specimen and, in the process, loses an amount of energy that is characteristic of the elemental identification of the atom in question as well as the energy level of the ejected electron. This signal can be utilized for quantitative compositional determination at very high spatial resolutions. It can also reveal information related to the chemical structure and bonding of the atoms in the material because the fine structure of the spectral features provides information about their density of states.

In addition to the core loss EELS features, the low-loss region of the energy loss spectrum (about 0–100 eV) is dominated by the excitation of bulk plasmons in the material due to the collective oscillation of valence electrons in response to the incident beam.[143] For carbonaceous materials, a plasmon peak is often observed near the 23 eV loss, but this position is sensitive to changes in density and electronic structure. Indeed, several groups have exploited these subtle differences to generate contrast in multiphase organic films.[144–148] As the differences in spectral output for many organic materials are so similar, difficulties often arise in extracting meaningful data from the collected spectra. For example, as shown in Figure 7.25, Yakovlev and Libera[144] used a routine based on multiple least-squares fitting of EELS reference spectra to analyze a nanocolloid of poly(dimethyl siloxane) (PDMS) and acrylate. By fitting the collected data using the reference spectra, they were able to generate phase distribution maps for the PDMS, acrylate and the surrounding ice matrix.

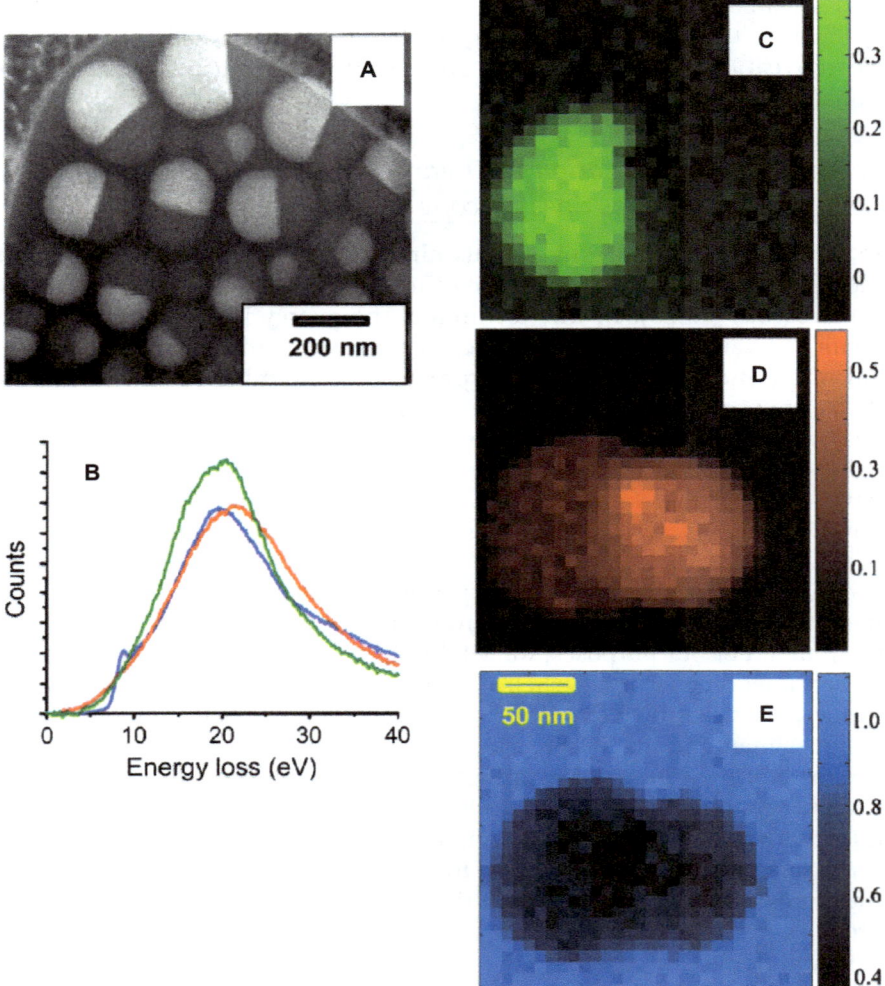

Figure 7.25 STEM–EELS analysis of a frozen section of a PDMS:acrylate nanocolloid. STEM image (A) shows the distribution of particles with the ice matrix. The reference EELS spectra (B) were collected from pure phases of the three components: PDMS, acrylate, and ice. (C) Phase maps extracted from a hyperspectral image by multiple least-squares fitting showing the spatial distribution of the PDMS (green), acrylate (red), and ice (blue).
Reprinted from S. Yakovlev and M. Libera, Dose-limited spectroscopic imaging of soft materials by low-loss EELS in the scanning transmission electron microscope, *Micron*, **39**, 734–740, Copyright (2008), with permission from Elsevier.

7.4.3.5 Three-dimensional Characterization via Electron Tomography

Despite the power of the TEM techniques discussed here, one important caveat must be considered: each of these approaches results in a two-dimensional projection through the specimen and contains no directly interpretable information about the through-thickness dimension. One way of retrieving this missing information is through electron tomography, which involves acquiring projection images of the specimen over a range of orientations and then reconstructing a model of the three-dimensional structure.[149] This has been particularly useful for the characterization of BHJ organic solar cell devices[148,150,151] because the morphological features of the three-dimensional structure are critical for charge transport and collection. In examples where contrast generation between the donor and acceptor phases has proved difficult, tomographic techniques have been combined with energy-filtered TEM imaging to more robustly assess the morphology.[147,148] Tomographic characterization of cryogenically stabilized BHJ films has also been accomplished using low-dose exposure techniques, resulting in a three-dimensional structural determination of the distribution of crystalline and amorphous domains.[152]

7.5 Grazing Incidence Scattering

7.5.1 Introduction to X-ray Scattering

X-ray scattering measurements are by far the most commonly used method to examine the extent of long-range order in organic semiconductors, the size of domains, and the coherence of order within domains. X-ray measurements typically involve a narrow and collimated incident beam that passes through a sample. Via Bragg[153,154] processes and as a result of heterogeneities in the electron density within the sample, the X-ray beam is scattered, exiting the sample at angles other than the incident angle. The intensity of the scattered light and its angle relative to the incident beam deliver information about the internal structure of the sample. Although laboratory-based X-ray sources (filament, rotating anode) are relatively ubiquitous, organic semiconductors often require a synchrotron source for low-noise measurements. The higher intensity is required because (1) organic semiconductors typically exhibit a low extent of order, (2) the films are relatively thin, so little material is in the beam, and (3) the measurement is often made with a fixed low-angle incidence and the scattering is off-specular.

7.5.2 Grazing Incidence X-ray Diffraction

Grazing incidence X-ray diffraction (GIXD) has become a common measurement technique to quantify the extent of regular molecular packing in organic semiconductor thin films. In large crystals, or powders containing reasonably large crystals, it is often possible to determine an explicit structure factor that can be used to extract the molecular arrangement within the unit cell. This practice is common in the analysis of small molecule and oligomeric semiconductors. After synthesis, a macroscopic single crystal is grown using solution methods or a zone sublimation furnace and the crystal is then subjected to a single-crystal synchrotron diffraction experiment with subsequent computer structure refinement to determine the explicit unit cell packing.

In thin films of organic semiconductors, however, it is often not possible to determine the explicit unit cell arrangement using GIXD. The packing, particularly the π–π stacking of polymer semiconductors, within the unit cell can be paracrystalline,[155,156] lacking long-range order and with a peak intensity pattern that is more diagnostic of the low extent of order rather than the molecular arrangement within the crystal. The expectation should be that GIXD will provide information about interplanar d-spacings within the unit cell (perhaps some, perhaps all) and the orientation distributions of those diffracting planes relative to the film normal. The relative extents of order can be compared by the diffraction strength when the two films are composed of the same material packed in the same polymorph.

Because the application of GIXD to organic semiconductors has been covered authoritatively by a detailed review article,[157] and we have ourselves treated this topic in textbooks[158,159] and review articles,[160–162] we will only discuss it briefly here, with an emphasis on the important and commonly missed details of its application.

The GIXD measurement is typically performed at a synchrotron in a grazing arrangement (Figure 7.26). The incident angle α is typically about 0.08–0.18°. A lower angle will measure only the film surface, whereas a larger

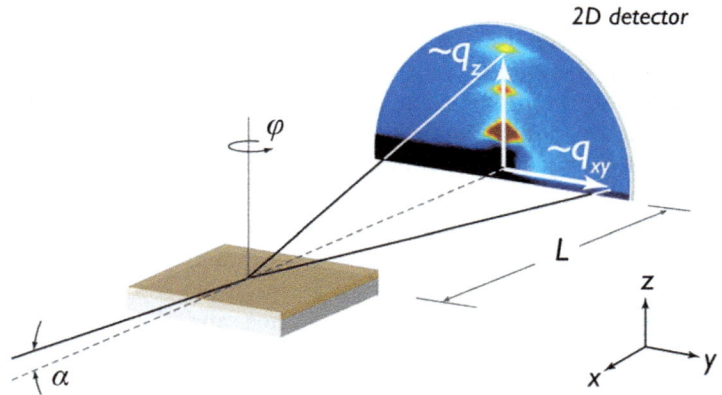

Figure 7.26 Typical GIXD experimental arrangement.
Adapted with permission from J. Rivnay, S. C. B. Mannsfeld, C. E. Miller, A. Salleo and M. F. Toney, *Chem. Rev.*, 2012, **112**, 5488–5519. Copyright 2012 American Chemical Society.

α will measure the whole film; the transition from surface- to bulk-sensitive occurs at a critical angle that varies depending on the material being measured. The pattern is typically measured on a two-dimensional detector and is collected as an image for analysis.

Although it is possible to measure GIXD with a point detector, it is difficult to cover the entire qz–qx space and therefore it is difficult to measure the full distribution of possible orientations. As a result, it is not usually possible to draw conclusions about the relative crystallinity from GIXD data collected with a point detector. Similarly, specular X-ray diffraction (*e.g.* the classic θ–2θ measurement in a powder diffractometer) cannot be used for relative crystallinity measurements of organic semiconductor thin films with an anisotropic distribution of crystal orientations (essentially all organic semiconductor films). The specular measurement covers only a very small part of the potential orientation distribution in which the crystal planes are oriented nominally parallel to the substrate plane.

Relative crystallinity measurements, and an assessment of the orientation distribution, can be made from the two-dimensional GIXD data by constructing a full pole figure. This can be done by combining the two-dimensional GIXD data with the narrow specular component described earlier.[163] Important corrections include a recalculation of the qz–qx coordinates to describe the real scattering angle, film thickness correction, and a correction based on the thin film geometry to account for sampling different amounts of the in-plane crystal orientation distribution at different scattering angles.[162,164] This last correction is one of the most significant alterations of the raw data; it can change the shape of the perceived orientation distribution tremendously.

The corrected pole figure can be used to assess the relative diffraction strength of the peak among various samples, provided that the material and crystal polymorph are identical and all of the appropriate normalizations have been performed. The relative diffraction strength is given by the ratio of the integrated curve areas of the corrected pole figures. Although this measurement does describe the relative crystallinity, there is currently no robust means by which to extract the absolute crystallinity from GIXD for the majority of organic semiconductors because the structure factors are not known. Relative crystallinities can, however, be compared with the absolute crystallinities established by methods based on other physical principles such as DSC.[28]

7.5.3 Grazing Incidence Small-angle X-ray Scattering

X-ray scattering techniques can also be used to determine the structure in organic semiconductors at larger length scales. If a material has significant compositional heterogeneity, such as a BHJ blend for OPVs, there may be sufficient electron contrast for X-ray scattering experiments to reveal the length scale of domains within the composite.[165,166] In compositionally homogeneous samples, however, there is unlikely to be sufficient contrast for the technique to detect domains unless the contrast is between the sample and air, which typically describes only surface features.

Grazing incidence small-angle X-ray scattering (GISAXS) measurements are typically performed in the geometry shown in Figure 7.26. The sample to detector distance is, however, far extended with a vacuum flight tube to mitigate air scattering. The resultant smaller q values can describe domain sizes typically in the 10–100 nm range. Unlike a GIXD pattern, which often shows significant structure, GISAXS patterns of organic semiconductors are typically dominated by scattering decay, within which a shoulder or peak may indicate a dominant length scale. A more complex scattering pattern is only expected in colloidal crystals or block copolymers that have domain structures arranged with extensive long-range order.

Similar information can be collected with small-angle neutron scattering (SANS). SANS has found special application in OPVs[165,167,168] because there is a natural contrast between the proton-deficient fullerene and other more proton-rich organic semiconductors. Unlike GISAXS, SANS measurements are performed in a transmission geometry, which can require a multi-film stack to achieve sufficient signal to evaluate the scattering pattern. As with GISAXS, there is little expectation that conventional SANS will be useful in the analysis of compositionally homogeneous organic semiconductor films.

7.5.4 Polarized Resonant Soft X-ray Scattering

A new small-angle scattering technique has been developed in recent years that may have significant application in the study of organic semi-conductors. This emerging method is resonant soft X-ray scattering (RSoXS),[169–171] which uses a transmission geometry similar to SANS, but derives its contrast in a unique way. RSoXS is performed with incident energies in the soft X-ray region of the spectrum, near the resonant edge of elements within the sample. Contrast can be derived by fluctuations in the X-ray scattering length density that occur near peaks in the near-edge X-ray absorption fine structure (NEXAFS) spectrum.[172] The technique thus combines the ability of small-angle scattering to measure domain sizes with the fine chemical element and bond specificity of NEXAFS spectroscopy.

In chemically heterogeneous samples, such as a BHJ, RSoXS provides a straightforward means to "dial in" contrast between phases. However, the real power of the technique shines when it is applied to compositionally homogeneous samples. When the measurement is applied with a polarized incident beam, in certain cases it can derive contrast from fluctuations in orientation within the sample. In this mode the technique is referred to as polarized RSoXS (P-RSoXS). Figure 7.27 shows the application of P-RSoXS to films of pBTTT prepared by different methods.[173] The technique has been shown not to be sensitive to the well-known terrace sizes of pBTTT,[174] but rather the true underlying length scale of the in-plane crystal orientation fluctuation that was earlier revealed by DF TEM.[138] The technique has

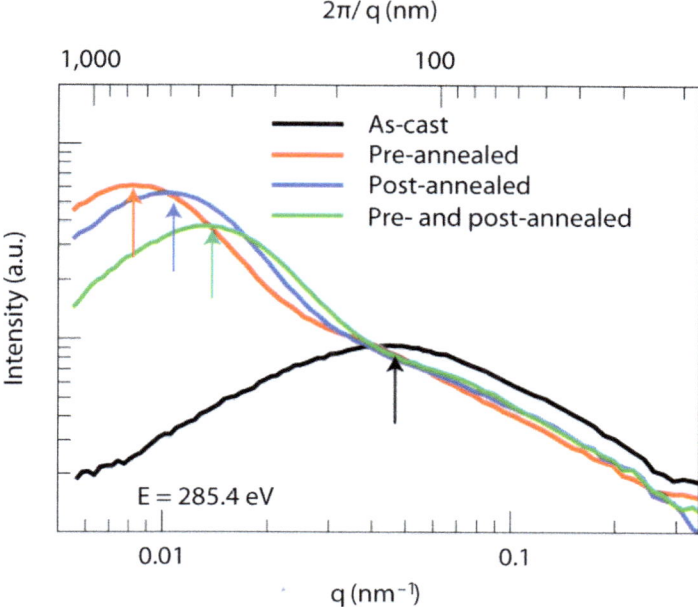

Figure 7.27 Application of P-RSoXS to pBTTT thin films to derive domain length scales.
Adapted with permission from B. A. Collins, J. E. Cochran, H. Yan, E. Gann, C. Hub, R. Fink, C. Wang, T. Schuettfort, C. R. McNeill, M. L. Chabinyc and H. Ade, *Nat. Mater.*, 2012, **11**, 536–543. Copyright © 2012 Macmillan Publishers Limited.

promise for a range of structure measurement issues, including assessing the orientation fluctuations in materials that exhibit little to no scattering in GIXD, and as a means for determining orientation at key interfaces in organic semiconductor-based devices.

7.5.5 *In situ* X-ray Scattering Studies of Structure Evolution

A modern theme in the application of X-ray scattering techniques to organic semiconductors is the advent of *in situ* studies that reveal the structure evolution of organic semiconductor films during a casting and drying process. Hard X-rays are particularly amenable to *in situ* X-ray scattering studies because they can pass through air for considerable distances. It may be some time before soft X-ray techniques such as P-RSoXS can be adapted to measure structure evolution. The typical approach is to perform the casting process at a high-flux synchrotron source while the beam is on, with continuous data collection using a fast two-dimensional imaging detector (about 100 ms per frame or faster).

The application of *in situ* GIXD and GISAXS has been especially powerful in the investigation of organic semiconductor-based systems.[175–192]

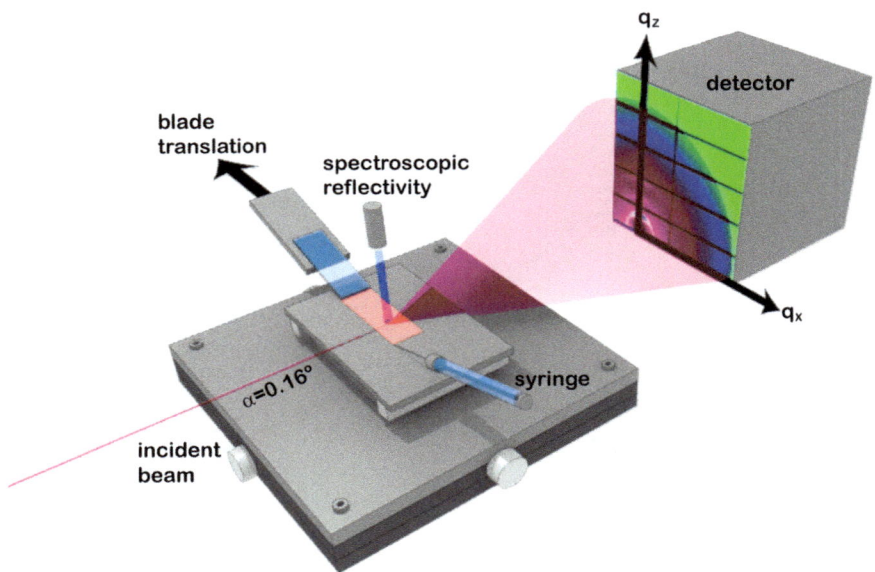

Figure 7.28   *In situ* measurement of GIXD from a blade-coated film.
Adapted with permission from L. J. Richter, D. M. DeLongchamp, F. A.
Bokel, S. Engmann, K. W. Chou, A. Amassian, E. Schaible and
A. Hexemer, *Adv. Energy Mater.*, 2015, **5**, 1400975. Copyright © 2014
WILEY-VCH Verlag GmbH & Co. KGaA, Weinheim.

A particular emphasis in recent years has been on the structure evolution of
BHJ blends for OPVs. An example of an *in situ* measurement scheme for
blade-based coating methods for BHJ OPVs is shown in Figure 7.28. The
solution is dispensed and cast remotely inside the X-ray hutch. The fluid
film thickness is collected simultaneously using spectroscopic reflectivity.

In situ X-ray scattering methods such as that described in Figure 7.28 can
reveal complex structural evolution routes during solidification. In OPV BHJ
films based on P3HT, for example, the technique revealed how formulation
additives affect the P3HT crystallization by acting as either direct or indirect
plasticizers.[189] The rich crystallization behavior of pure small molecule organic semiconductors has also been revealed by *in situ* X-ray scattering
studies.[193,194] The application of *in situ* methods to evaluate structure evolution is not only a fascinating frontier of materials science, it also has
immense practical value as it can directly inform process design and control
strategies for the manufacturing of organic semiconductor-based products
and devices.

References

1. R. Noriega, J. Rivnay, K. Vandewal, F. P. V. Koch, N. Stingelin, P. Smith,
   M. F. Toney and A. Salleo, *Nat. Mater.*, 2013, **12**, 1038–1044.
2. Y.-K. Lan and C.-I. Huang, *J. Phys. Chem. B*, 2009, **113**, 14555–14564.

3. S. A. Mollinger, B. A. Krajina, R. Noriega, A. Salleo and A. J. Spakowitz, *ACS Macro Lett.*, 2015, 708–712.
4. W. W. Wendlandt, P. K. Gallagher and E. A. Turi, *Thermal Characterization of Polymeric Materials*, Academic Press, New York, 1981, pp. 3–90.
5. B. Wunderlich, *Thermal Analysis of Polymeric Materials*, Springer, New York, 2005.
6. S. M. Sarge, E. Gmelin, G. W. H. Höhne, H. K. Cammenga, W. Hemminger and W. Eysel, *Thermochim. Acta*, 1994, **247**, 129–168.
7. E. Gmelin and S. M. Sarge, *Pure Appl. Chem.*, 1995, **67**, 1789–1800.
8. S. M. Sarge, W. Hemminger, E. Gmelin, G. W. H. Höhne, H. K. Cammenga and W. Eysel, *J. Therm. Anal.*, 1997, **49**, 1125–1134.
9. S. M. Sarge, G. W. Höhne, H. K. Cammenga, W. Eysel and E. Gmelin, *Thermochim. Acta*, 2000, **361**, 1–20.
10. G. Vanden Poel and V. B. F. Mathot, *Thermochim. Acta*, 2006, **446**, 41–54.
11. ASTM Standard D3418, 2012e1, Standard Test Method for Transition Temperatures and Enthalpies of Fusion and Crystallization of Polymers by Differential Scanning Calorimetry, 2012.
12. C. M. Guttman and H. Flynn Joseph, *Anal. Chem.*, 1973, **45**, 408–410.
13. V. B. F. Mathot and M. F. J. Pijpers, *J. Therm. Anal.*, 1983, **28**, 349–358.
14. V. B. F. Mathot and M. F. J. Pijpers, *Thermochim. Acta*, 1989, **151**, 241–259.
15. W. F. Hemminger and S. M. Sarge, *J. Therm. Anal.*, 1991, **37**, 1455–1477.
16. M. Alsleben, C. Schick and W. Mischok, *Thermochim. Acta*, 1991, **187**, 261–268.
17. B. Wunderlich and W. Chen, in *Liquid-Crystalline Polymer Systems – Technological Advances*, ed. A. I. Isayev, T. Kyu and S. Z. D. Cheng, 1996, vol. 632, pp. 232–248.
18. J. L. Feijoo, G. Ungar and V. Percec, *Mol. Cryst. Liq. Cryst. Sci. Technol., Sect. A*, 1993, **231**, 129–135.
19. C. R. Snyder, R. J. Kline, D. M. DeLongchamp, R. C. Nieuwendaal, L. J. Richter, M. Heeney and I. McCulloch, *J. Polym. Sci., Part B: Polym. Phys.*, 2015, **53**, 1641–1653.
20. A. Toda, K. Taguchi, K. Nozaki and M. Konishi, *Polymer*, 2014, **55**, 3186–3194.
21. T. Yamamoto, K. Sanechika and A. Yamamoto, *J. Polym. Sci., Polym. Lett. Ed.*, 1980, **18**, 9–12.
22. B. Wunderlich, *Thermochim. Acta*, 1999, **341**, 37–52.
23. E. Hempel, M. Beiner, T. Renner and E. Donth, *Acta Polym.*, 1996, **47**, 525–529.
24. M. Pyda, A. Boller, J. Grebowicz, H. Chuah, B. V. Lebedev and B. Wunderlich, *J. Polym. Sci., Part B: Polym. Phys.*, 1998, **36**, 2499–2511.
25. M. Pyda, E. Nowak-Pyda, J. Heeg, H. Huth, A. A. Minakov, M. L. Di Lorenzo, C. Schick and B. Wunderlich, *J. Polym. Sci., Part B: Polym. Phys.*, 2006, **44**, 1364–1377.
26. B. Wunderlich, *J. Appl. Polym. Sci.*, 2007, **105**, 49–59.

27. B. Crist and F. M. Mirabella, *J. Polym. Sci., Part B: Polym. Phys.*, 1999, **37**, 3131–3140.
28. C. R. Snyder, R. C. Nieuwendaal, D. M. DeLongchamp, C. K. Luscombe, P. Sista and S. D. Boyd, *Macromolecules*, 2014, **47**, 3942–3950.
29. P. J. Flory, *J. Chem. Phys.*, 1947, **15**, 684.
30. P. J. Flory, *J. Chem. Phys.*, 1949, **17**, 223–240.
31. C. R. Snyder, J. S. Henry and D. M. DeLongchamp, *Macromolecules*, 2011, **44**, 7088–7091.
32. I. C. Sanchez and R. K. Eby, *Macromolecules*, 1975, **8**, 638–641.
33. W. Chen, Y. Fu, B. Wunderlich and J. Cheng, *J. Polym. Sci., Part B: Polym. Phys.*, 1994, **32**, 2661–2666.
34. J. D. Hoffman, G. T. Davis and J. I. Lauritzen Jr., in *Crystalline and Noncrystalline Solids*, ed. N. B. Hannay, Plenum Press, New York, 1976, vol. 3, pp. 497–614.
35. M. G. Broadhurst, *J. Chem. Phys.*, 1962, **36**, 2578–2582.
36. J. D. Hoffman and J. J. Weeks, *J. Res. Natl. Bur. Stand.*, 1962, **66A**, 13–28.
37. B. Wunderlich, *Macromolecular Physics, Vol. 3: Crystal Melting*, Academic Press, 1980.
38. T. Y. Cho, B. Heck and G. Strobl, *Colloid Polym. Sci.*, 2004, **282**, 825–832.
39. D. C. Bassett and R. Davitt, *Polymer*, 1974, **15**, 721–728.
40. S. H. Chen, C. H. Su, A. C. Su, Y. S. Sun, U. Jeng and S. A. Chen, *J. Appl. Crystallogr.*, 2007, **40**, s573–s576.
41. M. Canetti, F. Bertini, G. Scavia and W. Porzio, *Eur. Polym. J.*, 2009, **45**, 2572–2579.
42. C. Müller, N. D. Zhigadlo, A. Kumar, M. A. Baklar, J. Karpinski, P. Smith, T. Kreouzis and N. Stingelin, *Macromolecules*, 2011, **44**, 1221–1225.
43. B. Crist, *Polymer*, 2003, **44**, 4563–4572.
44. F. P. V. Koch, M. Heeney and P. Smith, *J. Am. Chem. Soc.*, 2013, **135**, 13699–13709.
45. N. Bekkedahl and L. A. Wood, *J. Chem. Phys.*, 1941, **9**, 193.
46. L. A. Wood and N. Bekkedahl, *J. Appl. Phys.*, 1946, **17**, 362–375.
47. L. A. Wood and N. Bekkedahl, *J. Res. Natl. Bur. Stand.*, 1946, **36**, 489–510.
48. H. Marand, J. Xu and S. Srinivas, *Macromolecules*, 1998, **31**, 8219–8229.
49. J. Xu, S. Srinivas, H. Marand and P. Agarwal, *Macromolecules*, 1998, **31**, 8230–8242.
50. P. Juhász, J. Varga, K. Belina and H. Marand, *J. Therm. Anal. Calorim.*, 2002, **69**, 561–574.
51. G. Strobl, *Prog. Polym. Sci.*, 2006, **31**, 398–442.
52. V. Causin, C. Marega, A. Marigo, L. Valentini and J. M. Kenny, *Macromolecules*, 2005, **38**, 409–415.
53. S. Malik and A. K. Nandi, *J. Polym. Sci., Part B: Polym. Phys.*, 2002, **40**, 2073–2085.
54. B. S. Beckingham, V. Ho and R. A. Segalman, *Macromolecules*, 2014, **47**, 8305–8310.
55. L. Mandelkern, A. L. Allou and M. R. Gopalan, *J. Phys. Chem.*, 1968, **72**, 309–318.

56. L. Mandelkern and P. J. Flory, *J. Am. Chem. Soc.*, 1951, **73**, 3206–3212.
57. P. J. Flory, *Principles of Polymer Chemistry*, Cornell University Press, Ithaca, New York, 1953.
58. L. E. Alexander, *X-Ray Diffraction Methods in Polymer Science*, Robert E. Krieger Publishing Company, 1979.
59. D. R. Kozub, K. Vakhshouri, L. M. Orme, C. Wang, A. Hexemer and E. D. Gomez, *Macromolecules*, 2011, **44**, 5722–5726.
60. J. Borrajo, C. Cordon, J. M. Carella, S. Toso and G. Goizueta, *J. Polym. Sci., Part B: Polym. Phys.*, 1995, **33**, 1627–1632.
61. G. Elicabe, C. Cordon and J. Carella, *J. Polym. Sci., Part B: Polym. Phys.*, 1996, **34**, 1147–1154.
62. C. S. Lee and M. D. Dadmun, *Polymer*, 2014, **55**, 4–7.
63. A. J. Müller and M. L. Arnal, *Prog. Polym. Sci.*, 2005, **30**, 559–603.
64. A. J. Müller, R. M. Michell, R. A. Pérez and A. T. Lorenzo, *Eur. Polym. J.*, 2015, **65**, 132–154.
65. R. L. Danley, *Thermochim. Acta*, 2002, **395**, 201–208.
66. R. L. Danley, P. A. Caulfield and S. R. Aubuchon, *Am. Lab.*, 2008, **January**, 9–11.
67. V. B. F. Mathot, B. Goderis, R. L. Scherrenberg and E. W. van der Vegte, *Macromolecules*, 2002, **35**, 3601–3613.
68. M. F. J. Pijpers and V. B. F. Mathot, *J. Therm. Anal. Calorim.*, 2008, **93**, 319–327.
69. E. A. Olson, M. Y. Efremov, M. Zhang, Z. Zhang and L. H. Allen, *J. Microelectromech. Syst.*, 2003, **12**, 355–364.
70. E. Zhuravlev and C. Schick, *Thermochim. Acta*, 2010, **505**, 14–21.
71. M. Peo, H. Forster, K. Menke, J. Hocker, H. A. Gardner and S. Roth, *Solid State Commun.*, 1981, **38**, 467–468.
72. C. S. Yannoni and T. C. Clarke, *Phys. Rev. Lett.*, 1983, **51**, 1191–1193.
73. K. Holczer, F. Devreux, M. Hechtschein and J. P. Travers, *Solid State Commun.*, 1981, **39**, 881–884.
74. M. Mehring, H. Weber, W. Muller and G. Wegner, *Solid State Commun.*, 1983, **45**, 1079–1082.
75. T. C. Clarke, J. C. Scott and G. B. Street, *IBM J. Res. Dev.*, 1983, **27**, 313–320.
76. M. Mehring and J. Spenger, *Phys. Rev. Lett.*, 1984, **53**, 2441–2444.
77. R. A. Wind, H. Lock and M. Mehring, *Chem. Phys. Lett.*, 1987, **141**, 283–288.
78. D. Kongeter and M. Mehring, *Phys. Rev. B: Condens. Matter Mater. Phys.*, 1989, **39**, 6361–6369.
79. W. Stocklein, H. Seidel, D. Singel, R. D. Kendrick and C. S. Yannoni, *Chem. Phys. Lett.*, 1987, **141**, 277–282.
80. F. Creuzet, C. Bourbonnais, L. G. Caron, D. Jerome and K. Bechgaard, *Synth. Met.*, 1987, **19**, 289–294.
81. A. C. Kolbert, S. Caldarelli, K. F. Their, N. S. Sariciftci, Y. Cao and A. J. Heeger, *Phys. Rev. B: Condens. Matter Mater. Phys.*, 1995, **51**, 1541–1545.
82. D. L. VanderHart, W. L. Earl and A. N. Garroway, *J. Magn. Reson.*, 1981, **44**, 361–401.

83. L. M. Ryan, R. E. Taylor, A. J. Paff and B. C. Gerstein, *J. Chem. Phys.*, 1980, **72**, 508.

84. R. C. Nieuwendaal, H. W. Ro, D. S. Germack, R. J. Kline, M. F. Toney, C. K. Chan, A. Agrawal, D. Gundlach, D. L. VanderHart and D. M. DeLongchamp, *Adv. Funct. Mater.*, 2012, **22**, 1255–1266.

85. D. Dudenko, A. Kiersnowski, J. Shu, W. Pisula, D. Sebastiani, H. W. Spiess and M. R. Hansen, *Angew. Chem., Int. Ed.*, 2012, **51**, 11068–11072.

86. E. Salager, R. Stein, S. Steuernagel, A. Lesage, B. Elena and L. Emsley, *Chem. Phys. Lett.*, 2009, **469**, 336–341.

87. W. K. Rhim, D. D. Elleman and R. W. Vaughan, *J. Chem. Phys.*, 1973, **58**, 1772.

88. E. O. Stejskal, J. Schaefer and J. S. Waugh, *J. Magn. Reson.*, 1977, **28**, 105.

89. K. Yazawa, Y. Inoue, T. Yamamoto and N. Asakawa, *Phys. Rev. B: Condens. Matter Mater. Phys.*, 2006, **74**, 094204.

90. K. Yazawa, Y. Inoue, T. Shimizu and N. Asakawa, *J. Phys. Chem. B*, 2010, **114**, 1241–1248.

91. O. F. Pascui, R. Lohwasser, M. Sommer, M. Thelakkat, T. Thurn-Albrecht and K. Saalwächter, *Macromolecules*, 2010, **43**, 9401–9410.

92. A. E. Tonelli, *Macromolecules*, 1978, **11**, 565–567.

93. W. L. Earl and D. L. VanderHart, *Macromolecules*, 1979, **12**, 762–767.

94. R. C. Nieuwendaal, C. R. Snyder and D. M. DeLongchamp, *ACS Macro Lett.*, 2014, **3**, 130–135.

95. E. R. deAzevedo, W. G. Hu, T. J. Bonagamba and K. Schmidt-Rohr, *J. Am. Chem. Soc.*, 1999, **121**, 8411.

96. J. Schaefer, E. O. Stejskal, R. A. McKay and W. T. Dixon, *Macromolecules*, 1984, **17**, 1479.

97. N. Bloembergen, E. M. Purcell and R. V. Pound, *Phys. Rev.*, 1948, **73**, 679–712.

98. F. Bloch, *Phys. Rev.*, 1956, **102**, 104–135.

99. F. Bloch, *Phys. Rev.*, 1957, **105**, 1206–1222.

100. R. K. Wangsness and F. Bloch, *Phys. Rev.*, 1953, **89**, 728–739.

101. A. G. Redfield, *IBM J. Res. Dev.*, 1957, **1**, 19–31.

102. A. G. Redfield, *Adv. Magn. Reson.*, 1965, **1**, 1–32.

103. A. Abragam, *The Principles of Nuclear Magnetism*, Oxford University Press, Oxford, U. K., 1961.

104. C. P. Slichter, *Principles of Magnetic Resonance*, Springer-Verlag, Berlin, Germany, 3rd edn, 1990.

105. R. R. Ernst, G. Bodenhausen and A. Wokaum, *Principles of Nuclear Magnetic Resonance in One and Two Dimensions*, Oxford University Press, Oxford, U. K., 1987.

106. R. Kimmich, *NMR Tomography, Diffusometry, Relaxometry*, Springer-Verlag, Berlin, Germany, 1997.

107. C. K. Hall and E. Helfand, *J. Chem. Phys.*, 1982, **77**, 3275–3382.

108. E. Helfand, *Physica*, 1983, **118**, 123–135.

109. J. Skolnick, D. Perchak and R. Yaris, *J. Magn. Reson.*, 1984, **57**, 204–220.

110. B. B. Pant, J. Skolnick and R. Yaris, *Macromolecules*, 1985, **18**, 253–259.
111. F. Martini, S. Borsacchi, S. Speera, C. Carbonera, A. Cominetti and M. Geppi, *J. Phys. Chem. C*, 2013, **117**, 131–139.
112. E. R. deAzevedo, W. G. Hu, T. J. Bonagamba and K. Schmidt-Rohr, *J. Chem. Phys.*, 2000, **112**, 8988–9001.
113. A. C. Bloise, E. R. deAzevedo, R. F. Cossiello, R. F. Bianchi, D. T. Balogh, R. M. Faria, T. D. Z. Atvars and T. J. Bonagamba, *Phys. Rev. B: Condens. Matter Mater. Phys.*, 2005, **71**, 174202.
114. G. C. Faria, E. R. deAzevedo and H. von Seggern, *Macromolecules*, 2013, **46**, 7865–7873.
115. R. F. Cossiello, E. Kowalski, P. C. Rodrigues, L. Akcelrud, A. C. Bloise, E. R. deAzevedo, T. J. Bonagamba and T. D. Z. Atvars, *Macromolecules*, 2005, **38**, 925.
116. E. R. deAzevedo, K. Saalwachter, O. Pascui, A. A. de Souza, T. J. Bonagamba and D. Reichert, *J. Chem. Phys.*, 2008, **128**, 104505.
117. K. Schmidt-Rohr and H. W. Spiess, *Multidimensional NMR and Polymers*, Academic Press Limited, London, U. K., 1994, pp. 402–439.
118. D. L. VanderHart and G. B. McFadden, *Solid State Nucl. Magn. Reson.*, 1996, **7**, 45–66.
119. R. Mens, F. Demir, G. Van Assche, B. Van Mele, D. Vanderzande and P. Adriaensens, *J. Polym. Sci., Part A: Polym. Chem.*, 2012, **50**, 1037–1041.
120. R. Mens, P. Adriaensens, L. Lutsen, A. Swinnen, S. Bertho, B. Ruttens, J. D'Haen, J. Manca, T. Cleji, D. Vanderzande and J. Gelan, *J. Polym. Sci., Part A: Polym. Chem.*, 2008, **46**, 138–145.
121. X. Jia, X. Wang, A. E. Tonelli and J. L. White, *Macromolecules*, 2005, **38**, 2775–2780.
122. J. Clauss, K. Schmidt-Rohr and H. W. Spiess, *Acta Polym.*, 1993, **44**, 1–17.
123. R. C. Nieuwendaal, C. R. Snyder, R. J. Kline, E. K. Lin, D. L. VanderHart and D. M. DeLongchamp, *Chem. Mater.*, 2010, **22**, 2930–2936.
124. C. Yang, J. Hu and A. J. Heeger, *J. Am. Chem. Soc.*, 2006, **128**, 12007–12013.
125. N. C. Miller, E. Cho, M. J. N. Junk, R. Gysel, C. Risko, D. Kim, S. Sweetnam, C. E. Miller, L. J. Richter, R. J. Kline, M. Heeney, I. McCulloch, A. Amassian, D. Acevedo-Feliz, C. Knox, M. R. Hansen, D. Dudenko, B. F. Chmelka, M. F. Toney, J. L. Bredas and M. D. McGehee, *Adv. Mater.*, 2014, **24**, 6071–6079.
126. K. R. Graham, C. Cabanetos, J. P. Jahnke, M. N. Idso, A. El Labban, G. O. N. Ndjawa, T. Heumueller, K. Vandewal, A. Salleo, B. F. Chmelka, A. Amassian, P. M. Beaujuge and M. D. McGehee, *J. Am. Chem. Soc.*, 2014, **136**, 9608–9618.
127. X. Jia, J. Wolak, X. Wang and J. L. White, *Macromolecules*, 2003, **36**, 712–718.
128. X. L. Yao, K. Schmidt-Rohr and M. Hong, *J. Magn. Reson.*, 2001, **149**, 139–143.
129. R. Sondergaard, M. Hosel, D. Angmo, T. T. Larsen-Olsen and F. C. Krebs, *Mater. Today*, 2012, **15**, 36.

130. A. B. Barnes, G. De Paepe, P. C. A. van der Wel, K.-N. Hu, C.-G. Joo, V. S. Bajaj, M. L. Mak-Jurkauskas, J. R. Sirigiri, J. Herzfeld, R. J. Temkin and R. G. Griffin, *Appl. Magn. Reson.*, 2008, **34**, 237–263.
131. R. Egerton, P. Li and M. Malac, *Micron*, 2004, **35**, 399–409.
132. R. Egerton, *Ultramicroscopy*, 2012, **127**, 100–108.
133. M. Isaacson, *Ultramicroscopy*, 1979, **4**, 193–199.
134. M. Isaacson, *Princ. Tech. Electron Microsc.*, 1977, 1–78.
135. D. T. Grubb, *J. Mater. Sci.*, 1974, **9**, 1715–1736.
136. P. H. Geil, *J. Polym. Sci.*, 1960, **44**, 449–458.
137. R. Geiss, G. Street, W. Volksen and J. Economy, *IBM J. Res. Dev.*, 1983, **27**, 321–329.
138. X. Zhang, S. D. Hudson, D. M. DeLongchamp, D. J. Gundlach, M. Heeney and I. McCulloch, *Adv. Funct. Mater.*, 2010, **20**, 4098–4106.
139. J. C. Spence, *Experimental High-resolution Electron Microscopy*, Oxford University Press, 1988.
140. D. C. Martin and E. L. Thomas, *Polymer*, 1995, **36**, 1743–1759.
141. L. F. Drummy, R. J. Davis, D. L. Moore, M. Durstock, R. A. Vaia and J. W. P. Hsu, *Chem. Mater.*, 2011, **23**, 907–912.
142. R. F. Egerton, *Electron Energy-loss Spectroscopy in the Electron Microscope*, Springer Science + Business Media, New York, 3rd edn, 2011.
143. D. Pines, *Rev. Mod. Phys.*, 1956, **28**, 184.
144. S. Yakovlev and M. Libera, *Micron*, 2008, **39**, 734–740.
145. K. Varlot, J. M. Martin and C. Quet, *Polymer*, 2000, **41**, 4599–4605.
146. F. Allen, M. Watanabe, Z. Lee, N. Balsara and A. Minor, *Ultramicroscopy*, 2011, **111**, 239–244.
147. M. H. Gass, K. K. K. Koziol, A. H. Windle and P. A. Midgley, *Nano Lett.*, 2006, **6**, 376–379.
148. A. A. Herzing, L. J. Richter and I. M. Anderson, *J. Phys. Chem. C*, 2010, **114**, 17501–17508.
149. J. Frank, in *Electron Tomography: Methods for Three-dimensional Imaging of Visualization in the Cell*, ed. J. Frank, Springer, New York, 2nd edn, 2006, pp. 1–16.
150. S. S. van Bavel, E. Sourty, G. de With, S. Veenstra and J. Loos, *J. Mater. Chem.*, 2009, **19**, 5388.
151. S. S. Van-Bavel, E. Sourty, G. de With and J. Loos, *Nano Lett.*, 2009, **9**, 507–513.
152. M. J. Wirix, P. H. Bomans, H. Friedrich, N. A. Sommerdijk and G. de With, *Nano Lett.*, 2014, **14**, 2033–2038.
153. W. H. Bragg and W. L. Bragg, *Proc. R. Soc. London, Ser. A*, 1913, **88**, 428–438.
154. B. D. Cullity and S. R. Stock, *Elements of X-ray Diffraction*, Prentice Hall, Upper Saddle River, NJ, 2001.
155. J. Rivnay, R. Noriega, R. J. Kline, A. Salleo and M. F. Toney, *Phys. Rev. B: Condens. Matter Mater. Phys.*, 2011, **84**, 045203.
156. J. Rivnay, R. Noriega, J. E. Northrup, R. J. Kline, M. F. Toney and A. Salleo, *Phys. Rev. B: Condens. Matter Mater. Phys.*, 2011, **83**, 121306.

157. J. Rivnay, S. C. B. Mannsfeld, C. E. Miller, A. Salleo and M. F. Toney, *Chem. Rev.*, 2012, **112**, 5488–5519.
158. B. A. Collins, F. A. Bokel and D. M. DeLongchamp, in *Organic Photovoltaics*, ed. C. Brabec, U. Scherf and V. Dyakonov, Wiley-VCH Verlag GmbH & Co. KGaA, 2014, pp. 377–420.
159. R. J. Kline and D. M. DeLongchamp, *Organic Electronics II*, Wiley-VCH Verlag GmbH, Boschstr. 12, Weinheim, -1, 69469, Germany, 0.
160. A. Salleo, R. J. Kline, D. M. DeLongchamp and M. L. Chabinyc, *Adv. Mater.*, 2010, **22**, 3812–3838.
161. D. M. DeLongchamp, R. J. Kline, D. A. Fischer, L. J. Richter and M. F. Toney, *Adv. Mater.*, 2011, **23**, 319–337.
162. D. M. DeLongchamp, R. J. Kline and A. Herzing, *Energy Environ. Sci.*, 2012, **5**, 5980–5993.
163. J. L. Baker, L. H. Jimison, S. Mannsfeld, S. Volkman, S. Yin, V. Subramanian, A. Salleo, A. P. Alivisatos and M. F. Toney, *Langmuir*, 2010, **26**, 9146–9151.
164. M. R. Hammond, R. J. Kline, A. A. Herzing, L. J. Richter, D. S. Germack, H.-W. Ro, C. L. Soles, D. A. Fischer, T. Xu, L. Yu, M. F. Toney and D. M. DeLongchamp, *ACS Nano*, 2011, **5**, 8248–8257.
165. D. Chen, F. Liu, C. Wang, A. Nakahara and T. P. Russell, *Nano Lett.*, 2011, **11**, 2071–2078.
166. D. R. Kozub, K. Vakhshouri, L. M. Orme, C. Wang, A. Hexemer and E. D. Gomez, *Macromolecules*, 2011, **44**, 5722–5726.
167. W. Yin and M. Dadmun, *ACS Nano*, 2011, **5**, 4756–4768.
168. J. W. Kiel, A. P. R. Eberle and M. E. Mackay, *Phys. Rev. Lett.*, 2010, **105**, 168701.
169. S. Swaraj, C. Wang, H. Yan, B. Watts, J. Lüning, C. R. McNeill and H. Ade, *Nano Lett.*, 2010, **10**, 2863–2869.
170. B. A. Collins, J. R. Tumbleston and H. Ade, *J. Phys. Chem. Lett.*, 2011, **2**, 3135–3145.
171. B. A. Collins, Z. Li, J. R. Tumbleston, E. Gann, C. R. McNeill and H. Ade, *Adv. Energy Mater.*, 2013, **3**, 65–74.
172. J. Stöhr, *NEXAFS Spectroscopy*, Springer, Berlin, New York, Corrected edn, 2003.
173. B. A. Collins, J. E. Cochran, H. Yan, E. Gann, C. Hub, R. Fink, C. Wang, T. Schuettfort, C. R. McNeill, M. L. Chabinyc and H. Ade, *Nat. Mater.*, 2012, **11**, 536–543.
174. D. M. DeLongchamp, R. J. Kline, E. K. Lin, D. A. Fischer, L. J. Richter, L. A. Lucas, M. Heeney, I. McCulloch and J. E. Northrup, *Adv. Mater.*, 2007, **19**, 833–837.
175. J. Perlich, M. Schwartzkopf, V. Körstgens, D. Erb, J. F. H. Risch, P. Müller-Buschbaum, R. Röhlsberger, S. V. Roth and R. Gehrke, *Phys. Status Solidi RRL*, 2012, **6**, 253–255.
176. E. R. Meshot, E. Verploegen, M. Bedewy, S. Tawfick, A. R. Woll, K. S. Green, M. Hromalik, L. J. Koerner, H. T. Philipp, M. W. Tate, S. M. Gruner and A. J. Hart, *ACS Nano*, 2012, **6**, 5091–5101.

177. R. Li, H. U. Khan, M. M. Payne, D.-M. Smilgies, J. E. Anthony and A. Amassian, *Adv. Funct. Mater.*, 2013, **23**, 291–297.
178. M. Sanyal, B. Schmidt-Hansberg, M. F. G. Klein, A. Colsmann, C. Munuera, A. Vorobiev, U. Lemmer, W. Schabel, H. Dosch and E. Barrena, *Adv. Energy Mater.*, 2011, **1**, 363–367.
179. T. Wang, A. J. Pearson, A. D. F. Dunbar, P. A. Staniec, D. C. Watters, H. Yi, A. J. Ryan, R. A. L. Jones, A. Iraqi and D. G. Lidzey, *Adv. Funct. Mater.*, 2012, **22**, 1399–1408.
180. E. Verploegen, C. E. Miller, K. Schmidt, Z. Bao and M. F. Toney, *Chem. Mater.*, 2012, **24**, 3923–3931.
181. S. Hong, A. Amassian, A. R. Woll, S. Bhargava, J. D. Ferguson, G. G. Malliaras, J. D. Brock and J. R. Engstrom, *Appl. Phys. Lett.*, 2008, **92**, 253304.
182. Y. Gu, C. Wang and T. P. Russell, *Adv. Energy Mater.*, 2012, **2**, 683–690.
183. S. Kowarik, a Gerlach, S. Sellner, F. Schreiber, L. Cavalcanti and O. Konovalov, *Phys. Rev. Lett.*, 2006, **96**, 125504.
184. K. W. Chou, B. Yan, R. Li, E. Q. Li, K. Zhao, D. H. Anjum, S. Alvarez, R. Gassaway, A. Biocca, S. T. Thoroddsen, A. Hexemer and A. Amassian, *Adv. Mater.*, 2013, **25**, 1923–1929.
185. B. Schmidt-Hansberg, M. F. G. Klein, K. Peters, F. Buss, J. Pfeifer, S. Walheim, A. Colsmann, U. Lemmer, P. Scharfer and W. Schabel, *J. Appl. Phys.*, 2009, **106**, 124501.
186. B. Schmidt-Hansberg, M. F. G. Klein, M. Sanyal, F. Buss, G. Q. G. de Medeiros, C. Munuera, A. Vorobiev, A. Colsmann, P. Scharfer, U. Lemmer, E. Barrena and W. Schabel, *Macromolecules*, 2012, **45**, 7948–7955.
187. Y. Zheng, R. Wu, W. Shi, Z. Guan and J. Yu, *Sol. Energy Mater. Sol. Cells*, 2013, **111**, 200–205.
188. T. Agostinelli, S. Lilliu, J. G. Labram, M. Campoy-Quiles, M. Hampton, E. Pires, J. Rawle, O. Bikondoa, D. D. C. Bradley, T. D. Anthopoulos, J. Nelson and J. E. Macdonald, *Adv. Funct. Mater.*, 2011, **21**, 1701–1708.
189. L. J. Richter, D. M. DeLongchamp, F. A. Bokel, S. Engmann, K. W. Chou, A. Amassian, E. Schaible and A. Hexemer, *Adv. Energy Mater.*, 2015, **5**, 1400975.
190. S. Engmann, F. A. Bokel, A. Herzing, H. W. Ro, C. Girotto, B. Caputo, C. V. Hoven, E. Schaible, A. Hexemer, D. Delongchamp and L. Richter, *J. Mater. Chem. A*, 2015, **3**, 8764–8771.
191. J. J. van Franeker, M. Turbiez, W. Li, M. M. Wienk and R. A. J. Janssen, *Nat. Commun.*, 2015, **6**, 6229.
192. F. Liu, S. Ferdous, E. Schaible, A. Hexemer, M. Church, X. Ding, C. Wang and T. P. Russell, *Adv. Mater.*, 2015, **27**, 886–891.
193. D.-M. Smilgies, R. Li, G. Giri, K. W. Chou, Y. Diao, Z. Bao and A. Amassian, *Phys. Status Solidi RRL*, 2013, **7**, 177–179.
194. G. Giri, R. Li, D.-M. Smilgies, E. Q. Li, Y. Diao, K. M. Lenn, M. Chiu, D. W. Lin, R. Allen, J. Reinspach, S. C. B. Mannsfeld, S. T. Thoroddsen, P. Clancy, Z. Bao and A. Amassian, *Nat. Commun.*, 2014, **5**, 3573.

Subject Index

aromatic polyamide, 4
aryllithium monomers, Murahashi coupling polymerization of, 29–30

back-scattering spectrometers, 179
block copolymers (BCPs), 63–66
    all-conjugated block copolymers, 137–139
    all-conjugated donor–acceptor block copolymers, 139–142
    chain-growth synthesis of fully conjugated block copolymers, 66–71
    semi-conjugated block copolymers, 142–146
boronic acid (ester) monomer, Suzuki–Miyaura coupling polymerization of
    catalysts, 20–22
    discovery, 16–17
    initiators, 20–22
    monomers, 17–20
    polymer ends, functionalization of, 22–24

catalyst transfer polymerization (CTP), 85–86
chain-growth synthesis
    of alternating copolymers, 56–63
    block copolymers, 63–66
    of conjugated homopolymers, 44–55
    of fully conjugated block copolymers, 66–71
    of other copolymers, 71–78

condensative chain polymerization, 5
conjugated and non-conjugated blocks
    synthesis via macromolecular coupling reactions, 99–102
    synthesis via one-pot polymerization, 107–109
    synthesis with macromolecular initiators, 102–107
conjugated gradient copolymers, 97–99
conjugated polymer brushes, 109–112
conjugated polymers
    controlled chain growth in, 41–44
controlled chain growth mechanism of, 41–44
controlled chain polymerization, 39, 40
CTP. See catalyst transfer polymerization (CTP)

dark field techniques, 255–257
degree of branching (DB), 154
differential scanning calorimetry (DSC)
    equilibrium melting temperature, 225–228
    fusion per repeat unit, 228–231
    qualifying and quantifying crystallinity, 221–225
    self-nucleation and successive annealing, 231–232
    thin film samples, 232–235

electron energy loss spectroscopy
(EELS), 259–260

fully conjugated block copolymers
donor–acceptor type, 91–97
donor–donor type conjugated
block copolymers, 86–91

GIXD. *See* grazing incidence X-ray
diffraction (GIXD)
grazing incidence scattering
grazing incidence small-angle
X-ray scattering, 263–264
grazing incidence X-ray
diffraction (GIXD), 261–263
in situ X-ray scattering,
265–266
polarized resonant soft X-ray
scattering, 264–265
X-ray scattering, 261
grazing incidence small-angle X-ray
scattering, 263–264
grazing incidence X-ray diffraction
(GIXD), 261–263
Grignard monomers, Kumada-Tamao
coupling polymerization of
catalysts, 11–13
discovery, 4–6
initiators, 13–16
mechanistic studies, 6–9
monomers, 9–10
polymer ends,
functionalization of, 16

HBPs. *See* hyperbranched polymers
(HBPs)
hydroxymethylphenyl moiety, 154
hyperbranched polymers (HBPs),
154–155

in situ X-ray scattering, 265–266
intramolecular catalyst transfer
cross-couplings, 111

Kohlrausch–Williams–Watts (KWW)
fits, 180

Kumada–Tamao coupling
polymerization, Grignard
monomers
catalysts, 11–13
discovery, 4–6
initiators, 13–16
mechanistic studies, 6–9
monomers, 9–10
polymer ends,
functionalization of, 16

light-emitting properties
blending, 200–207
chemical design, 189–195
metal-enhanced fluorescence,
207–212
physical and physicochemical
methods
chemical cross-linking,
198
confined spaces, 199–200
controlled aggregation,
199
dewetting, 196–197
pressure, 196
solvent vapour and
solvent quality, 195–196
stretchable structures,
200

molecular weight and dispersity,
124–126

neutron scattering, thin films
grazing incidence small-angle
neutron scattering, 175–177
neutron reflectometry,
174–175
transmission SANS, 172–174
neutron spin echo (NSE)
spectrometers, 179
nitroxy-mediated radial
polymerization (NMRP), 130
non-metallated monomers,
Mizoroki–Heck coupling
polymerization of, 30

organo-stannanes. *See* Stille
    coupling polymerization,
    tin-containing monomers

P3HT and polyaniline (PANI), 180
poly(3-hexylthiophene) (P3HT)
    absorption spectra for, 205
    chemical structures of, 198
    controlled synthesis of, 39
    SANS of, 169
poly(N-isopropylacrylamide)
    (PNIPAM), 210
poly[2,5-bis(3′,7′-dimethyloctyloxy)-
    1,4-phenylenevinylene]
    (BDMO-PPV), 212
poly[2,5-bis(3-tetradecylthiophen-2-
    yl)thieno[3,2-b]-thiophene]
    (pBTTT), 202
poly[2,6-(4,4-bis(2-ethylhexyl)-4H-
    cyclopenta[2,1-b;3,4-b′]-
    dithiophene)-alt-4,7-(2,1,3-
    benzothiadiazole)] (PCPDTBT),
    172
poly[2-methoxy-5-(2-ethylhexyloxy)-
    1,4-phenylenevinylene]
    (MEH-PPV), 171, 189
poly[3-(6-hydroxyhexyl)thiophene]
    (P3HHT), 126
poly[4,8-bis(5-(2-ethylhexyl)-
    thiophen-2-yl)benzo[1,2-b;4,5-
    b′]dithiophene-2,6-diyl-alt-(4-
    (2-ethylhexyl)-3-fluorothieno-
    [3,4-b]thiophene-)-2-carboxylate-2-
    6-diyl)], 40
poly[9,9-dioctylfluorene-co-
    benzothiadiazole] (F8cBT), 199
poly[9,90-bis(600-(N,N,N-
    trimethylammonium)-
    hexyl)fluorene-2,7-ylenevinylene-
    co-alt-1,4-phenylene dibromide]
    (PFV), 210
poly[9,99-bis(60-(N,N,N-
    trimethylammonium)-
    hexyl)fluorene-co-alt-2,5-
    dimethoxy-1,4-phenylene
    dibromide] (PFPMO), 210

poly[(9,9-dioctylfluorenyl-2,7-diyl)-
    alt-(4,7-bis(3-hexylthiophen-5-yl)-
    2,1,3-benzothiadiazole)-2′,2″-diyl],
    202
poly(2,6-bisthien-5-yl)naphthalene-
    1,4,5,8-tetracarboxylic-N,N′-
    bis(2-alkyl) diimide, 40
poly(3-alkylthiophene)s (P3ATs), 122
polythiophenes
    all-conjugated block
        copolymers, 137–139
    all-conjugated block
        copolythiophenes, 134–136
    all-conjugated donor–acceptor
        block copolymers, 139–142
    chain end functional, 128–134
    controlled synthesis of,
        122–123
    dehydrogenative synthesis of,
        126–128
    graft copolymers, 146–149
    hyperbranched polymers
        (HBPs), 154–155
    semi-conjugated block
        copolymers, 142–146
    star-branched polymers,
        149–154

quasi-elastic neutron scattering
    (QENS)
    instrumentation and methods
        for analysis, 179–180
    of polymer semiconductors,
        180–184
    theory, 177–179

regio-regularity, 123–124

SANS. *See* small-angle neutron
    scattering (SANS)
scattering length density (SLD), 167
self-nucleation and successive
    annealing (SSA), 226
silylated monomers, nucleophilic
    substitution polymerization of,
    30–31

SLD. *See* scattering length density
(SLD)
small-angle neutron scattering (SANS)
of colloidal polymer
nanostructures, 168–171
of dissolved conjugated
polymers, 167–168
solid-state NMR spectrometry
crystallinity and order, 236–238
donor/acceptor blends, 248–251
relaxation and dynamics,
238–248
SSA. *See* self-nucleation and
successive annealing (SSA)
star-branched polymers,
polythiophene segments, 149–154
Stille coupling polymerization,
tin-containing monomers, 28–29
Suzuki–Miyaura coupling
polymerization, boronic acid
(ester) monomer
catalysts, 20–22
discovery, 16–17
initiators, 20–22
monomers, 17–20
polymer ends,
functionalization of, 22–24

tin-containing monomers, Stille
coupling polymerization of,
28–29
transmission electron microscopy
(TEM)
characterizing polymeric
materials in, 253–254
dark field techniques,
255–257
electron energy loss
spectroscopy (EELS),
259–260
low-dose high-resolution
imaging, 257–259
three-dimensional
characterization, 261
X-ray diffraction techniques,
255

ultra-small-angle neutron scattering
(USANS), 166

X-ray diffraction techniques, 255

Zn-containing monomers, Negishi
coupling polymerization of,
24–28